HEAT TRANSFER
IN ENERGY PROBLEMS

HEAT TRANSFER IN ENERGY PROBLEMS

Edited by

Tokuro Mizushina

Kyoto University

and

Wen-Jei Yang

The University of Michigan

⊙ **HEMISPHERE PUBLISHING CORPORATION**

Washington   New York   London

DISTRIBUTION OUTSIDE NORTH AMERICA

SPRINGER–VERLAG

Berlin   Heidelberg   New York   Tokyo

HEAT TRANSFER IN ENERGY PROBLEMS

1 2 3 4 5 6 7 8 9 0   B C B C   8 9 8 7 6 5 4 3

Library of Congress Cataloging in Publication Data
Main entry under title:

Heat transfer in energy problems.

   Proceedings of a seminar sponsored by the United States–Japan Cooperative Science Program, on Sept. 29–Oct. 2, 1980, in Tokyo, Japan.
   Includes bibliographical references and index.
   1. Heat–Transmission–Congresses. 2. Power (Mechanics)–Congresses. I. Mizushina, Tokuro, date. II. Yang, Wen–Jei, date. III. United States–Japan Cooperative Science Program.
TJ260.H394   621.402'2   82-910
                           AACR2
ISBN 0-89116-251-8 Hemisphere Publishing Corporation

DISTRIBUTION OUTSIDE NORTH AMERICA
ISBN 3-540-12202-8 Springer-Verlag Berlin

Contents

HIGH-FLUX HEAT TRANSFER: APPLICATIONS

HIGH-TEMPERATURE HEAT TRANSFER: APPLICATIONS

NOVEL HEAT TRANSFER TECHNIQUES FOR ENERGY UTILIZATION, HEAT STORAGE, AND RECOVERY

HEAT TRANSFER IN NONCONVENTIONAL ENERGY (POWER AND PROPULSION) SYSTEMS

ROUNDTABLE DISCUSSION

Preface

With limited supplies and soaring prices, energy has become a matter of worldwide concern. An urgent need for both the conservation of energy and the development of new energy resources is well recognized today. Heat transfer plays an important role in fulfilling the need, as it is directly involved in the processes of energy conversion, transfer, disposal, and storage. Both the United States and Japan represent high-energy-consuming countries and are technically advanced in heat transfer research. Therefore, it is very timely for Japanese and U.S. scientists in heat transfer engineering to meet to exchange first-hand information, to identify key problems, to discuss research strategies, and to promote areas of future cooperative heat transfer research. For this purpose a joint seminar was held September 29–October 2, 1980, in Tokyo, under the financial support of the U.S.–Japan Cooperative Science Program. This program is jointly administered by the U.S. Division of International Programs (INT)/National Science Foundation and the Japan Society for the Promotion for Science (JSPS).

This volume, the edited version of the proceedings, begins with an introduction to current Japanese heat transfer research in energy problems, including basic research, the Sun-shine project, the Moon-light project, and the Green project, and to new research endeavors in the United States. The major thrust of the volume lies in four important areas of heat transfer that are related to energy utilization and development:

1. *High-temperature heat transfer.* Film and impingement cooling of high-temperature surfaces, radiation, and radiative properties of metals and alloys at high temperature are presented as the subjects under basic phenomena. Topics dealing with applications are centered around nuclear-process heat exchange in the High-Temperature Gas-Cooled Reactor (HTGR) system, He–He intermediate heat exchanger, and nuclear power safety-reflooding and core uncovery phenomena.

2. *High-flux heat transfer.* Boiling and condensation are the focal points of the studies on basic phenomena, including critical heat flux, metal vapor condensation, and droplet evaporation on heated surfaces. Papers on applications cover power and propulsion systems together with gaseous liquid flows.

3. *High-performance heat transfer surfaces.* This area is divided into two parts, single flows and phase changes. Augmented surfaces for nucleate boiling are treated in great detail.

4. *Novel heat transfer techniques.* Included in this subject are both novel heat transfer techniques for energy utilization, heat storage, and recovery, and heat transfer in nonconventional energy systems. The former deals with heat transfer considerations in the use of new energy sources including solar energy, ocean thermal energy conversion, aqueous solutions of inorganic salts, and geothermal energy together with sensible heat storage and phase change storage. The latter introduces heat transfer problems in some advanced power systems including solar-thermal receiver technology, coal-fired MHD, and nuclear fusion power system concepts. Heat transfer problems in mist-cooled condensers for geothermal power plants are also discussed.

The work concludes with brief notes on nuclear reactor safety, energy research in Japan, a paper heat exchanger, combustion, advanced power systems, and cryogenic heat transfer.

We would like to express our sincere gratitude to the JSPS and INT/NSF for their sponsorship of the meeting. Our thanks are also extended to all the participants for their individual contributions and especially to the Japanese organizing committee members for their hard work and kindness. We also greatly appreciate the effort and devotion of Professor M. Akiyama of the University of Tokyo and the assistance of Miss Ling H. Yang, an engineering student at the University of Michigan, who checked and corrected the manuscripts. It is to the future U.S.-Japan cooperative efforts in heat transfer research that this volume is dedicated.

Tokuro Mizushina
Wen-Jei Yang

PRESENT STATUS
OF HEAT TRANSFER RESEARCH

Heat Transfer Research in Japan

TOKURO MIZUSHINA
Kyoto University, Kyoto, Japan

Quite an amount of money is being spent for the Sunshine Project and Moon-light Project by the Ministry of International Trade and Industry and Green Energy Project by the Ministry of Agriculture and Fishery all of which are the projects for developing the practical processes.

On the other hand, the basic sciences relating to those problems are being studied in universities. A special research project of energy problems has started in the frame of the Grant in Aid for Scientific Research of the Ministry of Education, Science and Culture.

Although many heat transfer problems must be solved in each research topic, it is a difficult task for me to extract all of those problems. In the following, therefore, each subject of the Programmed Research Groups of the Special Research Project with the name of its promoter will be described.

1. Energy Analysis and Economic Analysis of Energy Problems.

(1) Energy Problems and Economic Analysis.
    Prof. C. Moriguchi, Inst. of Economic Research, Kyoto Univ.

This research group has been organized to explore various possibilities of economic analysis in application to the contemporary energy problems and to play a role of coordinator/assessor overviewing competitive researchers for energy conservation and alternative energy sources. In this relation the objective of this group can be described as (A) providing a common ground for understanding and evaluating benefits and costs of energy researches within an economic framework and (B) providing a long-run scenario of the world economic growth under the constraints of energy supply.

(2) Holistic Approach to Evaluation of Energy Supply and Utilization Systems.
    Prof. Y. Kaya, Dept. of Electrical Engineering, Univ. of Tokyo.

The objective of this research group is to establish the assessment methodology of the contemporary energy system to give sound basis to the planning of the research and development program of new energy in Japan from a long-term and holistic standpoint, and then to envisage scenarios of the future energy system of Japan.

(3) Quantitative Analysis of Energy Flow in Biological Production Systems.
    Prof. T. Okuno, Dept. of Mathematical Engineering, Univ. of Tokyo.

The objective of this research group is to develop and establish the measurement and data analysis methods for evaluating quantitatively energy inputs and returns in several biological production systems; e.g. in cultivated lands, grass-lands and forests.

2. Effective Use of Energy.

The development of a technique for making effective use of energy depends on the basic research on the enhancement of efficiency of conversion among such various forms of energy as thermal, chemical, nuclear, kinetic and electrical energy. The objective of this research project is to obtain a basic understanding on the effective use of energy in cooperation of mechanical engineers, chemical engineers, applied chemists, electrical engineers and nuclear engineers. The research is conducted in the seven research groups.

(1) Utilization and Conversion of Thermal Energy at Temperatures Moderately Higher than Ambient.
    Prof. Y. Mori, Dept. of Physical Engineering, Tokyo Inst. of Technology.

This research aims at carrying out fundamental and systematic studies on efficient utilization and conversion of thermal energy at temperatures moderately higher than ambient from thermal engineering standpoints. The research is divided into the following four research items to cover the essential regions from basic to applied.
(A) Thermo-physical properties, transport properties and dynamical performances and losses of organic fluids which are used for electric power generation by use of Rankine cycle. — Formulation of equation of state and correlation equation of vapor pressure, measurements of heat capacity and thermal conductivity of various Freon compounds, experimental work on the effect of vibrational relaxation on heat transfer, and the measurement of the losses due to vibrational relaxation.
(B) Augmentation of heat transfer rate. — Performance of wavy duct heat exchanger, pipe flow boiling near the critical pressure, and the performance and optimum design criteria of shell and tube condenser.
(C) Optimum performance of direct contact heat exchanger, and performance and design criteria of heat pipes.
(D) Performance of various heat transfer systems. — performance of energy recovery systems utilizing dropwise condensation, performance of fluidized bed type energy recovery system, and the utilization of vacuum ejector in the energy conversion system from concentration difference.

(2) Extension of Technical Limits for Much More Effective Use of High Temperature Energy.
    Prof. Y. Katto, Dept. of Marine Engineering, Univ. of Tokyo.

The purpose of this study is to raise the efficiency for the production of available energy as high as possible by elevating the technological limits involved in the conversion of high temperature energy into the mechanical energy.
The research of this group can be divided into four items. (A) Conversion technology of high temperature energy. — The determination of optimum condition for the full coverage film cooling, and the experimental study for promoting heat transfer rate

3

in heat exchangers by means of thermal radiation. (B) Heat transfer technology with very high heat flux. — Studies on boiling and condensation heat transfer of potassium, and the relation between the flow patterns of two-phase flow and the critical heat flux. (C) Technology for controlling critical heat flux. — Experimental and analytical studies on critical heat flux in steady and unsteady heating systems, and experimental study on the prevention of the onset of critical heat flux phenomenon by using the extended surface. (D) Effective utilization system of high temperature energy. — Storage and transport of thermal energy, combined cycle of liquid metal vapor and steam, and the dynamic characteristics of a regenerator of a power plant using carbon-dioxide as a medium. Most of the results of this group contribute not only to the urgent problems of energy today and in the near future, but also to the peripheral technologies necessary in the future for realizing the energy production by means of nuclear fusion.

(3) Advanced Combustion Technique for Effective Use of Energy.
    Prof. H. Tsuji, Inst. of Space and Aeronautical Science, Univ. of Tokyo.

The objective of the present research project is to obtain basic data necessary to establish some advanced combustion techniques for the effective use of energy, namely, (A) the combustion technique of hydrogen which is known as a clean, future fuel, (B) the combustion technique of low calorific fuels and coal, and (C) the combustion technique of lean nonflammable, or barely flammable mixtures. The experiment of hydrogen combustion is being carried out by using a counterflow diffusion flame, a fuel-jet flame, and a large scale diffusion flame to make clear the ignition and combustion mechanisms, flammability limits, explosion limits, flame structure and stability of hydrogen and hydrogen-hydrocarbon mixtures.
The combustion experiment of low calorific fuels, such as methanol, lignite and turf, is being carried out, and the combusion characteristics of these low calorific fuels are being examined in detail. The experiment of fluidized bed combustion of coal is also being conducted and combustion characteristics of coal in this new technique are being investigated.
The combustion experiment of lean nonflammable, or barely flammable mixtures is being made by using a catalyzer and in an arc-discharge combustor. The mechanism of surface oxidation of lean mixtures on the catalyzer and the reaction-promoting mechanism by the addition of active species are being investigated. The flame-stabilization mechanism and turbulent combustion of lean mixtures are also studied.

(4) Advanced Utilization of Carbonaceous Fossil Resources.
    Prof. Y. Takegami, Dept. of Hydrocarbon Chemistry, Kyoto Univ.

Until nuclear fusion is established as commercial technology, the carbonaceous fossil resources such as coal, oil sand, and oil shale are considered to be the substitute of petroleum.
The principal objective of this research group is to develop new technology to change these fossil energy resources into a clean energy source by liquefaction and gasification of coal, and transformation of heavy oil.

(5) Development of Efficient Production of Energy Carrier.
    Prof. M. Takahashi, Research Inst. of Energy Materials, Yokohama National Univ.

Hydrogen can be considered to be a good energy carrier. The objective of this research group is to develop the new chemical processes to obtain hydrogen from water. To decompose water directly, the heat of high temperature of 4,000 $^{\circ}$C is required, but it is not practical to use such a high quality heat. Therefore, the research is being conducted to develop the multistage thermochemical water decomposing processes of high efficiency, and to extend the technical limits for water electrolysis.

(6) Development of New Batteries Suitable for an Effective Energy Conversion.
    Prof. S. Yoshizawa, Dept. of Industrial Chemistry, Kyoto Univ.

The objective of this research group is to develop the fuel cell, photogalvanic cell and other new effective batteries. The fuel cell type reactor may be an interesting candidate for energy saving in electrochemical industry, because the production of chemicals with a simultaneous recovery of electric power is possible. The photogalvanic cell is investigated for solar energy utilization and an effective battery is highly required to store electric energy in an off-peak period.

(7) Development of New Materials for Conversion and Storage of Energy.
    Prof. H. Tamura, Dept. of Applied Chemistry, Osaka Univ.

This research group intends to develop materials used in a hydrogen energy system, and effective membrane materials for energy saving in various separation processes and also for getting energy from the concentration difference.
Some of the investigators of this group are studying the fundamental characteristics of new materials and heat transfer media.

3. Photoconversion and Photosynthesis.

The goal of this research is an efficient utilization of solar radiation. The research is being conducted by four research groups working on the development of not only materials but also systematic method necessary for converting solar energy to storable and manageable types of energy like chemical and electrical energies.

(1) Photochemical Production of High Energy Compounds.
    Prof. I. Tanaka, Dept. of Chemistry, Tokyo Inst. of Technology.

Photochemical reactions like a valence isomerization between dienes and cage compounds are the suitable systems to store the solar radiation as chemical energy. The practical photochemical reactions must be developed to build up the efficient storage system of solar energy. Photon energy of solar light is not enough to excite most molecular systems and therefore an ultraviolet light source should be developed by the upconversion of solar light to the ultraviolet region.

(2) Chemical Conversion of Solar Energy by Means of Simulating Photosynthesis.
 Prof. T. Matsuo, Dept. of Organic Synthesis, Kyushu Univ.

In order to establish technical procedures for energy conversion from light into chemical potentials, attempts are being made to construct man-made chemical systems which mimic various processes involved in the photosynthesis. The following studies are being carried out for that purpose: (A) charge separation of the photoproduced redox species, (B) many-electron redox reactions in combination with photoinduced electron transport, (C) synthesis and alignment of dyes for light-harvesting systems, and (D) photochemical preparation of high-energy compounds suited for storage and transportation. The results, thus obtained, will be organized to design the best molecular system for solar energy conversion.

(3) Conversion and Storage of Solar Energy by the Photochemical Cell Systems.
 Prof. K. Honda, Dept. of Synthetic Chemistry, Univ. of Tokyo.

For the aim of conversion and storage of solar energy by means of chemical processes, the photochemical cells for light energy conversion are being studied in this project. The photochemical cell is based on the electron transfer process arising from the interaction between the photoexcited species and the substrate. Therefore, the simultaneous conversion of solar energy into both electric and chemical energies will be possible.

(4) Solid-State Solar Energy Conversion.
 Prof. M. Aoki, Dept. of Electronic Engineering, Univ. of Tokyo.

Solar energy is believed to be one of the promising substitutional energies. Among the various energy conversion systems, the solid-state direct conversion system is considered to be very hopeful. The researchers of this research group investigate suitable materials, preparation process, device structures and the method of characterization of the photovoltaic cell in order to clarify the utility of the solid-state conversion system and to improve its cost performance. This group also aims to determine the suitable method to construct the energy conversion system containing the battery system.

4. Natural Energy.

Emphasis of the research on natural energy is put on the fundamental studies for utilization of solar, geothermal, wind, hydro-, ocean and other types of natural energy from the new standpoint, and on the system studies for use of the combination of more than two types of the natural energy. The research project is to be carried out by the three research groups.

(1) Utilization of Solar Heat.
 Prof. K. Kimura, Dept. of Architecture, Waseda Univ.

Research on the utilization of solar energy covers the areas of solar energy utilization mainly in the form of low temperature heat. The objective of this research group is to develop the technique for effective use of solar energy as heat and analytical and experimental works are being made to obtain fundamental data for various systems of solar houses.

(2) Exploration and Development of Geothermal Energy.
 Prof. T. Yamasaki, Research Inst. of Industrial Science, Kyushu Univ.

The purpose of this study is to make clear the distinctive characteristics of geothermal reservoirs by conducting the following studies. (A) The geological measurement of fission track ages of igneous rocks from geothermal areas. (B) The geophysical study on the development of Natural electromagnetic induction method by measuring minute magnetic field and induced current of ELF domain. (C) The improvement of geochemical prospecting methods using mercury and other trace elements. (D) The research on thermal and hydrological structure of reservoirs by means of seismological observations. (E) Study on the polymerization of silica and mechanism of scale formation in geothermal hot waters.

(3) Utilization of Energy of Fluids in Nature.
 Prof. T. Ueda, Dept. of Mechanical Engineering, Univ. of Tokyo.

It is an important and pressing problem to develop systems converting the kinetic energy and thermal energy of fluids in Nature to electric power effectively. Two systems, one is low-head hydro-electric systems and the other ocean thermal energy conversion (OTEC), are studied in this project. These energy sources available are so diluted that it is necessary to solve many related problems in new technical development. The purpose of this project is to conduct fundamental studies in fluid dynamics and heat engineering fields pertinent to the technical subjects, and acquire basic information and data for realizing the effective power systems.

5. Bioenergy.

Green plants are considered to be conversion products of solar energy and are characterized as renewable energy sources. This research is concerned with the utilization of bioenergy, such as oily crops, various agro-waste, forest industry waste, and various useful chemicals produced from chemical, microbiological or enzymatic processing.

(1) High-Energy Plants.
 Prof. Y. Murata, Dept. of Agrobiology, Univ. of Tokyo.

Recently, several species of plants called "petroleum plants" have been discovered which produce hydrocarbon-like materials as product of photosynthesis. The extraction, identification and comparison of the constituents among species are now being made. Traditional oil crops can be used for the same purpose. Starch and sugar crops may also be used after changing their products into alcohol. In all these cases the focal problem is the final yield and overall efficiency. Thus, it is aimed in this study to obtain basic information necessary for the utilization of these high-energy plants.

(2) Conversion of Forest Resources into Energy.
Prof. T. Koshijima, Wood Research Inst.,
Kyoto Univ.

This research aims to obtain fuel material and starting material for chemical industry such as liquefied wood, methane, phenols, sugars, and organic acids by liquefaction, pyrolysis, and biological degradation of wood waste from both factory and forest.

(3) Production of Fuels, Solvents, and Other Useful Chemicals from Various Agrowastes through Microbial Activities.
Prof. T. Yamamoto, Dept. of Biology, Osaka City Univ.

This research team performs the fundamental study to find the methods practically effective for obtaining ethanol, butanol, acetone, organic acids, methane, hydrogen gas, etc., using various agrowastes as the raw materials, and also using microorganisms and enzymes as the catalysts of chemical reactions.

6. Thorium Fuel for Fission Reactor.

The present research topic is thorium fuel for fission reactor. Thorium is an important candidate for replacement of uranium as a nuclear reactor fuel, but only a few researches and developments have ever been made on thorium fuel compared with uranium fuel.

(1) Formation of Nuclides in the Neutron Irradiated Thorium by Mass Spectrometry.
Prof. M. Okamoto, Research Inst. of Nuclear Engineering, Tokyo Inst. of Technology.

As a basic and important technique, the mass spectrometry for the analysis of the nuclides formed in a neutron irradiated thorium should be accomplished for the use of thorium as a fertile material in nuclear reactors. There are many nuclides related to the thorium fuel cycle and most of them are the isotopes of thorium, protactinium and uranium. Therefore, it is necessary to accomplish a quantitative analysis method for the determination of isotopic composition of the nuclides formed in thorium by neutron irradiation. One of the techniques, which may be available, is the mass spectrometry using a mass spectrometer of high resolution power and high sensibility. In this project, a new computerized mass analysis system is designed and constructed, and the mass spectrometric analysis of the nuclides is being investigated.

(2) Biological Effect of Thorium to the Human Being.
Prof. S. Hatakeyama, School of Medicine, Tokyo Medical and Dental Univ.

One of the more serious potential problems involved in the use of thorium as a nuclear fuel is the long-term effects of the thorium daughter nuclides depositing in the human body. In Japan, the thorium dioxide-based angiographical contrast medium "Thorotrast" was used from 1928 to 1954 on tens of thousands of patients. Among this patient population traced to date, over 1,000 cases have been found to have developed carcinomas and other malignancies as late effects of Thorotrast administration 30 to 50 years ago. It is the objective of the present research program to obtain dose-response relationships of thorium series nuclides in Japanese by amplifying on the Thorotrast-related studies thus far conducted.

7. Efficient Utilization of Electric Energy and Making Electricity High Density.

(1) Design of Superconducting Generator and Its Characteristics in Power Systems.
Prof. T. Okada, Dept. of Electrical Engineering, Kyoto Univ.

Since the superconducting synchronous generator has many advantages, i.e. high efficiency, light weight and so on, it is expected to be put into practice. For the practical applications, the following subjects must be researched. (A) The developments of highly reliable machines. (B) The understanding of its electrical characteristics. Several tests under many kinds of load-conditions must be carried out. Therefore, in this research, a superconducting synchronous generator of a small size (20KVA) is designed and produced, and the following problems on the above subjects are investigated analytically and experimentally. (A) Design of highly reliable generators. (B) The characteristics in power systems.

(2) Power Transmission by Cryogenic Cable.
Prof. T. Kouno, Dept. of Electrical Engineering, Univ. of Tokyo.

The cryogenic cable is one of the new promising ways of electric power transmission in the future. However, there are many problems that must be solved to realize it. In this project, a desirable insulation system of a cryogenic cable is being mainly investigated.

(3) Development of Thin Film Electrodes for Photo-Electrochemical Cells and Fuel Cells.
Prof. T. Takagi, Dept. of Electronics, Kyoto Univ.

The objective is to develop thin film electrodes available for photo-electrochemical cells and fuel cells by using the ionized cluster beam (ICB) deposition technique and ion-based technology. This work will be conducted through an efficient use of technical installations at the Ion Beam Engineering Experimental Laboratory founded in 1979.

(4) Structures of Electrodes in a MHD Power Generation.
Prof. S. Shioda, Dept. of Energy Science, Tokyo Inst. of Technology.

MHD power generation has a potential to increase the overall thermal efficiency of the power plant over 50%. In recent years, considerable efforts have been made on the development of MHD power generator in many countries, particularly, in USSR and USA. As a result, progresses have been made in the study of electrical power output and enthalpy extraction. But, in order to commercialize the MHD power plant we must solve several problems.

One of the most important problems is to reduce the damage of electrodes in a MHD channel and develop an electrode that has a long life time. Thus, objectives in the present study are as follows. (A) Experimental study of electrode phenomena in both Faraday and Hall generators, (B) Research for the electrode configuration that yields a long life time for an electrode, (C) Experimental study of the discharge phenomena near electrodes with relatively large MHD channels.

(5) Load Factor Improvement of Power System and Its Evaluation.
   Prof. Y. Sekine, Dept. of Electrical Engineering, Univ. of Tokyo.

A promising means for the efficient utilization of electric energy supply installations and the effective use of electric energy is to improve the load factor of electric consumption.
This project aims at making clear the practical means for leveling electric energy consumption and their effects on the planning and operation of electric energy system.

(6) Energy Storage by Magnetic Coil.
   Prof. M. Nishimura, Dept. of Electrical Engineering, Osaka Univ.

The Superconductive Energy Storage (SCES) in electrical power systems is being studied. Utility of SCES will be discussed and estimated and a model plant of SCES is to be developed for conducting experimental research. The practical system will be designed based on those studies.

8. Effective Utilization of Energy in Agriculture.

This research topic is an effective utilization of energy in agriculture and is being conducted by four research groups. The objective of this research project is to promote the activities of agriculture, forestry and fishery and to secure a food supply through the effective use of energy in these industries in which a vast amount of petroleum is now directly or indirectly consumed.

(1) Energy Saving in Biological Production Systems.
   Prof. O. Kitani, Dept. of Agricultural Engineering, Univ. of Tokyo.

New technology is now required to save the auxiliary energy through production inputs such as fertilizer, chemicals, machines and facilities. The aim of this project is to develop some basis for this energy saving technology in agriculture, forestry and fishery.

(2) Efficient Energy Utilization in Food-Processing, Preservation and Distribution.
   Prof. T. Yano, Dept. of Agricultural Chemistry, Univ. of Tokyo.

The present methods of food processing, preservation and distribution have been developed based on the implicit assumption of sufficient energy supply. The aim of this project is to reexamine those processes from the viewpoint of efficient energy utilization.

(3) R & D and Utilization of Energy Source Materials and Unutilized Materials in Agriculture.
   Prof. H. Ezaki, Inst. of Agricultural Engineering, Univ. of Tsukuba.

We may find out many kinds of materials which will be useful in our life, but have not yet been used. These agricultural resources may be important for organic matters, as well as for energy resources.

(4) Alternative by Natural Energy in Agriculture.
   Prof. N. Kawamura, Dept. of Agricultural Engineering, Kyoto Univ.

The objective of this research is to investigate the alternative usage of natural energies, such as wind, water, solar and geothermal energies in agriculture. Natural energies have already been applied to cooling and heating in green house and drying grains and so forth.

The budget for the research projects described above is 800 million Yen for this year. For the participants' information, the main parts of the national budget for researches on energy problems in 1980 is listed below.

A. The Ministry of Education, Science and Culture.

   a. Grants in Aid for Special Research Projects of Energy Problems including the Research Projects described above.

      ¥ 1,400 million ($ 6,100 thousand)
      (¥ 600 million is for basic research on nuclear fusion.)

   b. Research Institutes of Nuclear Fusion in Universities.

      ¥ 6,200 million ($ 27,000 thousand)

   c. Miscellaneous Energy Research Institutes in Universities.

      ¥ 4,100 million ($ 17,800 thousand)

   d. Japan-US Cooperation in Energy Problems.

      ¥ 1,500 million ($ 6,500 thousand)

B. The Ministry of International Trade and Industry.

   a. Sunshine Project.

      ¥ 7,122 million ($ 31,000 thousand)

   (1) Solar energy.

      ¥ 1,769 million ($ 7,700 thousand)

   (2) Geothermal energy.

      ¥ 2,599 million ($ 11,300 thousand)

   (3) Gasification and liquefaction of coal.

      ¥ 860 million ($ 3,740 thousand)

   (4) Hydrogen energy

      ¥ 951 million ($ 4,130 thousand)

   (5) Others including the utilization of natural energy.

      ¥ 943 million ($ 4,100 thousand)

   b. Moon-light Project.

      ¥ 3,121 million ($ 13,600 thousand)

(1) MHD power generation.

 ¥ 1,322 million ($ 5,750 thousand)

(2) Utilization of waste heat.

 ¥ 173 million ($ 750 thousand)

(3) High-efficiency gas turbine.

 ¥ 777 million ($ 3,400 thousand)

(4) Energy storage by electric cell.

 ¥ 51 million ($ 220 thousand)

(5) Basic research on energy savings including the superconduction, new-type battery, heat transfer problems, and so on.

 ¥ 202 million ($ 880 thousand)

(6) Energy savings on air-conditioner, refrigerator and other miscellaneous projects.

 ¥ 596 million ($ 2,600 thousand)

C. The Ministry of Agriculture and Fishery.

a. Green Energy Project.
(Effective use of natural energy in agriculture)

 ¥ 964 million ($ 4,200 thousand)

D. The Science and Technology Agency.

a. Japan Atomic Energy Research Institute.

 ¥ 65,507 million ($ 285,000 thousand)

(1) Nuclear fusion.

 ¥ 51,719 million ($ 225,000 thousand)

(2) Research on safety.

 ¥ 9,867 million ($ 42,900 thousand)

(3) Multiobjective high temperature gas furnace.

 ¥ 3,921 million ($ 17,000 thousand)

b. Power Reactor and Nuclear Fuel Development Corporation.

 ¥ 81,465 million ($ 354,000 thousand)

(1) Power reactor.

 ¥ 44,640 million ($ 194,000 thousand)

(2) Nuclear fuel.

 ¥ 32,880 million ($ 143,000 thousand)

(3) Treatment of used fuel.

 ¥ 3,945 million ($ 17,200 thousand)

c. Research on Atomic Energy in the Institute of Physical and Chemical Research.

 ¥ 911 million ($ 3,960 thousand)

d. Japan-US Cooperation in Energy Problems.

 ¥ 3,880 million ($ 16,900 thousand)

Heat Transfer Research in the United States

WIN AUNG
Heat Transfer Program, National Science Foundation, Washington, District of Columbia, U.S.A.

1. INTRODUCTION

Heat transfer research is concerned with a broad range of thermal phenomena and processes in energy transport and conversion. A great number of studies in the United States deal with the development of new energy sources and of schemes to effect the efficient utilization of existing resources. There are also efforts directed at understanding the natural processes that occur in the atmosphere, in natural bodies of water, and underground, in order to facilitate the safe and efficient operation of engineering systems in these environments.

Traditionally, the chemical process industry places a continuing need for new heat transfer knowledge needed to deal with new materials and processes. Much of the heat transfer research conducted in this country's academic and industrial laboratories continues to be relevant to the process industry, although there is an increasing tendency to extend heat transfer research into other industries, such as the manufacturing industry. Here research must provide the basis for the design of heat transfer equipment with adequate thermal control and with the minimum material and energy requirements. This necessitates a good understanding of the behavior of materials under heat transfer conditions, and of the phenomena occurring in heat exchangers and other thermal systems, in the design of which there is now an obvious trend in moving away from the past practice of large design tolerances. New design technologies are needed not only for the efficient, safe and reliable operation of industrial plants, but also for reducing operating costs in existing ones. For example, the recent rapid rise in fuel costs has made it economically feasible to consider improving the furnaces in ethylene plants where fuel costs have climbed to 70% or more of the total operating costs of the plants.

In the remaining parts of this paper, six ongoing fundamental research projects in heat transfer are briefly outlined. The diversity of the ongoing activities and the limitation imposed on the present paper renders it extremely difficult to select a final list. The aim here is to indicate the rationale for and character of current heat transfer activities, and the six projects described serve to accomplish this goal quite well. Finally, future research needs and areas of technology that can benefit from additional heat transfer studies are also pointed out.

2. CURRENT RESEARCH ACTIVITIES

2.1 Liquid Metal Heat Transfer

Liquid metals such as mercury, sodium and potassium are receiving increasing attention because of their importance as heat transfer media in energy-related applications. For example, in fast breeder and fusion nuclear reactors currently under various stages of development, liquid metal cooling is a crucial factor. It is also of importance to liquid metal Magnetohydrodynamic (MHD) power generators which recently have shown great promise as extremely efficient converters of thermal power to electricity without moving parts.

An important study aimed at furthering our understanding of the thermal behavior of liquid metals is being conducted by Professor Paul S. Lykoudis of Purdue University. The study deals with liquid metal heat transfer with and without magnetic field effects. In the non-magnetic aspect of the study, the emphasis is on describing turbulent velocity and temperature profiles in single phase liquids. This work has yielded a fairly complete picture of the temperature field in turbulent pipe flow and of the free convection influence on the velocity and temperature fields. Spectral data on velocity and temperature fluctuation have also been acquired and they have provided valuable information on the turbulence structure in liquid metals. The work is valuable for the correct design of the core of Liquid Metal Fast Breeder Reactors (LMFBR) among other energy related applications.

The work relating to the effect of a magnetic field has been focused on both nucleate boiling and single-phase heat transfer situations. In nucleate boiling, the behavior and growth of a liquid metal bubble in the presence of a spherical magnetic field has been analyzed. An analysis has shown that a magnetic field acts to inhibit boiling heat transfer, and the effect is stronger on alkali metals, such as sodium and potassium, than on mercury. The result is that the overall energy transport is reduced. This can have serious ramification in system operation, and must be taken into account in considering the cooling requirements. For this purpose, the quantitative relations generated by the Purdue group, giving the boiling rate as a function of various operating parameters, could be used as a guide. The work should be of importance to the design of lithium blankets in proposed fusion reactors that operate with magnetic confinement.

So far the study has been confined to a situation that is akin to the case of a horizontal magnetic field. In the case of a vertical magnetic field, it is speculated but has yet to be analyzed and confirmed, that even though the growth of the bubble will be restricted in the horizontal direction, the eventual rising of the bubbles due to the buoyant forces will be performed along the smooth and orderly vertical trajectories imposed by the magnetic field, thus enhancing heat transfer. On the other hand, when the magnetic field is horizontal, buoyancy will be restrained, and more bubbles will remain longer in their nucleation sites, thus leading to early transition from nucleate to film boiling, a less effective mode of heat transport. This is an area that requires further investigation.

The suppression of heat transfer by a magnetic field is also found to occur in the single phase turbulent flow situation. In this case the field acts to decrease velocity fluctuation and the corresponding momentum transport due to turbulence. In this study, a method for predicting the heat transfer in turbulent pipe flow with a magnetic field aligned with the flow, has been developed. This work is of general importance to fusion reactor technology and also to MHD power generators. The present study which is being continued, has provided useful and significant understanding of the phenomena involved in liquid metal heat transfer and how it is affected by a magnetic field. The quantitative predictive relations developed will be useful for guiding the design of cooling system utilizing liquid metals as heat transfer media.

2.2 Thermal Convection

Thermal convection, the process by which fluid motion is induced as a result of density difference arising from temperature variations, is important in many engineering and scientific problems. It is the basis for a large number of atmospheric, astrophysical and geophysical phenomena, including plate tectonics and the circulation on planetary atmospheres. This process has also assumed increased significance because of developments in fire detection, solar collectors, thermal energy storage, dry cooling towers, underground transmission of electricity, cooling of nuclear reactors, thermal insulation for buildings, and the manufacturing of semiconductor materials for the photovoltaic conversion of solar energy into usable elctricity and for data processing machines.

In spite of the great diversity of application, thermal convection remains not well understood. The lack of understanding stems mainly from a need for research instrumentation capable of measuring heat transfer and fluid flow rates with extreme accuracy. Recent advances in instrumentation have enabled Professor Richard J. Goldstein and his associates to undertake a significant study on thermal convection. This group of researchers at the University of Minnesota are utilizing the most advanced experimental techniques that include laser-doppler velocimetry, laser-interferometry, and electrochemical mass transfer. Their attention is directed at thermal convection in horizontal and inclined fluid layers and in cylindrical annuli. The immediate objective is to establish the condition leading to the onset of convection and to attain convection regimes beyond those achieved in the past. An attendant difficulty lies in accurately maintaining the thermal conditions of the test cells. This difficulty which has plagued thermal convection studies in the past has now been alleviated by a unique application of the analogy of heat and mass transfer. By this analogy, heat transfer measurements have been replaced by the equivalent mass transfer experiments using the electrochemical method. Results of great accuracy are now obtained for thermal convection so that in certain situations, it is now possible to determine the entire range of the heat transfer regimes such as conduction, steady laminar convection, unsteady laminar

convection and turbulent convection. In these situations, the conditions for the onset of convection and the criteria for laminar to turbulent transition have been determined.

The study by Professor Goldstein's group is continuing. Future efforts will focus on the structure of turbulnet transport in thermal convection, transient convection from discrete sources, and transport phenomena immediately following the onset of thermal convection. In addition to increasing our general descriptive capability in thermal convection, this program of research will also provide data that can be directly applied in the design of many engineering systems.

2.3 Radiative Properties and Transport in Gases

The research has just been concluded by Professor Ralph Greif at the University of California at Berkeley, and is concerned with a study of the radiative properties of gases and the use of these properties in determining the flow and energy transfer to absorbing gases and surfaces. The research compromises both an experimental and a theoretical determination has been made of both the spectral absorption and the total absorption properties of radiating gases. Results are based on the model of a rigid rotator for a diatomic gas and have been compared with the experimental data for carbon monoxide and nitric oxide. Good agreement has been obtained. These results are especially important because they are of a basic nature and may be used where experimental information is incomplete. Simple and accurate limiting expressions have also been obtained which are particularly useful in radiation transport calculations and analyses.

Experiments have been carried out in a shock tube and in an electrically heated tube. The shock tube measurements were made with a McPherson scanning polychromator to obtain spectral data for a radiating gas that is essentially stationary in the region near the end wall off the shock tube. Measurements have also been made of the temperature distribution in a gas that is flowing in a tube. Under laminar flow conditions the effects of buoyancy and thermal radiation are of importance and have been clearly demonstrated. Absorption measurements of carbon dioxide in the 2.7 micron region have been made at elevated temperatures, 2400K and 3500K. These data represent new results which were previously not available.

A theoretical determination of the absorption properties of diatomic gases has also been completed. The study uses a theoretical expression for the intensity of a radiating gas based on a rigid rotating molecule in conjunction with the Lorentz expression for the spectral line shape. The resulting series has been summed and both numerical and analytical results have been obtained for both the absorption coefficient and the total band absorptance. The results are given in terms of the basic spectroscopic variables, e.g., line spacing, line width, etc. and do not employ any arbitrary constants or coefficients. A comparision has been made with experimental results for carbon monoxide and nitric oxide and good agreement has been obtained over the range of data available. However, at very large path lengths there is a need to consider the effects of nonrigid rotation. The research has provided new and basic data for gas properties. The results, including the data and the analyses, may be applied to a variety of systems to determine the energy and mass transport and fluid flow. Such diverse systems as combustion chambers, furnaces, gas turbines, heat exchangers, flue ducts, chimneys and power plants would be appropriate illustrative systems. Furthermore, problems associated with the production and dispersion of such pollutants as carbon monoxide, sulfur dioxide, nitric oxide, etc. are relevant to the program.

2.4 Improvement of Film Boiling Heat Transfer

One of the methods frequently used to effect high heat transfer rates from solid surfaces is through the application of a boiling liquid. However, this mode of heat transfer has an inherent shortcoming in that at high heat rates, the process operates in a regime called Leidenfrost film boiling. In this regime, the bubble which absorbs heat and grows in the vicinity of the solid surface does so on a vapor film which adheres to the surface.

This reduces the effectiveness of heat transfer and leads to either low heat transfer rates or an undersirable high surface temperature necessary to effect a prescribed heat transfer. Inspite of its limitations, film boiling remains the most frequently encountered boiling process in industrial process heat exchangers such as steam generators, etc.

A novel approach to a possible solution to this limitation of boiling heat transfer effectiveness has been applied by Dr. Edward G. Keshock of the University of Tennessee. The principal objective of this experimental study is to enhance the normally low heat transfer effectiveness by means of macroscopic surface projections such as regularly spaced pins or machined protrusions. The basic idea here is that these protrusions would penetrate the vapor film to make contact with the bulk liquid thereby improving the heat transfer performance of the surface. Dr. Keshock is currently testing out this idea with a variety of liquids and boiling surfaces, including surfaces of high purity copper, aluminum, stainless steel and mild steel. The last material is chosen to permit the use of magnetic pins as surface protrusions.

Initial efforts in this relatively new project have been directed at obtaining the temperature at which film boiling is initiated. This has resulted in the surprising discovery that there is a large disparity between the measured film boiling temperature for water on a stainless steel surface (temperature difference between the surface and the bulk liquid of 185 C) and on an aluminum surface (100 C). In addition, an "aging" effect has been discovered on the aluminum surface such that the Leidenfrost temperature difference is 118 C for a newly prepared surface while 11 days later it becomes 100 C (all surfaces are cleaned prior to testing). The mechanisms causing these phenomena are being investigated in Dr. Keshock's laboratories.

2.5 Two Phase Thermal Convection in a Porous Media

The aim of this study is to provide basic understanding of a number of important heat transfer phenomena in earth and ocean sciences. The study consists of an experimental and a theoretical part.

The experimental measurements have been carried out in a "sand box" that is heated from below. The sand is covered with distilled water which is cooled from above to a prescribed temperature using cooling coils. The temperature distribution in the sand layer has been measured using arrays of thermocouples. The temperature measurements show a pattern of near steady-state thermal convection with a two-phase mixture of steam and water adjacent to the lower boundary of the two-phase region is elevated in regions of ascending convection and is depressed in regions of descending convection. These experiments are used to establish the meaningfulness of the theoretical calculations. The theoretical study is conducted through numerical calculations. Initial calculations were carried out for a saturated one-phase porous layer with uniform permeability and a semi-infinite half space with permeability decreasing exponentially as a function of distance from the surface. Both permeable and impermeable upper boundaries have been considered.

The results of these studies have been applied to thermal convection in the oceanic crust. Calculated surface heat flows have been used to explain the periodicities in measured sea-floor heat flow adjacent to ocean ridges. Hydrothermal circulation of sea water may be responsible for concentrating minerals leading to ore deposits. The present study has provided quantitative information on the volume of sea water circulated and its residence time, as well as distributions of temperature and velocity. This will lead to important information on such questions as whether minerals such as copper are extracted from the sea water or are leached from one part of the crust and are then deposited in another part as the temperature of the sea water changed. The results can be used to explain for example the porphyry copper deposits in Cyprus. Cyprus is exposed oceanic curst. The economic copper deposits in Cyprus are regularly spaced and may be related to the spatial periodicities of the convection cells predicted by the present theory. Work is continuing on this study.

2.6 Heat Transfer in Phase Change Materials

Phase change heat transfer such as melting and solidification is of practical importance in a wide range of technologies and geophysics. Problems of this type are of interest in purification of materials, processing of semiconductors, metallurgical processing, crystal growth from melts and solutions, nuclear reactor safety, processing of foodstuffs, atmospheric entry (ablation), latent heat-of-fusion thermal energy storage, thermal control of spacecraft using phase change materials, melting of ice and freezing of waters, melting and freezing of soils, and many other areas of practical applications. Heat transfer in melting and solidification falls among a general class of phase change problems referred to as moving boundary problems. These types of problems have been receiving considerable interest because they are relevant to numerous fields of application.

A common feature of the classical solid-liquid phase change heat transfer problems, also referred to as Stefan or Neumann problems, is the existence of a solid-liquid interface which moves either into the solid (melting) or into the liquid (solidification) region, depending on the relative magnitude of the temperature gradients on either side of the interface. Because of the interface motion, the moving boundary problems are nonlinear; an analytical understanding adequate for design, for example, of thermal energy storage systems has not been available. To overcome this, a group of researchers headed by Professor Raymond Viskanta of the School of Mechanical Engineering at Purdue University have recently mounted a major research effort aimed at elucidating the role of natural convection in liquids during phase transformation.

Results have shown that during melting, natural convection develops in the liquid and influences both the rate of melting and the melting profile. Through the use of a horizontal cylindrical heat source embedded in a melting solid it has been found that melting primarily occurs above the cylinder with very little melting occurring below the cylinder. The timewise variation of the cylinder surface temperature is found to deviate markedly from that which would have been predicted by a pure conduction model and is more characteristic of transient natural convection.

The timewise variation of the average Nusselt number also reveals that conduction dominates only during the early stages of the melting process. These findings indicate that the practice of applying heat conduction models to describe the process can lead to erroneous results. In the case of solidification, the Purdue study shows that the process can be arrested by natural convection even in a slightly superheated liquid. Thus, whereas natural convection accelerates the phase change process in melting, in freezing the opposite effect prevails. During melting, the heat transfer at the heated surface is smaller than that in transient natural convection in the absence of phase change and this is attributed to the recirculation of the liquid in a small volume melt region.

The research at Purdue University is being continued, with increasing attention focused on the solidification process, and on melting and solidification on semi-transparent materials.

3. FUTURE NEEDS IN RESEARCH

It is clear from the previous Section that heat transfer research, while fundamental in nature, is also strongly motivated by applications. As the field matures and the techniques and methodologies advance, new research opportunities are opening up. The scope of heat transfer research is expanding. In the following, areas where additional heat transfer research can make a significant impact are indicated. It is not an exhaustive list and no priority ranking is intended.

3.1 Vapor Explosion

This area has been recently studied in connection with nuclear reactor safety, as well as marine transport of liquefied natural gas. Explosion due to inadvertent mixing of molten metal and water have occurred in foundries, aluminum plants and steel mills. Numerous explosions have been recorded in mixing "smelt" and water in kraft paper manufacture. Volcanic lava interactions with water also can lead to enormous releases of energy. A number of unresolved problems exist in these applications. One of the key questions concerns the existence of a steady detonation wave propagating through what amounts to a fuel-coolant mixture at pressures well above the coolant critical pressure. An issue that has not been addressed is the role of fuel drop interfacial instabilities in vapor explosion.

3.2 Radiative Transfer

In recent years, radiative transfer studies have been extended to include a number of important areas such as cryogenic engineering, combustion, porous media and spectroscopic identification of planetary cloud, haze and frost compositions. There is a need to further our understanding of transfer processes in gases which contain scattering and absorbing particulates or aerosols. A better characterization of radiative properties of solid surfaces is also needed. The interaction of natural or forced convection with radiative transfer also needs further elucidation. Application for these processes occurs not only in the relatively high temperature environment of combustion but also in thermal insulations used in the lower temperature surroundings in industrial plants and in homes and in thermal energy transport in soils.

3.3 Multiphase Heat Transfer

Under this heading is grouped some of the least understood topics in heat transfer. They range from film boiling and dispersed flow heat transfer to fluidized bed heat transfer. In recent years, important steps have been taken towards the mechanistic modeling of the transfer processes in dispersed flow, but the process is not yet completely understood. For example, the role of drop deposition in heat transfer is not yet sufficiently characterized. Boiling phenomena in vertical channels and in gravity-driven liquid films occur in a variety of practical situations, such as advanced cooling equipment in boilers and nuclear reactors. In spite of the important industrial relevance, these phenomena are not yet put on a technology, one of the the key needs is to develop a fundamental theory that is capable of predicting the heat transfer to tubes immersed in a fluidized bed.

3.4 High Speed, High Temperature Flows

Although this topic has not attracted as much attention as during the era of rapid development in aerodynamics, there is a renewed urgency to study it because of a need to improve the performance of power generators driven by gas turbines. The increased cost of fuel for the jet engines has made it necessary to reduce the fuel consumption by operating the turbines at higher temperatures and pressures to increase the efficiency. The thermal design of the turbines requires an accurate estimate of the local gas-side heat flux loads on the blades and vanes. At present, our quantitative knowledge concerning the local heat transfer rates at the stagnation region for the turbine blades and vanes, and the heat fluxes in the laminar, transition and turbulent boundary layers over the blades and vanes, is not adequate. There continues to be a need in this area for carefully planned and executed theoretical and experimental studies. This is an area where some of the recently emerged theoretical and experimental research techniques could be used to advantage.

3.5 Sprays

The spray-jet problem in heat transfer is not a new one. It appears in liquid spray cooling, spray drying of food and of biochemical particles, and in waste discharges. The recent energy crisis has given new significance to this problem as research in this area hinges on our ability to incorporate the proper turbulence models of momentum and heat transport for a spreading high temperature, often multi-component, fuel jet. Also needed are the proper transport coefficients and a knowledge of the role of entrainment at the jet boundaries. In the case of the liquid fuel, the process of jet break-up into liquid droplets and the subsequent heating and evaporation of the droplets is important. All these are in need of better characterization, along with convection within and around the droplets. In addition, the question as to under what conditions the droplets completely vaporize prior to entering the flame appears not be settled.

3.6 Porous Media

This area has been alluded to earlier in connection with the discussion on radiative transfer. Interest in this topic is stimulated by the need to understand and characterize transport phenomena in the recovery of geothermal energy, the subterranean storage of solar energy, the improvement of thermal oil recovery methods, the gasification of coal, and the processing of certain chemical and biochemical materials such as food and agricultural products. A recent workshop sponsored by the NSF indicated that in this subject more efforts have been given to analytical studies than to experimental . Experimental studies for the most part have been concerned with obtaining correlations for the effective thermal conductivity which is defined as the gross heat flux divided by the gross temperature gradient. Existing data are usually restricted to a particular geometry. A key future need is in developing general experimental techniques for obtaining this important property given the geometric parameters and the thermal parameters of the constituent materials. This will apply to porous materials either partially or fully saturated with a stagnant fluid.

Recently, classical analytical solutions based on boundary layer approxiamtions have been applied to convection in saturated porous media situated adjacent to an impermeable wall. Thus, as in the classical theory of convective heat transfer, boundary layer approximations in the case of porous media also lead to analytical solutions. A mathematical theory for convection in porous media now exists, but it is restricted to free or combined free and forced convection adjacent to impermeable, flat surfaces at high Rayleigh numbers. It is also limited to specific types of velocity and temperature variations in the case of combined convection in porous media.

4. SUMMARY REMARKS

In the above discussion, an attempt has been made to indicate the "engineering" basis of heat transfer research by, as much as possible, putting the discussion into the technological context. The research is fundamental in nature, in that in each case the ranges and limits of natural phenomena and processes are fully explored without regard to the specific need of any given application. The emphasis is thus in the general applicability of the research results.

The topics indicated in this paper have been selected in part out of an interest to stimulate discussions, but mainly out of convenience. It is certain that additional topics will emerge in the future that will warrant further discussion.

HIGH-TEMPERATURE HEAT TRANSFER: BASIC PHENOMENA

Heat Transfer
in High-Temperature Gas Turbines:
Film Cooling and Impingement Cooling

R. J. GOLDSTEIN

Mechanical Engineering Department, University of Minnesota,
Minneapolis, Minnesota 55455, U.S.A.

ABSTRACT

Film cooling and impingement cooling as used in modern gas turbine systems are described. Both processes lead to three-dimensional turbulent flows which are complex and difficult to analyze. Different experimental techniques, including the use of mass transfer analogies, have been used in studying such flows to permit more accurate measurements and to avoid the necessity of going to a high-temperature hostile environment in the laboratory. The present paper primarily describes research that has been performed in the Mechanical Engineering Department of the University of Minnesota.

NOMENCLATURE

D = injection hole diameter
h = heat transfer coefficient
h_0 = heat transfer coefficient in absence of injection
I = momentum flux ratio
K = acceleration parameter
M = blowing rate $(\rho_2 u_2 / \rho_\infty u_\infty)$
Nu = Nusselt number
Re_D = Reynolds number $(u_\infty D / \nu_\infty)$
Re_j = $u_2 D / \nu_2$
S = spacing between injection holes
T = temperature
T_{aw} = adiabatic wall temperature
T_r = recovery temperature of mainstream
T_w = wall temperature
T_2 = temperature of injected fluid
T_∞ = temperature of mainstream
u_2 = velocity of injected (coolant) fluid
u_∞ = velocity of mainstream
X = distance downstream from trailing edge of injection hole
Y = distance normal to surface
Z = position along span, measured from center of injection hole
α = angle between injected flow and mainstream
δ^* = momentum thickness of boundary layer
ρ_2 = density of injected fluid
ρ_∞ = density of mainstream fluid
η = film cooling effectiveness
η_{CL} = film cooling effectiveness at Z=0
$\overline{\eta}$ = film cooling effectiveness averaged over the span
θ = $(T-T_\infty)/(T_2-T_\infty)$

ν_2 = viscosity of injected fluid
ν_∞ = viscosity of mainstream fluid

INTRODUCTION

Gas turbine engine designs have progressed to higher and higher temperatures at the inlet to the turbine section. In some systems, this inlet temperature is higher than the melting point of the turbine blades. Thus, some means must be provided to protect the blades and maintain them at temperatures not only below their melting point but sufficiently low to ensure long operating life in the face of extreme temperature gradients and the potential of stress rupture and creep.

On stationary gas turbine systems, various coolants may be employed to maintain moderate temperatures on the exposed surfaces. One design under consideration uses water cooling. If the blade is only cooled internally, however, the temperature drop across the blade skin —even with ideal cooling— may be so high that the external temperature of the blade surface may be at an unacceptable value. With water-cooling systems, a very thin skin is used and relatively low temperature water serves as the coolant.

For an aircraft engine, the only coolant generally available is the air that flows through the engine. This usually is taken off at some stage of the compressor, often at the last stage, and thus is already at a somewhat elevated temperature. In addition, because of the relatively poor heat transfer characteristics of air, internal convection cooling may not be sufficient. Often some means must be taken to reduce the heat flow from the hot gas to the outside surface of the blade itself.

Figure 1 shows a high-performance turbine blade for a modern engine [1]. Rather than a simple solid blade, the inside is a complex geometry of flow passages to provide different types of cooling. Near the leading edge region of the blade the external heat transfer is very high: an impingement system provides an array of jets which strike the inside surface in this region. These jets provide a relatively high internal convection heat transfer coefficient. The spent air from the jets can then flow out through film cooling holes. The injected (or coolant) air provides considerable heat transfer during its flow

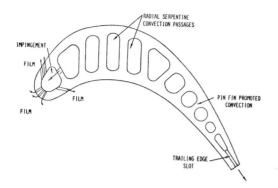

Figure 1   Typical High Pressure Turbine Blade Using a Combination of Film, Impingement, and Promoted Convection Cooling [1]

through the skin of the blade, particularly in the leading edge region. In terms of film cooling action upon injection to the external flow, the coolant should be kept close to the surface where it serves as an insulating layer between the surface and the hot gas flowing over it. In addition to the leading edge region, film cooling holes are often provided on both the pressure and suction sides of the blade. In the aft region of the blade, internal convection cooling includes the use of pin fins. These promote the heat transfer in the trailing edge region of the blade.

The air that is used for cooling is supplied by the compressor. If not used for cooling, it could flow through the compressor and turbine stages, providing more engine output. Design practice calls for using as little cooling air as possible to maximize the throughflow to the combustor and turbine. In addition, the air injected by film cooling or out the trailing edge of the blades should not decrease too greatly the aerodynamic efficiency of the turbine.

The work to be described primarily concerns film cooling research conducted at the Mechanical Engineering Department at the University of Minnesota. Developments to measure film cooling effectiveness and heat transfer following coolant injection are described. Research has also been done on impingement cooling in the presence of crossflow, either from a separate air stream or from an upstream set of injection holes.

FILM COOLING

If only internal air cooling is used, the temperature drop across the skin of a gas turbine blade may be too high. Thus, in many systems, some means of reducing the heat flow to the outside of the turbine blade must be provided. This can be done with film cooling, in which relatively cool air is injected close to the surface of the blade. This air flows downstream along the surface, acting as an insulating layer and separating the hot gas stream

from the turbine blade. Another way of looking at the film coolant is that it tends to decrease the temperature in the boundary layer and thus reduces the heat flow to the blade surface. The flow through injection holes also provides more internal cooling of the blade skin. This is particularly important in the leading edge holes which are often designed to have a fairly large length-to-diameter ratio.

Different injection systems have been used for film cooling [2]. In what has come to be called two-dimensional film cooling, a continuous slot (across the span of the system) is used to inject fluid. Various types of step-down slots, angled injection, and flow through porous regions along a wall have been used (Figure 2). These two-dimensional systems are usually quite effective, as the continuous flow of coolant across the span is, essentially, impressed down upon the wall by the mainstream flow, providing effective protection of the surface. However, such geometries are not feasible in many systems —such as in gas turbine blades. In these, a series of discrete injection holes is required for the proper structural characteristics; either a single row or multiple rows of holes can be used. With full surface film cooling there is a two-dimensional array of injection holes along the surface of the blade, a system that approaches transpiration cooling.

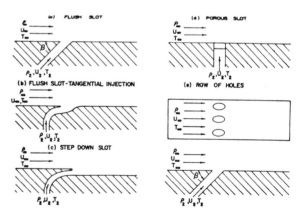

Figure 2   .Film Cooling Geometries

With discrete hole injection —for example, through a single row of angled injection holes— the film cooling performance is often far worse than with a continuous slot. The resulting three-dimensional flow (three-dimensional film cooling) can permit individual jets of injected fluid to penetrate through the boundary layer into the mainstream. The injected fluid, then, is not close to the surface which it is intended to protect. Mainstream flow can go around the jet and heat or overheat the surface which it is supposed to protect.

Figure 3 is a representation of the flow following injection at relatively high blowing rate, M —in this case, through a

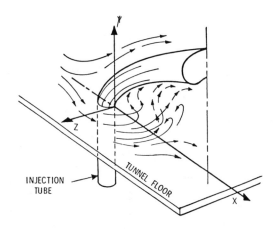

Figure 3   Sketch of Half-Jet Based on
Tuft Observations

single hole normal to the surface. The jet
takes on a kidney-like shape with mainstream flow going around and under the jet.

Figure 4 shows the temperature profiles
downstream of a single normal jet [3]. Note
that when the flowing rate is fairly small,
the jet appears to remain very close to the
surface. As the blowing rate increases, the
jet lifts away from the surface, with the
center of the jet considerably elevated and
rising continually as it goes downstream.
Figure 5 shows contours of constant temperature at various positions downstream of a
single injection hole [3]. Note the kidney-
shaped profile of the isothermal lines.
From these profiles, high blowing rate injection would not be expected to provide
good film cooling performance because the
coolant is away from the surface.

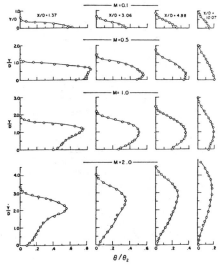

Figure 4   Temperature Profiles for 90 Deg
Injection Angle, Z/D=0

The interaction of a jet entering at some
angle to the mainstream can result in very
large eddies of the order of size of the
jet diameter or greater. This large-scale
interaction makes analytical prediction of
the flow and heat transfer along the wall
quite difficult. Figure 6 shows the flow
interaction of a jet entering normal to a

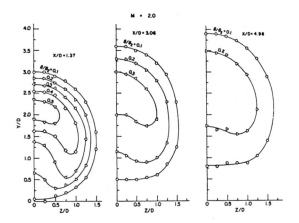

Figure 5   Constant-Temperature Contours
for 90 Deg Injection Angle,
M=2.0

fluid stream [4]. Greater penetration is
observed at the large blowing rates. The
short-time exposure photos show the large
scale of the eddy motion of the flow. It
is not apparent from the picture that
there is an actual "lift-off" of the jet
at high blowing ratio as is indicated in
Figs. 3-5. Lift-off does occur, although at times the eddies from the jet
do touch down upon the surface.

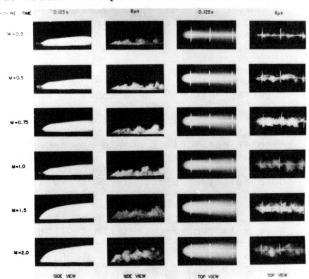

Figure 6   Photographs of Jet in Crossflow
with Normal Injection

Many of the figures shown previously represent flow from a single jet entering a
mainstream. In practice, of course, at
least a row of jets is used in film cooling. In addition, the injection is generally not normal but inclined as close to
the wall as possible. This tends to keep
the jet closer to the surface. Design
limitations, however, often preclude the
jets being at angles smaller than 30 or 35
degrees from the surface.

In determining the effectiveness of protection that film cooling affords, different
parameters can be used. A common and convenient one is a measure of the adiabatic

wall temperature, T_{aw}. This is the temperature of a wall downstream of film cooling, assuming the wall is truly adiabatic, without any heat passing through it. It is the limiting temperature that the wall would reach and permits the use of a simple and convenient dimensionless parameter, the "film cooling effectiveness",

$$\eta = \frac{T_{aw} - T_r}{T_2 - T_r} \qquad (1)$$

The adiabatic wall temperature provides a reference that can be used in defining a heat transfer coefficient,

$$q = h (T_w - T_{aw}) \qquad (2)$$

Far downstream the dimensionless film cooling effectiveness would generally approach zero while near injection, at least with two-dimensional film cooling, T_{aw} is close to the injection temperature and the effectiveness is close to unity.

The film cooling effectiveness is a function of many variables, including the geometry of the injection system, the properties of the mainstream and injection fluids (in particular, the density difference if the two are the same chemical composition), the position along the surface, and the blowing parameter, M, or the momentum flux ratio, I.

Though not the only possible definition for h, the one given by Equation 2 is convenient. When the wall temperature approaches the adiabatic wall temperature, h maintains its finite value. Far downstream the heat transfer coefficient should approach that due to the mainstream flow alone. This latter property has often been found to be the case, even relatively close to injection for small and moderate blowing rates.

Figure 7 shows the variation of the adiabatic wall temperature (film cooling effectiveness) with blowing rate for injection through tubes inclined at an angle of 30 degrees to the wall surface [5]. The effectiveness is determined along a line downstream of the center of an injection hole (Z=0). Note the maximum effectiveness occurs at a blowing parameter of approximately 0.5. There is relatively close agreement between the results for single hole injection and for a row (across the span) of holes, at least at low and moderate blowing rates and small distances downstream. As more mass is injected through the injection holes (M larger), one might expect more flow along the surface to act as an insulating layer or more mass to reduce the temperature in the boundary layer. However, at sufficiently high blowing rates the jet tends to leave the surface, resulting in poor effectiveness. With a row of holes at high blowing rates, the jets can merge together some distance downstream and be partially turned towards

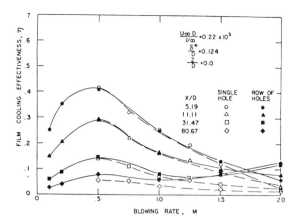

Figure 7   Effect of Blowing Rate on Centerline Film Cooling Effectiveness for Single Hole and Multiple Hole Injection at an Angle of 35 Deg with the Flow

the wall, resulting in an increase of effectiveness with M at large X.

Strictly speaking, the results shown in Fig. 7 are only valid for a density ratio, ρ_2/ρ_∞, close to unity. To study the influence of larger variations in the density of injected gas compared to that of the mainstream, an injection temperature very different from that of the main flow could be used. This is what occurs in many applications where the absolute temperature of an air-free stream may be 1-1/2 to 2-1/2 times that of injected coolant air, resulting in very large density differences. The use of a mass transfer analogy permits a study of the effect of density ratio on film cooling at moderate (room) temperature. With this system [6], instead of injecting a hotter or colder gas through the injection holes, air mixed with another gas —either a tracer, to have a density ratio of approximately unity, or a large concentration of the other gas, to have a density higher or lower than that of air— is used. Downstream of injection, a wall concentration, rather than a wall temperature, is measured. This wall concentration leads to an impermeable wall effectiveness which can be shown to be equivalent to the film cooling (thermal) effectiveness [6,7].

The effect of density on film cooling can be observed in Figure 8 [6]. For varying injection rate, the maximum of effectiveness appears to occur at a fixed ratio of the velocity of the injected fluid to that of the mainstream fluid. Note that the figure shown is specific to the injection geometry used, a single row of holes inclined at approximately 35 degrees to the mainstream.

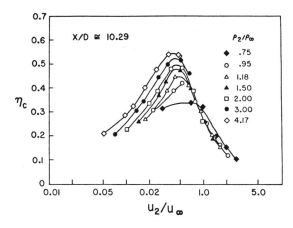

Figure 8 Effect of Injection Velocity On
the Center-Line Effectiveness
for Injection Through a Row of
Inclined Holes, Z/D=0

An important goal is reducing the effect of
blow-off on film cooling effectiveness.
Essentially this means preventing, or de-
laying, blow-off so that it does not occur
at the values of blowing rate or velocity
ratio used in many applications. At least
two designs to prevent or reduce blow-off
have been tried. One involves the use of
multiple rows of holes. With two staggered rows of holes, as shown in Figure 9,
the jets emanating from the upstream row
tend to fill the space between the jets
from the downstream row. This provides
more blockage across the span of the tunnel by the jets, affording less possibility
for the mainstream fluid to go around and underneath the jets. The jets are essentially
pressed down, maintaining the coolant flow
along the wall. The resulting improvement
in effectiveness is shown in Figure 10,
where the film cooling performance of a
single row of holes is compared to that of
two staggered rows [8].

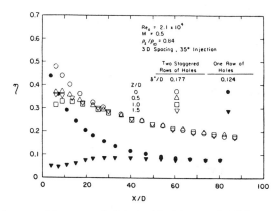

Figure 10 Local Distribution of Effectiveness for $M \cong 0.5$

Another means of having the injected jets
stay close to the wall and not penetrate
into the mainstream involves changing the
simple geometry of a round tube. The tubes
are widened near their exit and a fairly
straight lip is used as the downstream
edge of the hole. This is to simulate
the geometry that produces a two-dimensional
jet where, in the absence of a mainstream,
a Coanda-type effect might cause the injected flow to follow more closely the
surface downstream of injection [9]. Figure 11 shows flow following injection
through a round cylindrical hole and through
the special widened or shaped hole. Note
that both in the absence and presence of a
crossflow or mainstream flow, the jet leaving the shaped hole tends to stay closer to
the surface, which should give better film
cooling performance. Figure 12 shows that,
indeed, superior film cooling performance
occurs with the shaped hole, particularly
at the high blowing rates, where the jets
from the straight holes tend to lift off
the surface [9].

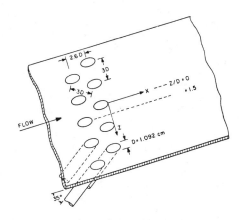

Figure 9 Film Cooling Injection with Two
Staggered Rows of Holes

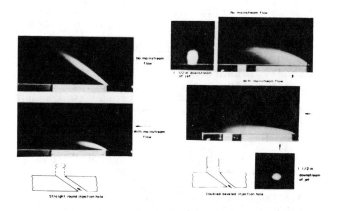

Figure 11 Flow Visualization Study of
Jets Leaving a Cylindrical or
Shaped Channel

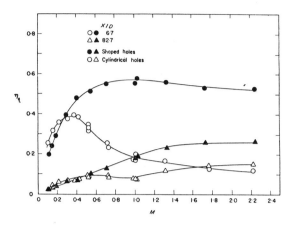

Figure 12   Centerline Effectiveness As a
Function of Blowing Rate For
Injection of Air Through Straight
Round Holes and Shaped Holes,
Row of Holes on 3 D Spacing

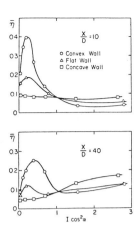

Figure 13   Average Film-Cooling Effectiveness As a Function of $I\cos^2\alpha$
for $\rho_2/\rho_\infty = 0.95$

The results for film cooling described
above refer to a flat surface at and downstream of injection. However, in many applications, including those on a turbine
blade, the mainstream flows over a curved
surface. There might be little difference
in the film cooling on such a surface following a two-dimensional slot. With flow
through discrete holes, however, film cooling is strongly affected by surface curvature.

Consider the flow of a jet along with a
mainstream around a curved surface. For
simplicity, assume the velocity of the jet
is in the same direction as the mainstream.
Above a convex surface, the pressure in the
mainstream flow increases with distance
from the wall. Thus, a jet of relatively
low momentum (compared to the mainstream)
might be pushed by this pressure gradient
towards the wall, tending to provide good
film cooling. If the jet had a high momentum flux relative to that of the free
stream, the jet might tend to leave the
surface on the convex wall, giving relatively poor film cooling. Just the opposite would be expected to occur on a concave wall, where, at relatively low velocity or momentum flux, the jet could be
"pushed" away from the surface by the
higher pressure near the surface. At high
momentum flux ratios on the concave surface, the jet might be expected to impinge
on the surface downstream of injection,
yielding good film cooling.

Experiments were performed with a turbine
cascade in which the geometry of the blades
corresponded to a current design [10]. The
mass transfer analogy was used to obtain
an impermeable wall effectiveness on the
pressure (concave) and suction (convex)
sides of a blade. A comparison of these
results with results obtained previously
for film cooling on a flat surface is given
in Figure 13. Note that at low blowing
rates (actually, momentum flux ratio in

mainstream direction), there is considerably better film cooling performance on
the convex surface than on the flat surface, while at higher blowing rates, where
the jet would tend to move away from the
surface, the convex surface gives the
poorer film cooling performance. Just the
opposite occurs on the pressure (concave)
side of the blade, where, at a low momentum flux ratio, the jet is "pushed" by the
pressure gradient away from the surface,
giving relatively poor film cooling effectiveness.

The influence of curvature on film cooling
is very important with a single jet or a
single row of jets. This has not always
been considered when predicting cooling
performance on turbine blades. Curvature
would not be expected to have as great an
influence on film cooling from two staggered rows of holes where the jets tend to
merge and act more like a continuous slot.

Other parameters that have been studied
relative to their influence on film cooling include freestream acceleration [11],
freestream turbulence [12], and the laminar versus turbulent nature of both the
injected coolant flow and the boundary
along the surface [13]. Modest acceleration of the mainstream flow does not have
a great effect on film cooling. At low
blowing rates, high turbulence mixing decreases somewhat the effectiveness along
the center line of the injection hole.
Freestream turbulence intensity does not
significantly alter the lateral distribution of effectiveness, while changing the
turbulence scale does affect it considerably. When the coolant flow in the injection hole is laminar, the effectiveness
is lower than that following turbulent injection. This lower effectiveness, particularly at blowing rates near where lift-
off tends to occur, is apparently due to
the peaked nature of the velocity profile
with laminar flow.

The influence of coolant injection on the heat transfer coefficient is shown in Figure 14, where the heat transfer coefficient is compared to that which would occur in the absence of blowing [14]. Note that at low blowing rates, the heat transfer coefficient is close to that found for zero blowing, as is also true some distance downstream, even at quite high blowing rates. The results on this figure were taken with actual heating of the surface downstream of injection. With heating, it is difficult to get accurate values of the surface heat flux close to the injection hole through which the coolant flows. Thermal conduction through the wall can lead to significant errors in measurement of the local heat transfer at small distances downstream of injection.

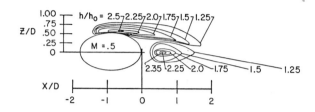

Figure 15   Contours of Constant Values of h/h_0 for M = 0.5

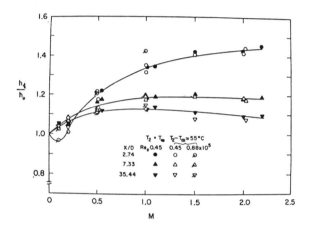

Figure 14   Variation of the Center-Line Heat Transfer Coefficient With Blowing Rate

As an alternate technique, a mass transfer analogy using the sublimation of naphthlene was used [15]. Measurements of the relative mass transfer rate close to injection can then be taken. Figure 15 shows the high mass transfer coefficients that occur in the vicinity of the injection hole. Values many times that obtained in the absence of injection occur both immediately adjacent to the hole and immediately downstream of the hole. It is important to know the heat transfer coefficients in this region because of their influence on the thermal stresses that are produced by temperature gradients along the surface.

IMPINGEMENT COOLING

In many applications, very high local heat transfer can be obtained with impingement of one or more jets onto a surface. Such impingement heat transfer is used in cooling of many systems, including the leading edge region of a turbine blade (cf. Fig. 1).

The high heat transfer in the impingement region is primarily due to the high velocity and thin boundary layer that occur there. Much of the early work on impinge-ment heat transfer was concerned with flow from a single round jet or from a two-dimensional slot into a still atmosphere. The influence of crossflow on impingement heat transfer, however, is important in many applications. Figure 16 shows the flow of an impinging jet with various crossflows at right angles to the jet [16]. At low values of the crossflow (high values of the flow parameter, M), the jet impinges almost directly across from its exit hole. As the jet flow relative to the main flow decreases, it is turned in the downstream direction until at sufficiently low jet velocity it appears not to impinge at all on the opposing surface.

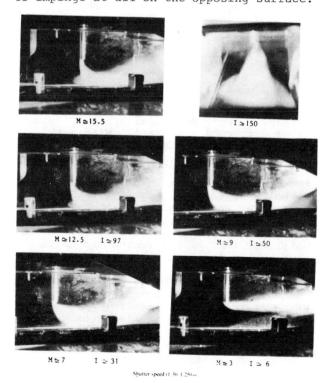

Figure 16   Impinging Jet flow for Jet-to-Wall Spacing of Six Diameters

23

In measuring heat transfer with an impinging jet, care must be taken because of the
high rates of heat transfer and, therefore,
small temperature differences that occur in
the region of impingement. In addition,
the recovery temperature along the wall is
not easy to predict. Recovery factors,
both greater and less than unity, are measured in the impinging region. The heat
transfer coefficient should be defined
using the temperature of the wall minus
the recovery temperature.

Figure 17 shows the influence of crossflow
on impingement heat transfer [16]. At relatively low crossflow, or high jet velocity, the maximum heat transfer is directly
opposite the center of the injection hole.
As the injection velocity decreases for a
fixed mainstream flow, the peak heat transfer decreases and the location of the peak
moves in the downstream direction.

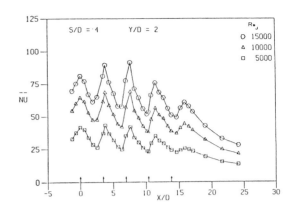

Figure 18   Spanwise Average Nusselt
Number For Impingement From
An Array of Jets

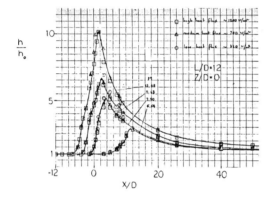

Figure 17   Heat Transfer Coefficient For
Jet Impingement With Crossflow

In many applications, a row of jets, or
perhaps a two-dimensional array of jets
is used to cool a surface. With a two-
dimensional array there is often a crossflow from the spent air of the upstream
jets. The heat transfer from one such
two-dimensional array made up of five rows
of holes is shown in Figure 18 [17]. With
this array the crossflow is produced by the
upstream rows of jets. The maximum heat
transfer at the first row of holes is somewhat less than at the second and third
rows, while the peak heat transfer for the
fourth or fifth row is considerably smaller. In addition, there is a displacement
of the maxima in the downstream direction
which is readily apparent for the downstream holes. There is also considerable
variation of the heat transfer between the
holes of an individual row.

CONCLUDING REMARKS

We have briefly surveyed studies on film
cooling and impingement heat transfer that
have been conducted at the University of
Minnesota. Flow phenomena related to the
interaction of discrete jets with a mainstream are quite complex and far from
fully understood. Because of the large
structure of the eddies and the resulting
complex interactions, it is difficult to
develop good analytical or numerical predictions for such flows. Several semi-
empirical techniques using turbulent eddy
diffusivities have been used and are convenient for developing correlations. These
are described in earlier works. Considerably more experimental and numerical work
is necessary to fully take into account
the influence of such things as surface
curvature and injection jet geometry on
adiabatic wall temperature and local heat
transfer.

ACKNOWLEDGMENT

Support from the Office of Army Research
during the writing of this survey is grate-
fully acknowledged.

REFERENCES

1. Elovic, E. and Koffel, W. K., Proceed-
   ings of the 1979 International Joint
   Gas Turbine Congress and Exhibition,
   Paper No. G-1(97) (1979).

2. Goldstein, R. J., Advances in Heat and
   Mass Transfer, vol. 7, pp. 321-379
   (1971).

3. Ramsey, J. W. and Goldstein, R. J.,
   J. Heat Transfer, 93, 365 (1971).

4. Goldstein, R. J., Eriksen, V. L., and
   Ramsey, J. W., Heat Transfer 1978,
   vol. 5, pp. 255-260 (1978).

5. Goldstein, R. J., Eckert, E.R.G., Eriksen, V. L., and Ramsey, J. W., Israel J Technology, $\underline{8}$, 145 (1970).

6. Pedersen, D. R., Eckert, E.R.G., and Goldstein, R. J., J. Heat Transfer, $\underline{99}$, 620 (1977).

7. Ito, Sadasuke, University of Minnesota Ph.D. Thesis, December 1976.

8. Jabbari, M. Y. and Goldstein, R. J., J. Engr. Power, $\underline{100}$, 303 (1978).

9. Goldstein, R. J., Eckert, E.R.G., and Burggraf, F., Int. J. Heat Mass Transfer, $\underline{17}$, 595 (1974).

10. Ito, S., Goldstein, R. J., and Eckert, E.R.G., J. Engr. Power, $\underline{100}$, 476,(1978).

11. Jabbari, M. Y. and Goldstein, R. J., $\underline{\text{Heat Transfer 1978}}$, vol. 5, pp. 249-254 (1978).

12. Kadotani, K. and Goldstein, R. J., J. Engr. Power, $\underline{101}$, 466 (1979).

13. Yoshida, Toyoaki, University of Minnesota M.S. Thesis, December 1977.

14. Eriksen, V. L. and Goldstein, R. J., J. Heat Transfer, $\underline{96}$, 239 (1974).

15. Taylor, J. R. and Goldstein, R. J., To Be Published.

16. Bouchez, J.-P. and Goldstein, R. J., Int. J. Heat Mass Transfer, $\underline{18}$, 719 (1975).

17. Behbahani, A. and Goldstein, R. J., To Be Published.

Studies of Full-Coverage Film Cooling

MASARU HIRATA and NOBUHIDE KASAGI
University of Tokyo, Tokyo 113, Japan

MASAYA KUMADA
University of Gifu, Gifu 504, Japan

ABSTRACT

The final target of this study is to develop a numerical prediction method for the temperature distribution within the full-coverage film-cooled wall. This paper consists of two fundamental experiments.

Firstly, the combined heat transfer/heat conduction tests were performed by adopting brass and acrylic resin as a material for the full-coverage film-cooled (FCFC) wall. The 30°-slant injection holes are distributed in the staggered array with the two kinds of hole pitches, which are five and ten hole diameters in the streamwise and lateral directions. The free stream velocity was varied as 10 and 20 m/s respectively, and the measurement of wall surface temperature was made with the help of liquid crystal as a temperature indicator. From these experiments, basic quantitative data are given for the local and averaged cooling effectiveness for each material tested.

Secondly, in order to obtain the information on the local heat transfer coefficients, a mass transfer simulation technique based on the naphthalene sublimation has been developed. The effects of the mass flux ratio and non-dimentional secondary injection concentration on the local mass transfer coefficient are made clear. It is confirmed from this experiment that the local Stanton number is a linear function of non-dimensional secondary injection temperature as expected from the analysis.

NOMENCLATURE

A = surface area associated with one hole, Fig. 19
A_H = backside and hole-inside area associated with one hole, Fig. 19
Bi = Biot number, hd/λ_B
C = concentration of naphthalene
c_p = specific heat at constant pressure
D = diffusion coefficient of naphthalene vapour
d = hole diameter
Feq = mass flux ratio at equivalent uniform blowing condition, $\pi d^2 M/4ps$
H = boundary layer shape factor or FCFC plate shape factor
$\overline{h}$ = heat transfer coefficient averaged over the area of A
h_d = mass transfer coefficient
$\overline{h}_H$ = heat transfer coefficient averaged over the area of A_H
k = proportional constant, $(St(1)-St(0))/St(0)$
M = mass flux ratio, $\rho_2 u_2/\rho_\infty u_\infty$
p = streamwise hole pitch
Re_d = Reynolds number, $u_2 d/\nu$
Sh = Sherwood number
Sc = Schmidt number
St = Stanton number, $\overline{h}/\rho_\infty u_\infty c_p$
St_H = Stanton number, $\overline{h}_H/\rho_2 u_2 c_p$
s = lateral hole pitch
T = temperature

t = thickness of FCFC plate
u = flow velocity
u_2 = injection velocity evaluated at the condition far upstream of injection holes
x = streamwise distance from the first row of holes
z = lateral distance
α = injection hole angle
η_w = local cooling effectiveness, $(T_w-T_\infty)/(T_2-T_\infty)$
θ = non-dimensional injection temperature, $(T_{20}-T_\infty)/(T_w-T_\infty)$ or non-dimensional injection concentration, C_2/C_w
θ_{20} = non-dimensional temperature at hole exit
λ = thermal conductivity of fluid
λ_B = thermal conductivity of FCFC wall material
ξ, η, ζ = coordinate system at hole exit
ρ = density
ν = kinematic viscosity

Superscripts

$(\overline{})$ = spanwise averaged value

Subscripts

w = evaluated at wall surface
∞ = evaluated at free stream condition
2 = evaluated far upstream of injection hole
20 = evaluated at hole exit
0 = value in the case of $\theta=0$ or value without injection
B = evaluated at backside, except where specially designated in nomenclature
T = averaged over all

INTRODUCTION

Increasing attention has been directed towards advanced technology for high temperature gas turbines to achieve high thermodynamic efficiency for energy conservation. Among these new technologies, development of more effective cooling techniques is essential to acceptable reliability and durability of gas turbine components. A feasible cooling scheme involves the secondary coolant injection through a number of discrete holes provided on the surface to be protected from the hot combustion gas. This has been termed as full-coverage film cooling (FCFC) and is currently under intensive studies [1-5].

Flow field and heat transfer in the near-hole region of FCFC are inevitably three-dimensional in contrast to most film cooling theories, which are well applicable to the region far downstream of injection. In addition, the heat transfer problem with FCFC becomes much more complicated with the heat conduction effect inside the film-cooled wall. Therefore further study should be needed to establish a reliable prediction method for FCFC.

A conventional procedure for treating film cooling heat transfer is based on the adiabatic wall temperature and the heat transfer coefficient

without fluid injection [6]. The heat transfer coefficient in FCFC, however, might be considerably
different from the value assumed, because the turbulent boundary layer on the film-cooled wall is
disturbed strongly by the coolant injection.
Moreover the heat conduction inside the wall would
change the temperature field dependent on the thermal properties of wall material.
An alternative procedure in the case of uniform
wall temperature has been proposed by the Stanford
group [5], in which the heat transfer coefficient
can be obtained for any value of coolant temperature ratio. They have reported detailed data of
the Stanton number averaged over each row of injection holes under the various experimental conditions. However, it is uncertain that the heat
transfer coefficient of uniform wall temperature
condition could be applicable to the general case
with heat conduction.
    In the present studies of FCFC, the final tar-
get is to develop the numerical prediction scheme
based on the finite element method for the calculation of temperature distribution within the full-
coverage film-cooled wall. This report presents
firstly the basic quantitative data obtained by
the combined heat transfer/heat conduction experiments with the adoption of acrylic resin and brass
as a material for the FCFC wall.
Secondly, in order to enable the numerical calculation of temperature distribution inside the full-
coverage film-cooled wall, the local Stanton number has been measured, not only on the surface
but also on the backside surface and the inside of
holes, by adopting the sublimation of naphthalene.
Varying the naphthalene vapour contents within the
secondary injection air, it is confirmed that the
local Stanton number is a linear function of non-
dimensional secondary injection temperature. Furthermore, flow visualization is performed by using
the oil film method in order to understand the
flow structure near the hole.

EXPERIMENTAL APPARATUS AND PROCEDURE

All the heat transfer experiments have been carried out in the test duct of 200mm x 400mm cross

Free Stream Velocity	u_∞ = 10.0, 20.0 (m/s)
Free Stream Turbulence	Tu = 0.2, 0.3 (%)
Pressure Gradient	zero, zero
Displacement Thickness #	δ^* = 2.83, 2.25 (mm)
Momentum Thickness #	θ = 2.07, 1.73 (mm)
Shape Factor #	H = 1.37, 1.30
Reynolds Number #	Re_θ = 1334, 2230

FCFC Plate

Hole Diameter	d = 12 (mm)
Injection Angle	= 30 (deg)
Streamwise Hole Pitch	p/d = 5, 10
Lateral Hole Pitch	s/d = 5, 10
Plate Thickness	t/d = 25/12

Material	Acrylic Resin	Brass
Thermal Conductivity λ_B (W/mK)	0.21	128
Specific Heat c_p (kJ/kgK)	1.47	0.381
Thermal Boundary Layer Thickness #$ δ_t (mm)	10.6	11.4
Enthalpy Thickness #$ θ^* (mm)	0.676	1.045

(# ; at the first row, $; u_∞ = 20.0 (m/s))

Table 1 Experimental conditions

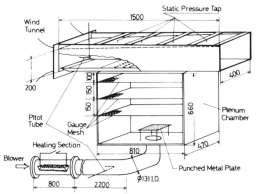

Fig. 1 Experimental setup for FCFC

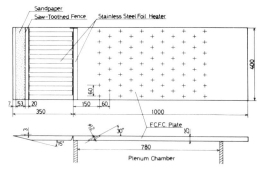

Fig. 2 FCFC plate tested ($p/d=s/d=5$)

section and 1500mm length provided at the exit of
a blowdown wind tunnel as shown in Fig. 1. The
secondary air is heated up to a prescribed temperature at the heating section and is injected into
the mainstream through the settling chamber and
the holes of FCFC plate. The height of a ceiling
plate of the test section has been adjusted so
that the pressure gradient in the streamwise direction should be eliminated.
As a FCFC plate material, acrylic resin as well as
brass have been adopted, of which thermal properties are summarized in Table 1. If the heat transfer coefficient of the turbulent boundary layer
without injection is assumed, the Biot number
based on the hole diameter is about 3 for acrylic
resin and 5×10^{-3} for brass respectively. The hole
diameter, d, is 12mm, while the thickness of the
flat FCFC plate, t, is 25mm. The injection angle,
α, is slant by 30° from the downstream wall surface. These values remain constant throughout the
experiments. The injection holes have been drilled in the staggered manner with the two kinds of
hole pitches, which are five and ten hole diameters in the streamwise and lateral directions,
i.e., $p/d = s/d$ =5,10.
Fig. 2 shows the FCFC plate of 5 diameters pitch
with the 350mm upstream preheating plate of acrylic resin. With turbulence promotors at the leading region, the turbulent boundary layer at the
first row of holes becomes almost fully-developed.
The preheating electric power input into the stainless steel foil has been regulated so that the wall
temperature changes continuously from the preheating plate to the FCFC plate.
The free stream velocity, u_∞, has been varied as
10 and 20 m/s. The mass flux ratio of secondary
injection, M, is ranged from 1.1 to 2.6 for p/d=10
and from 0.3 to 0.7 for p/d=5. These values of M
cover almost the same range of total secondary air
supply and are considered to be feasible in the
practical design of film-cooled components of gas

turbines.

The temperature measurements have been made by copper-constantan thermocouples of 70μm in diameter, except that of FCFC wall surface temperature obtained by making use of the selective light scattering of cholestelic liquid crystal [7,8]. The liquid crystal presently used consists of cholesteryl-oreil carbonate, cholesteryl chloride and cholesteryl nonanoate. It has been enclosed in capsules of 5–30μm in diameter and these capsules are pasted and fixed uniformly by 100μm in thickness on the FCFC plate. In the case of an acrylic resin plate, the wall surface covered with liquid crystal is illuminated by a sodium light source of 5890Å in wavelength. Then the isothermal contour, which is corresponding to a certain absolute temperature, can be recorded by monochromatic photography. The diagram of local cooling effectiveness later described has been composed by the various contours obtained with the temperature change of secondary air supplied. On the other hand, no appreciable spanwise variation of surface temperature has been observed in the case of a brass plate due to the remarkable effect of heat conduction. Therefore, under each experimental condition the location of green band, x/d, has been measured and decided through the iterated observation of naked eyes. After the precise calibration, the reproducibility of temperature with the liquid crystal has been estimated at least within $\pm0.2°C$ in comparison with the Cu-Co thermocouples.

All the experiments of the mass transfer simulation for the local heat transfer measurements have been carried out in the test duct of 300mm × 400mm cross section and 2000mm length provided at the exit of a blowdown wind tunnel. The geometrical shape of the FCFC plate is the same as that in the heat transfer experiments above explained. In this mass transfer experiment, the heaters are set at the inlet of both blowers used so that the temperature difference between the mainflow and the secondary injection flow is regulated within $\pm0.5°C$. Furthermore, in order to soak the secondary injection air with naphthalene vapour, the honeycombed duct with molded naphthalene is connected with the inlet of the injection blower and the concentration of naphthalene is kept under control by varying the duct length. The concentration of naphthalene is calculated from weighing the quantity of sublimation and it is further checked by a gas chromatography at three positions beneath the FCFC plate that the naphthalene vapour is distributed uniformly into each injection hole. Measurement of sublimation is performed within the accuracy of 1μm by using the electric micrometer which is attached to the automatic traversing mechanism.

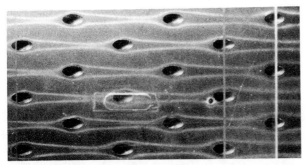

Fig. 3 Surface temperature measurement with the aid of cholesteric liquid crystal

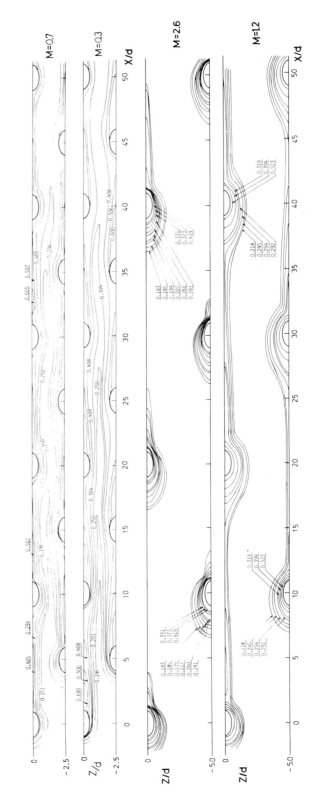

($p/d=s/d=5$, u =20m/s) ($p/d=s/d=10$, u =10m/s)

Fig. 4 Cooling effectiveness diagram of acrylic resin wall

EXPERIMENTAL RESULTS AND DISCUSSION

Cooling effectiveness

In the present studies, the cooling performances of FCFC are expressed in the form of the cooling effectiveness defined as follows:

$$\eta_w = (T_w - T_\infty)/(T_2 - T_\infty) \qquad (1)$$

Fig. 3 is the typical photograph of $5d$-pitch acrylic resin plate covered with liquid crystal which is illuminated by the sodium light source. The similar photographic records of various isothermal contours have been accumulated by changing the temperature of heated secondary air flow. Then by the superposition of these contours the cooling effectiveness diagram can be composed as shown in Fig. 4. As shown in these diagrams, immediately downstream of each injection hole, there are slender regions of higher cooling effectiveness which are covered well with the film flow of the secondary coolant. These regions, however, become narrower with the mass flux ratio increased.

Fig. 5 summarizes the distributions of spanwise averaged effectiveness, $\overline{\eta}_w$, along the streamwise distance, x/d, for $u_\infty = 10\text{m/s}$. It was found that the influence of free stream velocity is comparatively small as long as M is kept constant. But, as shown in these figures, the injection hole pitch gives an appreciable influence on the cooling effectiveness due to an abrupt change of the flow pattern on the FCFC wall. The $\overline{\eta}_w$ value of $p/d=10$ is not increasing with the downstream distance in contrast to that of $p/d=5$, although a somewhat high value can be recognized around each injection hole.

Throughout the experiment of brass FCFC plate, no appreciable spanwise distribution of η_w has been observed due to the strong heat conduction effect within the wall. As a result, the variation of η_w value in the lateral direction is estimated about $\pm 3\%$ at most at the x/d location examined.

The results of brass wall for $u_\infty = 20\text{m/s}$ are summarized and represented in Fig. 6. Generally speaking, the effectiveness of brass is considerably higher than that of acrylic resin and at the same time the streamwise distribution is smooth without any peaks at the injection. With M increased, the cooling effectiveness increases in contrast to the acrylic resin and seems to approach a certain saturated value. Again the mainstream velocity did not markedly affect the value and distribution of $\overline{\eta}_w$. But it is important to mention that the results of $p/d=10$ give considerably poor values compared with those of $p/d=5$ and that the former tends to be saturated in a shorter streamwise distance. The reason for the decrease of $\overline{\eta}_w$ might be the intense mixing of the coolant jet with the mainstream fluid as well as the decrease of the hole interior surface where the heat transfer is much enhanced. All the experimental data of $\overline{\eta}_w$ at the three streamwise positions are plotted versus the equivalent uniform blowing ratio, Feq, in Fig. 7. Additionally, the effectiveness obtained in the case of uniform blowing has been calculated by referring to Torii et al. [9] as shown in this figure. At first, the general tendency can be noted that the cooling effectiveness of brass is much higher than that of acrylic resin, so the suitable design process must take into account the thermal properties of film-cooled wall. Secondly, the injetion hole pitch feasible in practical design is considered to range about from $5d$ to $10d$, while the effect of hole pitch is distinguished over this range. Therefore, it must be noted that the smaller pitch should be preferable from the view point of heat transfer. In addition, it can be also said that the cooling performance of uniform blowing is extremely good with the same total secondary coolant supply, although the FCFC performance can be considerably improved in the case of a thermally conductive wall as described later.

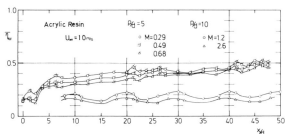

Fig. 5 Spanwise averaged cooling effectiveness of acrylic resin wall

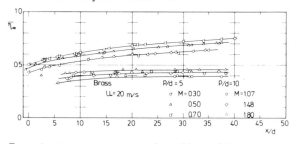

Fig. 6 Spanwise averaged cooling effectiveness of brass wall

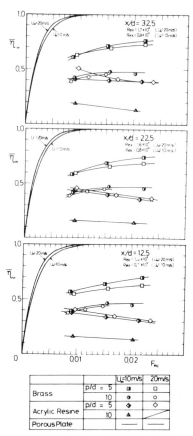

Fig. 7 Dependency of spanwise averaged effectiveness on equivalent uniform blowing ratio

Mass transfer experiments

For the case of the uniform wall temperature, Crawford et al. [5] applied the superposition theory on the linear thermal energy equation and proposed the following equation for the heat transfer coefficient in FCFC:

$$\frac{St(\theta)}{St(0)} = 1 + \frac{St(1)-St(0)}{St(0)}\,\theta = 1 + k\theta \qquad (2)$$

where $St(1)$ and $St(0)$ are the local Stanton number when $\theta=1$ and $\theta=0$, respectively. In order to calculate the heat transfer coefficient, h, for a given injection temperature, it is necessary to obtain the data of h at two θ values for the same value of M. Therefore, if the experiments are run with two different values of θ for a single value of M, the value of k is determined. Then it becomes possible to obtain the heat transfer coefficient for an arbitrary injection temperature ratio.

In this paper, the local heat transfer coefficient is determined by adopting the mass transfer simulation technique as naphthalene sublimation.

At first, in order to make clear the effect of M on the local Stanton number, measurements have been carried out by varying M under the condition that naphthalene vapour is not contained within the flow injected. The Stanton number obtained for this case is used as the reference value.

The spanwise distributions are typically shown in Fig. 8. It is shown that the effect of injection parameter M is not significant except at the region near the hole. The contour diagram of the local Stanton number is shown in Fig. 9. The noteworthy points from these diagrams are summarized as follows;

(1) The Stanton number varies drastically near the holes, while the pattern of contour is similar to each other.

(2) The higher values of Stanton number at the downstream of the holes are due to the reattachment of the secondary injection flow. As the value of M increases, this region expands and its value becomes higher.

(3) There is the considerably higher Stanton number region around the hole due to the induced flow.

These diagrams of St number distribution are very much complicated comparing with the cooling effectiveness diagrams shown in Fig. 4.

Fig. 10 shows typical spanwise distributions of lo-

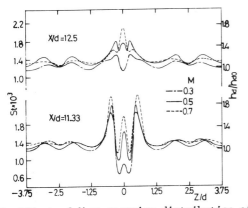

Fig. 8   Effect of M on spanwise distribution of the local Stanton number in the case of $\theta=0$

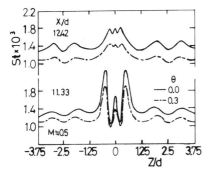

Fig. 10 Effect of θ on spanwise distribution of the local Stanton number in the case of $\theta\neq0$

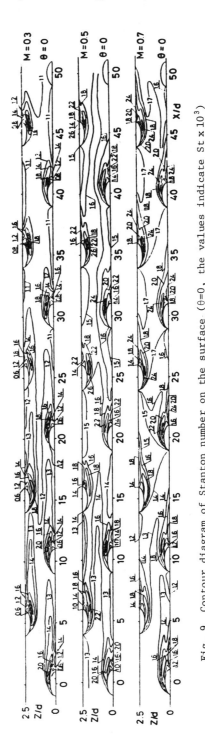

Fig. 9   Contour diagram of Stanton number on the surface ($\theta=0$, the values indicate St $\times 10^3$)

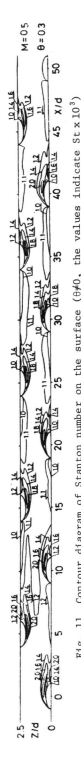

Fig. 11 Contour diagram of Stanton number on the surface ($\theta\neq0$, the values indicate St $\times 10^3$)

cal St number for $M=0.5$ and $\theta=0.3$, where the injected secondary flow contains naphthalene vapour. As is expected, the Stanton number in the case of $\theta=0.3$ is smaller than that of $\theta=0$ in every region. The contour diagram of the local Stanton number is shown in Fig. 11 for the case of $\theta=0.3$ and $M=0.5$. It is easily understood that the measured Stanton number has a general tendency to decrease with θ and that the rate of decrease becomes larger with an increase of the streamwise distance.

Spanwise averaged Stanton number on the surface, $\overline{St}$, has been calculated by the numerical integration of local variation of the Stanton number as shown in Fig. 12. Fig. 12 (a) represents the effect of M on $\overline{St}$ along the streamwise distance in the case of $\theta=0$. As might be expected from Fig. 8, $\overline{St}$ becomes higher as the value of x/d increases for all the values of M, in contrast with the Stanton number without injection, and at the same time, $\overline{St}$ increases appreciably with increase of M. In addition, $\overline{St}$ is relatively small in front of each hole and shows the peak value in the region downstream of injection, and this results in the jagged shape of streamwise distribution.

Fig. 12 (b) shows the dependency of $\overline{St}$ on the non-dimensional injection temperature parameter, θ, while the value of M is kept constant to be 0.3. Generally, $\overline{St}$ decreases with increase of x/d and with increase of θ. These characteristics are similar for the case of $M=0.5$ and 0.7.

The typical relationship between $St(\theta)/St(0)$ and θ is represented in Fig. 13. As shown in the figure, it seems the linear relationship between the local Stanton number and θ is well established. The value of k, which means the gradient of straight line in the figure, becomes larger as x/d increases and is also large at the region covered directly with the injection, for instance, at $z/d=0$.

Assuming that the linear relationship is established all over the surface, then it is easily shown that the spanwise averaged Stanton number comes into this relationship. The relation between $\overline{St}(\theta)/\overline{St}(0)$ and θ is shown in Fig. 14. As is expected, the linear relation for $\overline{St}(\theta)$ is again well confirmed. Therefore, based on the experimental data of k at various x/d, the spanwise averaged value, $\overline{k}$, have been calculated. Fig. 15 shows the results of typical 36 points in the streamwise direction, in which the symbol of ▲ denotes the value at the hole center position. The value of $\overline{k}$ decreases monotonically in the streamwise direction with x/d increased, indicating a step-like change immediately downstream of each injection hole.

In order to establish the numerical prediction method of the whole field of temperature for FCFC plate, it is also necessary to obtain the information on the heat transfer coefficient of backside surface of the FCFC plate. A typical example for the backside Stanton number distribution is shown in Fig. 16.

From the local Stanton number data shown in this figure, the dependency of the spanwise averaged Stanton number on x_B/d have been calculated as shown in Fig. 17. The profiles of $\overline{St}_B$ are similar

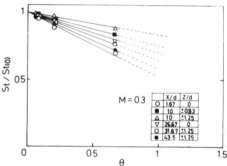

Fig. 13 Linear relation between the local Stanton number and θ for typical positions

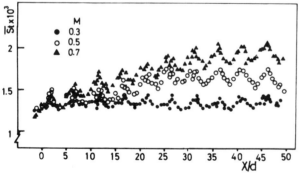

Fig. 12(a) Effect of M on distribution of spanwise averaged Stanton number

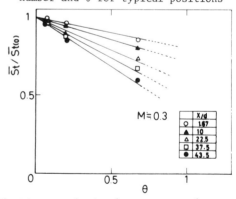

Fig. 14 Linear relation between spanwise averaged Stnaton number

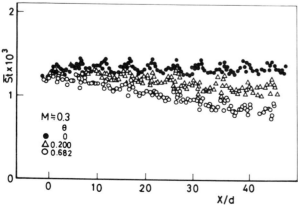

Fig. 12(b) Effect of θ on distribution of spanwise averaged Stanton number

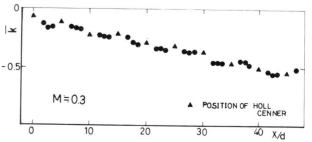

Fig. 15 Variation of $\overline{k}$ in the lateral direction

for every value of Re_d tested with a steep peak value at the edge of hole, while $\overline{St}_B$ decreases with Re_d increased.

The successive integration of these data gives the totally averaged Stanton number on the back surface, $\overline{St}_{BT}$, as shown in Fig. 18. The experimental data are well correlated by the straight line shown in this figure. This means the totally averaged heat transfer coefficient is proportional to the 2/3 power of u_2.

Improvement of FCFC effectiveness

According to some simplified analysis [10] for the temperature field of the FCFC plate based on the model as shown in Fig. 19, it can be concluded that the effectiveness can be remarkably improved if the backside and hole-inside heat transfer is augmented under the same conditions of M.

In the present study, the supplementary heat transfer experiment has been performed as shown in Fig. 20. The holes of d in diameter have been drilled on the plate of 3mm thickness in the staggered manner as same as the FCFC plate. This perforated plate is inserted by $3d$ upstream of the FCFC plate, while the placement of holes of the inserted plate and the FCFC plate are relatively staggered. Thus, the secondary coolant flow has an additional effect of high heat transfer performance as multi-impinging jets before flowing into the injection holes [11].

The results of $\overline{\eta}_w$ measurement are given in Fig. 21. As is easily understood, the cooling performance has been quite improved and the value reaches more than $\overline{\eta}_w=0.8$. From these results, it is concluded that the heat transfer augmentation on the backside surface should be of great importance to suitable design of FCFC.

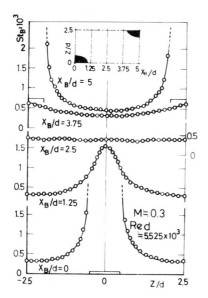

Fig. 16 Spanwise distribution of the local Stanton number on the backside surface

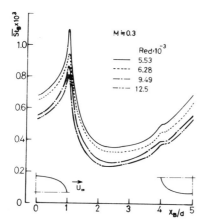

Fig. 17 Lateral distribution of spanwise averaged Stanton number on the backside surface

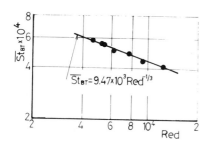

Fig. 18 Dependency of totally averaged Stanton number on the backside surface

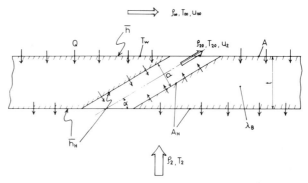

Fig. 19 Unit hole element model for FCFC heat transfer

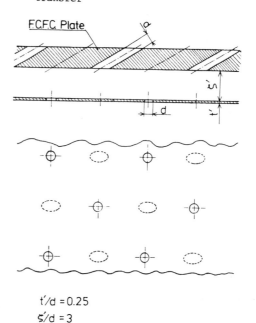

t′/d = 0.25

ς′/d = 3

Fig. 20 Jet orifice plate inserted

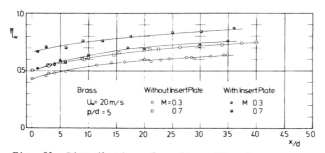

Fig. 21 Distribution of cooling effectiveness
improved by the jet orifice plate inserted

CONCLUSIONS

In order to develop a numerical prediction method
of the effectiveness of FCFC plate, two major experiments are carried out.
Firstly, combined heat transfer/heat conduction
experiments were performed by adopting the two
kinds of FCFC plate materials with particular
attention to the effect of thermal properties.
The surface temperature measurement has been successfully made with the aid of cholesteric liquid
crystal. The basic quantitative data are given
for the local and averaged cooling effectiveness
for each material tested. It is emphasized that
the thermal properties have remarkable influence
on the cooling effectiveness due to the heat conduction within the plate. Also the hole pitch and
the mass flux ratio affect appreciably the FCFC
performance.
Secondly, the basic measurement of the local
Stanton number both on the film-cooled surface and
on the backside of the FCFC plate has been performed successfully with the aid of the naphthalene sublimation technique. The effects of the
mass flux ratio and the non-dimensional secondary
coolant temperature on the local Stanton number are
experimentally clarified and discussed. It is
confirmed that the local and averaged Stanton number on the film-cooled surface is a linear function of non-dimensional injection temperature.
Further, the contribution of local heat transfer
on the backside to the cooling effectiveness is
measured.
Finally, the improvement of FCFC performance has
been discussed with a typical technique demonstrated experimentally.

ACKNOWLEDGEMENTS

The authors are grateful to Mr. N. Okuzumi, N.
Nishimura, Y. Kawano, T. Jozuka, T. Takahashi,
H. Kondo, M. Nakagawa, R. Noguchi and T. Mitsuya
for their cooperations in the present experiments.
This work was supported through the Grant-in-Aid
for Special Project Research by the Japanese Ministry of Education, Science and Culture.

REFERENCES

1. Metzger, D. E., Takeuchi, D. I., and Kuenstler,
    P. A., "Effectiveness and Heat Transfer With
    Full-Coverage Film Cooling," ASME Journal of
    Engineering for Power, Vol. 95, 1973, p. 180.
2. Mayle, R. E., and Camarata, F. J., "Multinole
    Cooling Film Effectiveness and Heat Transfer,"
    ASME Journal of Heat Transfer, Vol. 97, 1975,
    p. 534.
3. Bergeles, G., Gosman, A. D., and Launder, B.
    E., "Double-Row Discrete-Hole Cooling: An Ex-
    perimental and Numerical Study," in *Gas Turbine
    Heat Transfer 1978*, ASME, 1978, p. 1.
4. Sasaki, M., Takahara, K., Kumagai, T., and
    Hamano, M., "Film Cooling Effectiveness for
    Injection From Multirow Holes," ASME Journal
    of Engineering for Power, Vol. 101, 1979,
    p. 101.
5. Crawford, M. E., Kays, W. M., and Moffat, R.
    J., "Full-Coverage Film Cooling, Part 1 Com-
    parison of Heat Transfer Data for Three Injec-
    tion Angles," ASME Paper, 80-GT-43, 25th In-
    ternational Gas Turbine Conference, New
    Orleans, 1980.
6. Goldstein, R. J., "Film Cooling," in *Advances
    in Heat Transfer*, Vol. 7, Academic Press, New
    York, 1971, p. 321.
7. Fergason, J. L., "Cholesteric Structure-I, Op-
    tical Properties," Molecular Crystals, Vol. 1,
    1966, p. 293.
8. Fergason, J. L., "Liquid Crystals in Nonde-
    structive Testing," Applied Optics, Vol. 7,
    1968, p. 1729.
9. Torii, K., Nishiwaki, N., and Hirata, M., "Heat
    Transfer and Skin Friction in Turbulent Bound-
    ary Layer With Mass Injection," Proceedings of
    3rd International Heat Transfer Conference,
    Vol. 3, Chicago, 1966, p. 34.
10. Kasagi, N., Hirata, M., and Kumada, M.,
    "Studies of Full-Coverage Film Cooling (Part 1,
    Cooling Effectiveness of Thermally Conductive
    Wall)," to be presented at 26th International
    Gas Turbine Conference, Houston, 1981.
11. Hollworth, B. R., and Dagan, L., "Arrays of
    Impinging Jets With Spent Fluid Removal
    Through Vent Holes on the Target Surface,
    Part 1: Average Heat Transfer," ASME Paper,
    80-GT-42, 25th International Gas Turbine Con-
    ference, New Orleans, 1980.

Radiative Heat Transfer in the United States

C. L. TIEN and S. C. LEE
University of California, Berkeley, California, U.S.A.

ABSTRACT

The status of current research activities in radiative heat transfer in the United States is presented. The scope covers basic researches in the radiative properties of surfaces, gases and particulates, as well as the more complex multi-dimensional system analysis. Development of rigorous surface radiation analysis, along with more accurate surface properties, enables optimal design of energy systems such as solar collectors. Fundamental research in radiation properties of gases and soot particulates enables a more accurate prediction of radiative heat transfer at high temperatures of fires and combustion chambers. Although models in predicting radiative properties of gases have been rather successful, more simplified and convenient representations of the results should be developed for practical applications. Further research is also required in calculating radiation from gas-particle mixtures. For the prediction of radiative heat transfer from nonhomogeneous mixtures, techniques for solving the equation of transfer in an absorbing, emitting and scattering medium are of great interest. Various techniques, for example the spherical harmonics method, the differential approximation and the optical pathlength method etc., have been investigated. Extension of these techniques to treat multidimensional enclosures is also studied. Another aspect of major concern is the interaction of radiation with other modes of heat transfer. New techniques have been developed to solve the integrodifferential equations of combined convection and radiation. The problem of radiative heat transfer with phase change is also under much consideration.

INTRODUCTION

One of the factors that accounts for the importance of heat transfer by radiation is due to its characteristic temperature dependence. Unlike energy transfer by conduction and convection which depends on the temperature difference of the locations to approximately the first power, radiative energy transfer depends on the difference of the individual body temperatrues raised to the power of 4 or 5. It is evident that radiation becomes intensified at high temperatures. Consequently, radiation is the dominant mode of heat transfer in many combustion systems.

The importance of radiation is realized in many physical systems. In high temperature systems such as industrial scale fires and coal fired combustors, radiation from gases (carbon dioxide and water vapor) and particulates (ash, char, coal and soot) must be properly accounted for.

The determination of the effective emittance and absorptance of gas-particle mixtures is central to the analysis of heat transfer in these systems. In other systems like the fluidized bed combustors, radiation contributes significantly to the total heat transfer between the fluidized medium and the submerged heat transfer tubes. Moreover, in the design and analysis of solar energy utilization systems, the interaction of radiation with other modes of heat transfer plays an important role in enhancing heat transfer, particularly in air operated solar collectors. Climate modelling is another area where atmospheric radiation is important. Various gaseous species in the atmosphere, along with water droplets and aerosols, absorb and scatter radiation. Radiative modelling of the atmospheric constituents and aerosols forms the basis of climate modelling. Another interesting aspect of radiative heat transfer is found in aquaculture -- algae growing. Algae absorb and emit solar radiation, thus creating a greenhouse effect in ponds. A careful radiative heat transfer analysis is then required to achieve optimal growth for algae.

Despite the many efforts which have been devoted to the various aspects of radiation heat transfer, numerous problems still remain. A workshop on radiation heat transfer was recently held in the United States which encouraged communications between University researchers and scientists in industries and U.S. Government laboratories with similar interests. The topics of the workshop cover basic research in radiation properties of surfaces, gases and particulates, as well as analysis of radiative heat transfer in various systems. Each of the topics will be discussed accordingly.

SURFACE RADIATION PROPERTIES

The ideal behavior of a blackbody serves as a standard with which real bodies can be compared. The radiative behavior of real bodies depends on many factors such as surface finish, material composition, temperature, the spectral distribution of the incident radiation and the direction of the incident and emitted radiation, etc. Accurate prediction of the radiative behavior of real materials is vital to their utilization.

In the selection of surfaces for use in the collection of radiant energy, it is desirable to maximize the energy absorbed by the collector surface, while minimizing the loss by emission. Therefore, it is important to obtain accurate radiation properties of surfaces for use in solar energy systems. The thermal radiative properties of glass used in solar collectors have been studied extensively. The scope of research includes experimental measurements of glass properties, prediction of properties of anti-reflecting glass and condensate covered glass plates. Optical properties of the glass have been determined using an interferometric technique. Computer programs for evaluating properties of multilayer-coated glass and glass systems have been developed. Much work is still needed in obtaining extensive property measurements of other selective surfaces at elevated temperatures as well as investigating the stability of surface properties.

The interest in solar energy conversion and the trends toward high temperature processes in materials technology have produced a need to predict radiant heat transfer through semi-transparent solids. However, appropriate radiative property data of these solids at elevated temperatures are not readily available. Presently, experimental studies on the total band absorptance of semitransparent solids such as fused quartz have been made. Approximate analysis on the band absorptance has been developed based upon the dispersion relations of the classical electron theory. Further studies to reveal the fundamental band properties for other solids as well as formulating more rigorous analysis are certainly needed.

SURFACE RADIATIVE TRANSFER

The analysis of surface radiative heat transfer is often very complicated, because real surfaces are far from being ideal. Thus, surface property variations with temperature, wavelength and direction introduce considerable complexities in the analysis. For example, the effects of spectral and directional variations of spectrally selective coatings for temperature control in solar energy systems must be properly accounted for in order to produce accurate heat transfer predictions. However, simplifying approximations must be made to circumvent the complexities in analyzing surface heat transfer by radiation.

Currently, the study of surface radiative heat transfer is mostly associated with solar collector systems. Thermal analysis of solar collectors using compound-parabolic-concentrators (CPC) has been investigated. Based on the observation that a CPC collector would behave like a flat plate collector when the concentration ratio is one, the Hottel-Whillier-Woertz-Bliss formulation for flat plate collectors is adapted for use in CPC collectors. Other more fundamental research works include the investigation of the effects of radiation scattering from surface defects and deposits as well as the geometry effect of the collector system on the rate of radiative transfer.

A complete treatment of radiative heat transfer of real surfaces including all variations, while possible in principle if all radiation properties are known, is seldom attempted or justified. Future research work should be oriented toward obtaining simple, but analytically well-founded formulation for engineering use.

GASEOUS PROPERTIES

Thermal radiation of gases has long been a subject of fundamental importance in understanding many phenomena, such as combustion and momentum and energy transport processes involving radiating gases. Over the past decades, great progress has been achieved in understanding the thermal radiation properties of gases. Extensive theoretical and experimental studies on the properties have enabled more accurate predictions of gaseous radiation.

Prediction of gaseous radiation in combustion systems is generally obtained using either Hottel's emissivity charts, the narrow band statistical model or the exponential wide band model. The emissivity charts, due to their relatively empirical basis, are often not well suited to nonhomogeneous situations, or for cases involving mixtures of radiating gases. The narrow band statistical model has led to convenient analytical substitutes of the charts. However, the massive amount of inputs and the large amount of computer time required make the model inappropriate in practical applications. On the other hand, the exponential wide band model, despite its relative simplicity in nature, has been quite successful in producing accurate results even in non-homogeneous situations. Gray gas models have also been developed to obtain the emissivities of carbon dioxide and water vapor. These models approximate the calculations of real gas emissivity as the emission from many gray gases with different weighing factors and progressively increasing absorption coefficients.

Although the thermal radiation properties of gases are quite well understood now, computations for the emissivities, especially in nonhomogeneous systems, are still rather involved. Further work is required in developing simple, user-oriented formulae and charts to render efficient use in complicated combustion systems.

PARTICULATES

The combustion products from burning hydrocarbon fuels are generally consisted of several gaseous constitutents and numerous small solid particulates called soot. It has long been recognized that soot plays an important role in the radiation from flames and smoke. In large scale fires, smoke and coal-fired combustors, large particulates may be present and scattering of radiation can be quite significant. In order to estimate the radiative contribution by soot in these combustion systems, detailed soot information must be known.

Soot particles are characterized by the optical properties, composition (Hydrogen/Carbon ratio), size and shapes. Assuming homogeneous and spherical soot, prediction of soot radiation and optical diagnostic of soot concentration and size distribution in flames can be obtained if the soot optical properties are known. The optical properties of soot have been explored in the past using the dispersion relations from classical electron theory based on reflectance measurements on various forms of carbon outside the flame. The results so obtained may deviate from those of soot in flames. Recently, better results have been obtained based on a more rigorous consideration of the dispersion theory and in-situ flame transmission data. Using these soot optical properties, agreement is observed between the predicted and experimental spectral extinction coefficients of soot from flames of different hydrocarbon fuels. Alternate approaches to determine the soot optical properties are presently being considered. One proposal is to measure the bidirectional spectral reflectivity of highly polished pellet surface. The measurement is extremely sensitive to the surface finish of the compressed pellets, because voidage on the surface would reduce the surface specularity, thus introducing uncertainties in the measurement. Another technique to deduce the refractive index uses transmission measurements through suspension of particles in liquids.

Soot particles are generally assumed to be spherical in all flame radiation analyses. However, it is well known that soot particles usually conglomerate into larger chunks and long chains during and subsequent to the combustion process. The effect of conglomeration is seldom taken into consideration due to the complexities involved. Dependent scattering can be important in densely packed systems such as the fluidized combustor. Therefore, much research is needed concerning the effect of dependent scattering and soot conglomeration on particulate radiation.

RADIATIVE TRANSFER ANALYSIS

Exact solution to the equation of radiative transfer and energy conservation to yield temperature distributions and heat flow in an absorbing and emitting medium requires considerable effort in most practical cases. Several approaches have been taken to circumvent this complexity. When it is desired to determine only the radiant energy from a mass of isothermal gas to the boundaries of the enclosure, application of the concept of mean beam length greatly simplifies the computation. In reality the gas inside an enclosure is seldom isothermal. Thus, a more elaborate scheme such as the zonning method must be employed. And in cases when variations in both the gas properties and the enclosure surface properties must be accounted for, the Monte Carlo approach is more appropriate.

Radiative heat transfer from an absorbing and emitting medium is relatively well understood; however, the problem becomes much more complicated when scattering is important. Present effort is directed toward developing more efficient methods to solve the equation of transfer in an absorbing, emitting and scattering medium. Solution techniques such as the Fourier transform, power series expansion and source function expansion have been developed assuming isotropic scattering. In order to approximate the scattering characteristics more closely, anisotropic scattering is assumed in many analyses. Other than the spherical harmonics method or the differential approximation, a method called the optical pathlength approach has been developed for solving general nongray and anisotropic scattering problems. The basis of this method is to determine all possible pathlengths of all the photons in a conservative scattering medium and then determine the heat transfer with absorption.

Rigorous solution of the equation of transfer is by far restricted to one dimensional cases. Extension of the analysis to multidimensional systems must be considered in order to render application to real systems. On the other hand, development of simple analytical and efficient numerical methods to evaluate the radiative heat transfer in engineering systems is needed.

INTERACTION OF RADIATION WITH OTHER HEAT TRANSFER MODES

The interaction of radiation with conduction and convection in absorbing-emitting media occurs in many practical cases such as atmospheric phenomena, shock problems, rocket nozzles and industrial furnaces, etc. The combined modes of heat transfer often introduce additional mathematical complications than that for radiation alone. A nonlinear integrodifferential equation generally results for the energy equation of such problems. Analytical solutions are usually not obtainable due to the complex nature of heat transfer by mixed modes. Results are therefore generally obtained numerically.

Radiative heat transfer is an important mode of energy transfer in combustion systems due to the high temperature of the media. Many applications of fluidized beds involve simultaneous radiative and conductive heat transfer between the fluidized medium and the submerged heat transfer tubes. Heat transfer by combined convection and radiation also occurs in many physical systems. Such examples include flame stabilization by plates and rods in premixed flowing gases, the burning of a solid or liquid fuel in an oxidizer stream, or the protection of re-entry vehicle by ablative heat transfer, which all involve boundary layer thermal reaction with radiation heat transfer. The integrodifferential nature of the combined convective/radiation equations makes the solution a formidable task. Recent development in the solution technique using the Spline-collocation method has been successful. Extension of the technique to solve the more complicated combustion-radiation gas dynamics in laminar boundary layers is currently under consideration.

The problem of radiative heat transfer with phase change is also of fundamental research interest. Typical applications include the melting of ice by solar energy, crystal growth and melting and solidification of uranium fuels etc. The identification of a two-phase zone with partial melting, lying between the old and new phase zones, is attributed to the internal melting or solidification induced by internal thermal radiation. The existence of the zone calls for a new mathematical formulation to a wide class of melting and solidification problems.

Other phase change problems involve the radiative controlled condensation such as the fog formation of Uranium dioxide vapor mixture due to radiative cooling in hypothetical reactor core disruption. It is found that the coupled effect of fog formation and radiation is a very effective mode of heat transfer from a high temperature condensable mixture to the cooler boundary. Therefore, many analytical and experimental studies on radiation with phase change are needed.

CONCLUSION

It has long been recognized that radiation is an important mode of heat transfer at high temperature. Considerable progress has been achieved in the understanding of fundamental radiative properties as well as formulating rigorous radiative heat transfer analysis. Extensive studies on the experimental and analytical formulations of surface radiative properties have been made. The information thus allows more exact analysis of surface radiative heat transfer to be made. However, further research is still needed in analyzing practical engineering systems, particularly those related to solar collectors. Thermal radiation in flames, smoke and combustion chambers constitutes an important aspect in engineering analysis. Radiation in these systems is greatly influenced by the presence of both gaseous products and soot particles. The thermal radiation properties of gases are rather well understood. Future research should be aimed at developing well founded, but simple and convenient representations of gaseous properties for engineering applications. As for soot particulates, major problems still exist in obtaining accurate soot characteristics such as the optical properties, size and shape. Radiative transfer in an absorbing, emitting and scattering medium has received a lot of attention. Many solution techniques have been developed for solving the equation of transfer in one dimension. Extension of the techniques to treat multidimensional enclosures is of fundamental research interest. It is realized that simultaneous modes of heat transfer are often present in many systems. Development of simple, yet rigorous analysis is of great concern for engineering applications.

APPENDIX

A Workshop on Radiation Heat Transfer was held on July 27, 1980 in Orlando, Florida in conjunction with the 19th AIChE/ASME National Heat Transfer Conference. The Workshop was stimulated by Win Aung, Heat Tansfer Program Director, and was jointly organized by Win Aung and Ray Viskanta of Purdue University. The purpose of the informal discussion was to encourage communication among researchers and scientists in academia, industry and U.S. Government laboratories with similar interests in radiation heat transfer. Approximately fifty invited specialists attended the Workshop. The following five specific topics were discussed and each topic was first reviewed briefly by an invited panelist (shown in parenthesis) in a joint panel session:

1. Radiation Properties of Gases and Gas Particle Mixture (C.L. Tien, University of California at Berkeley)

2. Analysis of Multidimensional Radiative Transfer in Participating Media (A.L. Crosbie, University of Missoui - Rolla)

3. Radiation Heat Transfer in Combustion Systems (R. Viskanta, Purdue University at West Lafayette)

4. Radiation Heat Transfer in Fires (J.R. Lloyd and K.T. Yang, University of Notre Dame)

5. Radiation Heat Transfer in Solar Energy Utilization Systems (D.K. Edwards, University of California at Los Angeles)

In addition to the five short presentations and the subsequent group discussions on each of the five topics, a set of short abstracts of research in progress were provided by the participants. The titles of these research activities and the associated investigators are listed below:

1. Radiation Heat Transfer in Solar Cavity-Type Receivers, M. Abrams, J.S. Kraabel and R. Greif, Sandia National Laboratories at Livermore.

2. Radiative Transfer Properties of Semitransparent Solids, E.E. Anderson, University of Nebraska.

3. Higher Order Differential Approximation of Radiative Transfer, Y. Bayazitoglu, Rice University.

4. Radiation Heat Transfer with Scattering and Non-Gray Absorption, R.O. Buckius, University of Illinois at Urbana-Champaign.

5. Radiative Transfer in Planetary Atmospheres and Climate Modelling, R.D. Cess, State University of New York at Stony Brook.

6. Radiative Heat Transfer with Phase Change, Thermal Radiation Properties of Nuclear Reactor Materials, Multi-Dimensional Radiative Transfer in Non-Gray Gases, S.H. Chan, University of Wisconsin at Milwaukee.

7. Application of Spline Collocation Method to Chemically Reacting Boundary Layer Flows with Radiation Heat Transfer, T.C. Chawla, Argonne National Laboratory.

8. Radiative Heat Transfer in Fluidized Beds, J.C. Chen, Lehigh University.

9. A General Formulation and Computer Code for Scattering Radiative Transfer, A.C. Cogley, University of Illinois at Chicago Circle.

10. Radiative Heat Transfer in Combustion Systems, I.H. Farag, University of New Hampshire.

11. Experimental Determination of the Complex Refractive Index of Hydrocarbon Soot Particles, J.D. Felske, State University of New York at Buffalo.

12. Radiant Heat Transfer in Enclosures with Participating Media, Radiant Heat Transfer in Fibrous Insulations, W.A. Fiveland, Babcock and Wilcox Research Center.

13. Radiation Heat Transfer from a Hot Surface to Flowing Steam and Water Droplets, R. Greif, University of California at Berkeley.

14. Thermal Radiation in Coal Combustion and Non-isothermal Fires, Emmissivity of Coated Thermocouple Wires, W.L. Grosshandler, Washington State University.

15. Radiation Heat Transfer in Solar Energy Systems and Infrared Scanning Research, C.K. Hsieh, University of Florida.

16. Radiation Absorption and Scattering in Liquids, F.P. Incropera, Purdue University.

17. Spectral Emissivity Measurements of Nitrous Oxide, L.A. Kennedy, State University of New York at Buffalo.

18. Solution of Radiation Transfer in Partici-
    pating Media, M.N. Ozisik, North Carolina
    State University.

19. Radiative Heat Transfer in Gas-Particulate
    Systems, M.F. Modest, University of Southern
    California.

20. Radiative Transfer in Particle Laden Combust-
    ing Flows, H.G. Semerjian, National Bureau
    of Standards.

21. Bidirectional Reflectance of One-Dimensional
    Rough Surfaces, Convective and Radiative
    Transfer for Turbulent Flow in Energy
    Systems, T.F. Smith, University of Iowa.

22. Radiation Heat Transfer as an Enhancement
    Mechanism in Solar Collectors, E.M. Sparrow,
    University of Minnesota.

23. Enhanced Radiating Heating Due to Use of
    Alternative Fuels in Aircraft Turbine
    Combustors, J.S. Toor, Science Applications,
    Inc.

24. Generalization of the P-1 Approximation to
    Non-Gray Radiative Heat Transfer, Multi-
    dimensional Radiative Transfer, Evaluation
    of Mean-Beam Length, Geometric Mean
    Absorptance Between 3-D Objects with a Simple
    Contour Interpretation, W.W. Yuen, University
    of California at Santa Barbara.

Radiative Properties of Metals and Alloys at High Temperatures

Kyoto University, Kyoto, Japan

ABSTRACT

In order to estimate the radiative properties of
heat resisting superalloys, the infrared spectra
of reflectivities and emissivities of the basic
elements, iron, nickel, cobalt and chromium were
measured at high temperatures. The experimental
results were well expressed by the two-conduction-
electrons and one-bound-electron type dispersion
model of optical constants. Electronic constants
of the model were given and formulated in order to
learn the dependence of the macroscopic properties
of radiative heat transfer on wavelength, temperature and phase of crystal. Also the infrared
spectra of reflectivities and emissivities of
stainless steels and superalloys were measured.
The data on carbon steels were taken as well.

NOMENCLATURE

c = velocity of light in vacuum, ms^{-1}
e = electron charge, C
Im = imaginary part of complex quantity
k = imaginary part of optical constant
m_k^* = effective mass of conduction electron k, Kg
N_k = number of conduction electron k per unit volume, m^{-3}
n = index of refraction
$\hat{n}$ = optical constant
R = (energy) reflectivity
S = intensity of interband quantum transition
T = temperature, K
δ = parameter of vibration damping of bound electron
ε = emissivity
ε_o = dielectric constant of vacuum, Fm^{-1}
λ = wavelength of light in vacuum, m
λ_k = relaxation wavelength of conduction electron k, m
λ_o = resonance wavelength of bound electron, m
ρ_i = resistivity due to impurities, Ωm
ρ_o = resistivity of bulk metal, Ωm
σ_{dc} = electric dc conductivity, Sm^{-1}
σ_o = optical dc conductivity, Sm^{-1}
σ_k = optical dc conductivity for conduction electron k, Sm^{-1}

Subscripts

CS = carbon steel
H = hemispherical
HS = hard steel
i = impurity
k = k-th electron (k=1,2)
MS = mild steel
N = normal
The number in the subscript of reflectivity R indicates the incident angle.

INTRODUCTION

In order to use our limited sources of energy efficiently and to develop the possibilities of new
sources of energy, persistent efforts have been
made to develop heat resisting materials. Ni or Co
based superalloys are used as structural materials
in high temperature industrial installations,
which require the use of temperatures beyond 1300
K. Knowledge of the thermal radiation properties
of these metals is important and indispensable

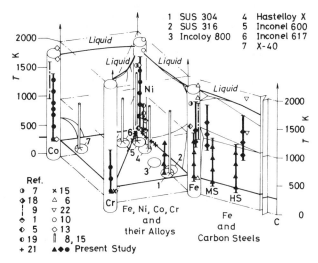

Fig.1 Temperature regions studied by the present author and the other authors

Table 1 Chemical compositions of the specimens (wt %)

	Fe	Ni	Co	Cr	C	Si	Mn	P	S	
Fe 99.9%	Bal.	0.01	–	<0.005	<0.002	0.01	0.02	0.002	0.003	0.006Cu
Ni 99.9%	0.0019	Bal.	0.014	–	0.001	<0.01	<0.01	0.001	<0.001	0.0010Cu, 0.0010Pb
Co 99.9%	0.002	0.027	Bal.	–	0.004	0.004	0.001	0.003	0.002	<0.01Cu
Cr 99.8%	0.093	–	–	Bal.	0.002	0.023	0.001	0.003	0.020	<0.003Cu, 0.010Al
Mild steel (SS 41)	Bal.	0.02	–	0.01	0.03	0.02	0.26	0.016	0.010	0.01Cu
Hard steel (S 55 C)	Bal.	0.02	–	0.02	0.56	0.23	0.74	0.029	0.011	0.02Cu
SUS 304	Bal.	9.3	–	19.	<0.08	<1.0	<2.0	<0.04	<0.03	
SUS 316	Bal.	12.	–	17.	<0.08	<1.0	<2.0	<0.04	<0.03	2.5Mo
Incoloy 800	Bal.	32.	–	20.5	0.04	0.35	0.75			0.3Cu
Hastelloy X	18.5	Bal.	1.5	22.	0.1	0.5	0.5	<0.04	<0.03	9Mo,0.6W
Inconel 600	7.2	Bal.	–	16.	0.04	0.2	0.2			0.1Cu
Inconel 617	–	Bal.	12.	20.	0.03					8.5Mo, 0.35Ti, 1.2Al
X-40 (Stellite 31)	1.5	10.	Bal.	25.	0.5	0.5	0.5	<0.04	<0.04	7.5W

41

when designing equipment for use at high temperatures. Most of the superalloys are mainly composed of one or more of the elements, Fe, Co and Ni and also Cr. However, the radiative properties of superalloys and their basic elements at high temperatures have scarcely been investigated. In the present paper we studied the radiative properties of the basic elements; Fe, Ni, Co and Cr and secondarily the radiative properties of carbon steels, stainless steels and superalloys. The macroscopic properties of radiative heat transfer were calculated to learn their dependencies on wavelength, temperature and phase of crystal. The temperature regions studied by the present author and the other authors are shown summarily in Fig.1 for each metal. The chemical compositions of the materials examined are listed in Table 1.

EXPERIMENTAL PROCEDURE

The surfaces of the specimens were buffed up to optically smooth surfaces with a maximum roughness less than 0.03 μm. The specimens of Fe, mild steel and hard steel were annealed in a vacuum at 660 °C for 1 hour. No preparatory heat treatment was performed on the other materials. The reflectivity measurements at room temperature were carried out in the three wavelength regions, 0.34∿0.70 μm (v.s.), 0.70∿2.5 μm (n.i.r.) and 2.5 ∿25 μm (i.r.). In the v.s. and n.i.r. regions a double beam spectrophotometer with a fused-quartz prism was used and R_{12N} was measured. In the i.r. region another double beam spectrophotometer with KBr prisms and AgCl polarizer was used and the spectra of R_{13N}, R_{60S}, R_{60P} and R_{70P} were measured.

As standard surfaces for the reflectivities, an Al evaporated film was used in the v.s. region and an Au evaporated film in the n.i.r. and i.r. regions. The experimental apparatus for the measurements of absolute reflectivities and high temperature reflectivities is shown schematically in Fig.2. The reflectivity attachment shown in the center of the figure has a KRS-5 window 5 and can be filled with gases. The wall of the chamber was painted with metal black and cooled by water to minimize the effect of stray light. High temperature reflectivities were measured by this attachment only for R_{70P} in the i.r. region. R_{70P} is sensitively affected by the difference of the optical indices. At high temperatures normal emissivities were measured by the relative method of comparing the monochromatic radiative intensi-

ties from the specimens with those from the blackbody. The wavelengths measured were from 0.95 to 12 μm. Although, in the emissivity measurements, the spectral data are limited to the wavelength region of intense emission, the emissivity measurement has the advantage of being insensitive to the macroscopic surface configuration, which might be accompanied by the phase transformations of crystals at high temperatures. The atmosphere for the reflectivity measurement was argon or hydrogen with oxgen at a partial pressure of 10^{-12} Pa. The emissivity was measured in a vacuum of less than 2×10^{-3} Pa.

DISPERSION EQUATION OF OPTICAL CONSTANTS

In order to obtain the macroscopic properties of radiative heat transfer systematically, it is desirable to express the wavelength dependence of the optical constant with a dispersion equation and to investigate the temperature dependencies of the parameters of the equation. The most general type of the optical dispersion equation based on the classical theory of electron is the following equation[1].

$$\hat{n}^2 \equiv (n-ik)^2 = 1 + \sum_j \frac{S_j \lambda^2}{\lambda^2 - \lambda_{oj}^2 + i\delta_j \lambda_{oj} \lambda} - \frac{\lambda^2}{2\pi c \varepsilon_o} \sum_k \frac{(N_k e^2 \lambda_k / 2\pi c m_k^*)}{\lambda_k - i\lambda} \quad (1)$$

The unity in the first term of the right hand side of the equation is the relative dielectric constant of vacuum. The second term is the contribution of the bound electrons corresponding to interband quantum transitions. The series of j=1,2,··· corresponds to each transition of an electron between quantum levels. The contribution of this term to the optical constant is important only at the wavelengths in the vicinity of λ_{oj}. The third term is the contribution of the conduction electrons and is dominant mainly in the infrared. The numerator $N_k e^2 \lambda_k / 2\pi c m_k^*$ is equal to the conductivity of the electron k. The values of m_k^* are related to the shape of the energy curves of electrons[2] and differ according to the particular direction of the Brillouin zone. Although the primitive Drude model supposes only one kind of electron, Roberts[1][3] introduced the additional terms of k≥2 to describe the optical dispersions of general metals and alloys. In the present study a modified formula of the general equation (1) is used as follows;

$$\hat{n}^2 = 1 + \frac{S\lambda^2}{\lambda^2 - \lambda_o^2 + i\delta\lambda_o\lambda} - \frac{\lambda^2}{2\pi c\varepsilon_o} \sum_{k=1}^{2} \frac{\sigma_k}{\lambda_k - i\lambda} \quad (2)$$

The dc conductivity σ_o optically determined is written as

$$\sigma_o = \lim_{\lambda\to\infty} \{-2\pi c\varepsilon_o \, Im(\hat{n}^2)/\lambda\} = \sigma_1 + \sigma_2 \quad (3)$$

The second term of the right hand side of eq.(2) expresses the interband transitions as a broad absorption band neglecting the fine structures of each transition. In the third term the contributions of conduction electrons are expressed by the two terms, k=1 and 2. The contribution of the conduction electron with the longer relaxation time, which dominates the absorption at the longer wavelengths, is expressed by k=1 and the contribution of the conduction electron with the shorter relaxation time, which dominates the absorption at the shorter wavelengths, is expressed by k=2.

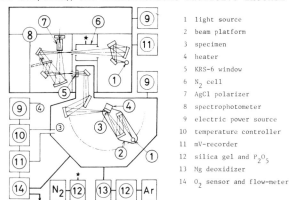

1	light source
2	beam platform
3	specimen
4	heater
5	KRS-6 window
6	N_2 cell
7	AgCl polarizer
8	spectrophotometer
9	electric power source
10	temperature controller
11	mV-recorder
12	silica gel and P_2O_5
13	Mg deoxidizer
14	O_2 sensor and flow-meter

Fig.2 Experimental apparatus for reflectivity measurement

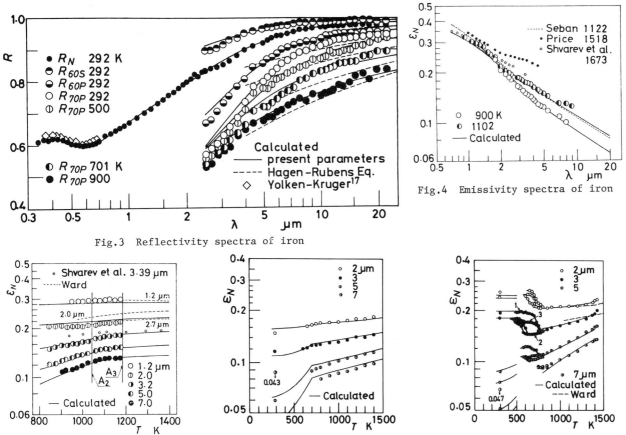

Fig.3 Reflectivity spectra of iron

Fig.4 Emissivity spectra of iron

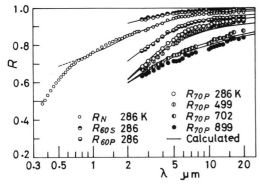

Fig.5 Temperature dependence of emissivity of iron

Fig.8 Temperature dependence of emissivity of nickel

Fig.11 Temperature dependence of emissivity of cobalt

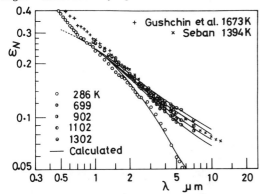

Fig.6 Reflectivity spectra of nickel

Fig.7 Emissivity spectra of nickel

The dispersion is described by the seven parameters of S, λ_o, δ, σ_1, σ_2, λ_1 and λ_2. The number of the parameters to be determined by the numerical analysis of the experimental spectra is, however, smaller. As will be shown, the temperature dependencies of the interband transitions are generally so small that some parameters can be presumed to be constant throughout the whole temperature range studied. As for the terms of conduction absorption, the wavelength region where the relaxation wavelength λ_2 is the dominant feature lies in the v,s. or the ultraviolet. In some metals which demonstrate intense interband transitions in the near-infrared, the minute values of λ_2 can not be specified, and in fact, no serious difficulty is encountered by assuming $\lambda_2 = 0$ μm. The values of σ_k/λ_k (k=1,2) depend on the shape of the energy curves of electrons and their temperature dependence is so small that the values are presumed to be independent of temperature also.

In view of the above discussion, the values of S, λ_o, δ and σ_k/λ_k may be presumed to be independent of temperature. And in the second stage only the temperature dependencies of the two kinds of σ_k (k= i,2) are to be analyzed. The first analysis of obtaining S, λ_o, δ and σ_k/λ_k were made generally for four kinds of reflectivity data at room temperature. In the computations of the optimization, the six- or four-dimensional hill-climbing method[4] was employed. In the second stage the direct searching method was employed using the data of reflectivity and emissivity at a specified temperature.

EXPERIMENTAL RESULTS AND DISCUSSIONS

Fe, Ni, Co and Cr

The experimental results of Fe are shown in Figs. 3∿5. R_N indicates the spectra of R_{12N} and R_{13N}. The maximum reflectivity is observed in the v.s. region and it is considered to be due to absorption by bound electrons. In the n.i.r. and i.r. regions, R_N increases as the wavelength increases, which indicates that absorption by conduction electrons plays a dominant role in these wavelength regions. The values of R_{60S}, R_{60P} and R_{70P} measured at room temperature are plotted in Fig.3. For the analysis of the parameters of the dispersion formula, these reflectivities are more useful than R_N, which is slightly affected by the differences of the parameters. The results of the spectral normal emissivity at three temperatures are shown in Fig.4 together with the results of other authors. The present results agree well with the values of Seban[5] and show the same tendency as that of the spectrum by Shvarev et al.[6]. Price[7] gave the considerably high emissivity. In Fig.5, the temperature dependencies of the emissivities at five wavelenghts are shown. The emissivities are continuous at all the transformation points, and none of the anomalies seen in some other thermophysical phenomena are observed. This behaviour of emissivity is similar to that of electric resistivity[8]. At the shorter wavelengths, the effect of temperature is small. Ward[9] showed the same tendency at wavelengths from 1.2 to 2.7 µm. These spectral behaviours suggest the existence of two kinds of conduction electrons, one of which has a longer relaxation time and responds to the infrared field in the same manner as to the dc field, sensitivity depending on temperature, and another of which has the shorter relaxation time and is not affected by temperature.[23][24]

The experimental results of Ni are shown in Figs. 6∿8, which show a similar tendency to that of Fe. The reflectivity changes continuously at the Curie temperature of 631 K and the spectrum is monotonous also in the paramagnetic crystal region. Nickel has no other transformation point except the Curie temperature, so that the qualitative behaviour may not change up to the neighbourhood of the melting point. The published optical constants at room temperature[3][5][10][11] give the normal emissivities which agree well with the present results. The values of Seban[5] and those of Gushchin et al[10] shown in Fig.8 agree qualitatively with present results.[25]

The experimental results of Co are shown in Figs. 9∿11. The optically smooth original surface prepared from the rolled plate by buffer polishing gives the R_N in Fig.9. There is a broad absorption band of quantum transition type in the n.i.r. region. At the longer wavelength the spectrum is of the conduction electron type. The published spectra at room temperature may be classified into two groups; the spectra of evaporated thin films are monotonous in the near infrared[11], however those of bulk specimens show a structure similar to R_N in Fig.9. The optical constants in the region of 0.3∿17 µm by Bolotin et al.[12] gives a R_N which agrees well with the present results at wavelengths longer than 0.5 µm. At temperatures up to 600 K, the emissivity increases with the rise of temperature (Fig.10, 11). In Fig.11, the highest emissivity curves such as curve 1 of $\lambda=3$ µm correspond to this heating process. The temperature dependencies are pronounced at the

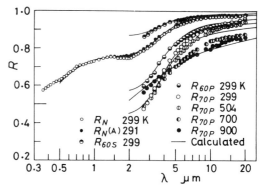

Fig.9   Reflectivity spectra of cobalt

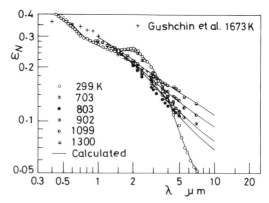

Fig.10   Emissivity spectra of cobalt

longer wavelengths and small at the shorter wavelengths. No evident change is observed in the absorption band of the interband transitions. At temperatures from 700 K to 800 K, a remarkable decrease of emissivity can be seen. This phenomenon is pronounced in the interband region and not so clear in the conduction absorption region. In Fig. 10 the absorption band dissappears and the spectra change to the conduction electron type. This is caused by the hcp ⇄ fcc martensitic phase transformation. The reconstruction of the energy band structure dissolves the narrow band gaps. The temperature dependence in the feromagnetic fcc region from 800 to 1400 K is of the two-electrons type. The ε_N-T curves of the present experiment and those of Ward[9] at 2 and 3 µm are continuous at the Curie temperature of 1400 K. The optical constants at 1673 K of Gushchin et al.[13] give the ε_N shown in Fig.10.

The most interesting characteristic of the optical properties of Co is the remarkable hysteresis phenomenon observed around the transformation temperature. In the ε_N curve at $\lambda=3$ µm in Fig.11, the buffed original specimen is heated along path 1. In the cooling process after the heating beyond 800 K, the ε_N takes path 2. Although the value of ε_N increases in the neighbourhood of 600 K, it does not return to the value observed in the first heating. The specimen when cooled to room temperature shows the reflectivity spectrum $R_N(A)$ in Fig. 9. Furthermore in the second heating the ε_N takes path 3. The values of ε_N in the second cooling agree with those in the first cooling. In the further repetitions, the ε_N follows the curves of the second scanning. The same tendency can be seen at the other wavelengths. The hysteresis phenomenon was also observed in the electrical resistivity[8]. The higher emissivity in the buffed original specimen is considered to be caused by the increased electron scattering due to the imperfectness of the crystal lattice introduced by the mechanical preparation of the surface. This imperfectness may be restored by the reconstruction of the lattice in the transformation temperature range. The electron diffraction experiment (E=75 KV, incident angle $\cong$ 90°) revealed that the original mechanically polished and buffed surface showed only the hallow of the non-crystalline layers, but that the annealed surface revealed weak diffraction rings of polycrystals in the hallow.[25]

The experimental results of Cr are shown in Figs. 12 and 13. Fig.12 shows the broad absorption band of interband transitions in the R_N at room temperature. Although the agreements of the published spectra are particularly poor in the n.i.r. region, the v.s. and the n.i.r. optical constants by Gorban et al.[15] and the i.r. optical constants by Lenham and Treherne[16] give R_N spectra which deviate less than 2 and 1%, respectively, from the present result. In the present experiment the data at high temperatures are limited to those below 700 K since a surface film considered to be Cr_2O_3 was grown at temperatures beyond 700 K. The spectra may be described by the same model as that for hcp Co. Since Cr experiences no transformation in the temperature range up to the melting point 2178 K, the results analyzed are considered to be useful in temperatures up to at least 1000 K.[25]

The results of the analysis are shown in Table 2 for the parameters S, λ_o, δ, σ_2 and σ_k/λ_k and in Figs.14~17 for the temperature dependencies of σ_o ($\equiv\sigma_1+\sigma_2$) and σ_1. The temperature dependencies of σ_o are described by the empirical expression,

$$\sigma_o^{-1} = AT^2 + BT + C \qquad (4)$$

The figures include the dc conductivities calculated from the published values of resistivities[8]. The values of σ_2, i.e. the difference between σ_o and σ_1, are nearly constant in the temperature range measured. This is caused by the fact that the free path of the electrons is in the order of interatomic distance even at room temperature. The values of σ_1 dominate nearly all the aspects of the temperature dependence of σ_o. The behaviour of σ_o is similar to that of σ_{dc}, but the absolute values of σ_o are different from those of σ_{dc}. The other parameters are estimated by the

following relations.

$$S, \lambda_o, \delta = const. \qquad (5)$$
$$\sigma_k/\lambda_k = const. \qquad (6)$$
$$\sigma_2 = const. \qquad (7)$$

By the relations of eq.(2) to eq.(7), the optical constants ($\hat{n}$) are determined and can be used to calculate various radiation properties. The calculated values of σ_o using eq.(4) and those of reflectivities and emissivities using eqs.(2)~(7) are shown by the solid lines in Figs.14~17 and in Figs. 3~13, respectively.

The four representative quantities of (spectral) hemispherical emissivity (ε_H), (spectral) normal emissivity (ε_N), total hemispherical emissivity

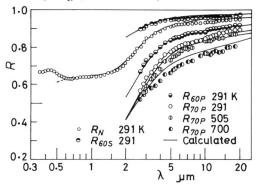

Fig.12 Reflectivity spectra of chromium

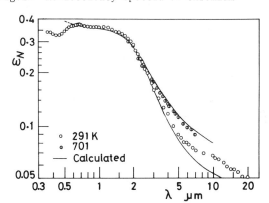

Fig.13 Emissivity spectra of chromium

Table 2 Dispersion parameters of eq.(2)

	T K	S	λ_0 m ×10^{-6}	δ	σ_1/λ_1 Sm^{-2} ×10^{11}	σ_2/λ_2 Sm^{-2} ×10^{12}	σ_2 Sm^{-1} ×10^6	A ΩmK^{-2} ×10^{-12}	B ΩmK^{-1} ×10^{-10}	C Ωm ×10^{-7}
Fe	200-1043				1.37	2.88	0.46	0.826	1.01	0.0
	1043-1185	0.0	-	-				0.0	9.82	0.690
	1185-1600				1.20	1.41	0.47		2.97	8.81
Ni	250- 700	0.0	-	-	2.23	2.42	0.77	1.52	-6.23	1.87
	700-1700							0.0	2.92	2.93
Co (hcp) buffed	250- 500	57.7	1.46	1.38	1.53	(λ_2=0)	0.27	1.28	-5.42	2.56
annealed	250- 500				1.75					2.17
(fcc)	800-1400	0.0	-	-	1.11	1.31	0.51	0.0627	10.2	-3.01
Cr	280-1000	37.1	0.82	1.70	1.87	(λ_2=0)	0.18	1.74	-9.88	5.79
Mild steel (SS 41)	200-1043				1.37	2.88	0.46	0.826	2.93	0.0
	1043-1135	0.0	-	-					21.8	-11.3
	1135-1600				1.20	1.41	0.47	0.0	2.97	10.11
Hard steel (S 55 C)	200-1043	0.0	-	-	1.37	2.88	0.46	0.826	4.14	0.0
	1043-1600				1.20	1.41	0.47		2.97	10.89

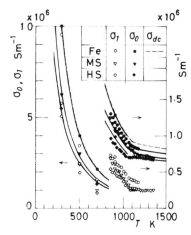

Fig.14 Temperature dependencies of σ_0 and σ_1 of iron, mild steel and hard steel

(ε_{Htotal}) and total normal emissivity (ε_{Ntotal}) are calculated and shown in Figs.18~21. The notations A and B in the figures indicate the mechanically buffed surface and the one annealed at temperatures beyond 800 K, respectively.

Carbon Steels

The experimental results of mild and hard steels are shown in Figs.22 and 23. The parameter obtained are shown in Table 2 and in Fig.14. In the case of carbon steels, since the concentrations of impurities are very small, the resistivity may be expressed by the following equation,

$$\rho_{oCS} = \rho_{oiron} + \rho_{oi}^* \tag{8}$$

Considering the above expression and the general temperature dependence of ρ_o, the same expression as eq.(4) was adopted. The calculated values of σ_o and those of reflectivity and emissivity are shown by the solid lines in Figs.14, 22 and 23. The calculated emissivities ε_H, ε_N, ε_{Htotal} and ε_{Ntotal} are shown in Figs.18 and 19.[23][24]

Stainless Steels and Superalloys

The experimental results of stainless steels and superalloys are shown in Fig.24 for room temperature. Although there is much difference in the chemical composition of the stainless steels and

the superalloys studied, the reflectivity spectra obtained show nearly the same absolute values and tendencies. The details of this research including high temperature condition will be published in the near future.

For the application to the thermal design of high temperature industrial installations, the effect of a thin coating of one superalloy or ceramic on another superalloy and the effect of ceramic dispersion in superalloy must be examined also. Further, it is desirable that radiative properties at high temperatures can be predicated by using

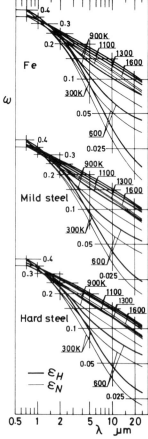

Fig.18 Calculated spectral emissivities ε_H and ε_N of iron, mild steel and hard steel

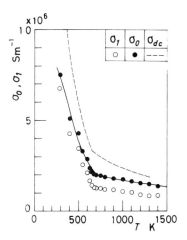

Fig.15 Temperature dependencies of σ_o and σ_1 of nickel

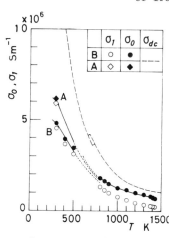

Fig.16 Temperature dependencies of σ_o and σ_1 of cobalt

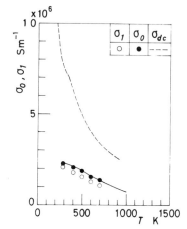

Fig.17 Temperature dependencies of σ_o and σ_1 of chromium

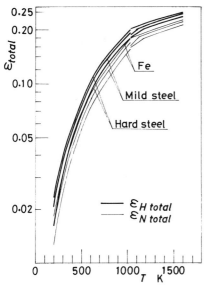

Fig.19 Calculated total emissivities ε_{Htotal} and ε_{Ntotal} of iron, mild steel and hard steel

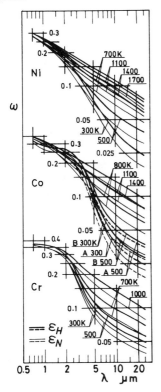

Fig.20 Calculated spectral emissivities ε_H and ε_N of nickel, cobalt and chromium

the room temperature data. For this purpose, the difference between σ_o and σ_{dc} should be treated and utilized skillfully since σ_{dc} can be measured very easily at high temperatures.

CONCLUSION

In order to obtain the thermophysical constants which determine the thermal radiative properties of heat resisting metals and alloys, reflectivities and emissivities were measured in the wave-length region from 0.34 to 25μm. The model parameters of the dispersion equation (2) of optical constants were obtained from the experimental data. The dependencies of the parameters on temperature were examined by the theory of electric conductivity. Spectral and total emissivities were calculated.

ACKNOWLEDGEMENT
The author was favored to have the assistance of Toshiro Makino who contributed sustained efforts to the accomplishment of this work.
This work was partially supported by the Grant-in-Aid for Fundamental Scientific Research of the Ministry of Education of Japan.

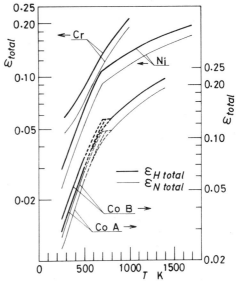

Fig.21 Calculated total emissivities ε_{Htotal} and ε_{Ntotal} of nickel, cobalt and chromium

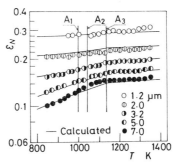

Fig.22 Temperature dependence of emissivity of mild steel

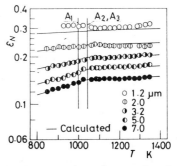

Fig.23 Temperature dependence of emissivity of hard steel

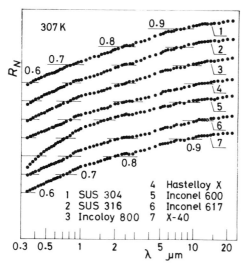

Fig.24 Reflectivity spectra of stainless steels and superalloys at room temperature

REFERENCES

1. Roberts, S., Phys. Rev., 114-101(1959).
2. Mott, N.F. and Jones, H., The Theory of the Properties of Metals and Alloys, 240(1936), Dover.
3. Roberts, S., Phys. Rev., 100-1667(1955).
4. Chestnut, H., Systems Engineering Tools, 437 (1965), John Wiley and Suns.
5. Seban, R.A., Trans. ASME. Ser.C, 87-173(1965).
6. Shvarev, K.M., et al., Teplophysica Byisokiif Temperatyp, 16-520(1978).
7. Price, D.J., Proc. Phys. Soc. Lond., 59-118 (1947).
8. Touloukian, Y.S., Thermophysical Properties of High Temperature Solid Materials, 1,2,3(1967), Macmillan.
9. Ward. L., Proc. Phys. Soc. Lond., 69-339(1956).
10. Gushchin, V.S. et al., Fiz. Tverd. Tela., 20-1637(1978).
11. Johnson, P.B., and Christy, R.W., Phys. Rev., 89-5056(1974).
12. Bolotin, G.A. et al., Fiz. Met. Metalloved., 35-699(1973).
13. Gushchin, V.S. et al., Dokl. Akad. Nauk. USSR, 240-320(1978).
14. Phys. Soc. Japan, Physical Properties of Metals (in Japanese), 394(1968), Shocabo.
15. Gorban, N.Ya. et al., Opt. Spectrosc., 35-687 (1973).
16. Lenham, A.P. and Treherne, D.M., J. Opt. Soc. Amer., 56-1137(1966).
17. Yolken, H.T. and Kruger, J., J. Opt. Soc. Amer., 55-842(1965).
18. Beattie, J.R. and Conn, K.T., Phil. Mag., 46-989(1955).
19. Edwards, D.K., Trans. ASME, Ser.C. 91-1(1969).
20. Touloukian, Y.S. and DeWitt, D.P., Thermophysical Properties of Matter, 7, (1970), IFI/Plenum.
21. Gorban, N.Ya. et al., Opt. Spektrosc., 35-508 (1973).
22. Shvarev, K.M. et al., Teplofiz. Vyis. Temp., 17-66(1979).

23. Makino, T. and Kunitomo, T., Bull. of the JSME, 20-1607(1977).
24. Makino, T. et al., Bull. of the JSME, 23-1835 (1980).
25. Makino, T. et al., Prepr. of Jpn. Soc. Mech. Eng., 800-18-36(1980).

HIGH-FLUX HEAT TRANSFER:
BASIC PHENOMENA

Needed Research in Boiling Heat Transfer

WARREN M. ROHSENOW
Massachusetts Institute of Technology,
Cambridge, Massachusetts, U.S.A.

ABSTRACT

Current research results will be presented for a few of our active research projects with suggestions for future research in these areas. Then, a more detailed presentation of research needs in mechanisms of boiling, pool boiling and flow boiling will be given.

NOMENCLATURE

B_M dimensional constant in Eq. (8)
C_p specific heat, $J \cdot kg^{-1} \cdot K^{-1}(Btu \cdot lbm^{-1} \cdot {}^\circ F^{-1})$
D_T tube diameter, m(ft)
D_d drop diameter, m(ft)
$F(X_{tt})$ parameter in Eq. (4)
F_2 parameter in Eq. (7)
G mass flux, $kg \cdot s^{-1} \cdot m^{-2}(lbm \cdot hr^{-1} \cdot ft^{-2})$
g gravitational acceleration, $m \cdot s^{-1}(ft \cdot hr^{-2})$
g_0 constant, $1\ kg\ m\ N^{-1} \cdot s^{-2}(4.17x10^8$
$\quad\quad lbm \cdot ft \cdot lbf^{-1} \cdot hr^{-2})$
h_{fg} enthalpy of vaporization, $J \cdot kg^{-1}(Btu \cdot lbm^{-1})$
h heat transfer coefficient, $W \cdot m^{-2}$
$\quad\quad (Btu \cdot hr^{-1} \cdot ft^{-2} \cdot {}^\circ F^{-1})$
k thermal conductivity, $W \cdot m^{-1}\ K^{-1}$
$\quad\quad (Btu \cdot hr^{-1} \cdot ft^{-1} \cdot {}^\circ F^1)$
n number density of drops
Pr Prandtl number
q heat flux, $W \cdot m^{-2}(Btu \cdot hr^{-1} \cdot ft^{-2})$
T temperature, $K({}^\circ F)$
ΔT temperature difference, $K({}^\circ F)$
x quality
x_b quality at dryout position
α void fraction
μ dynamic viscosity, $kg \cdot s^{-1} \cdot m^{-1}(lbm \cdot hr^{-1} \cdot ft^{-1})$
ρ density, $kg \cdot m^{-3}(lbm \cdot ft^{-3})$
σ surface tension, $N \cdot m^{-1}(lbm \cdot ft^{-1})$

Subscripts

B boiling
Bi value obtained from fully-developed boiling correlation at the incipient boiling point
FC forced convection
ib value at the incipient boiling point
ℓ liquid
sat saturation, or, with respect to saturation
v vapor
w wall

INTRODUCTION

The current research topics to be presented are: predictions of forced convection boiling data and of dispersed flow film boiling data, transient critical heat flux, CHF, blow-down heat transfer CHF in thin flowing films, direct contact heat transfer between immiscible fluids, and computer

calculation methods. Specific research needs are then presented. These resulted from a discussion at the Multiphase Symposion Workshop in Florida, April 1979.

FORCED CONVECTION FLOW BOILING

Many correlations have been proposed for predicting heat transfer for forced convection boiling tubes. The superposition of the effects of forced convection and boiling developed for the low quality region by Rohsenow [1] was extended to the higher quality region by Chen [2].

In Chen's correlation, F modifies the forced convection Reynolds number, and S accounts for the suppression of boiling as the incipient boiling point is approached. These two factors are empirically determined.

The correlations proposed here [3] for the subcooled and low quality region and for the high quality region obtain the suppression factor directly from the incipient boiling criterion of Bergles and Rohsenow [4] and calculate the convective component from non-boiling models in each region. Only one empirically determined coefficient is needed for the boiling contribution. This coefficient is determined here for water data.

Higher Quality Region

Annular flow exists for void fractions greater than about 80 percent. In this region, the superposition technique to be used consists of adding the forced convection heat flux, q_{FC}, to the fully developed boiling heat flux, q_B, and adjusting the result to force the effect of boiling to be zero at the incipient boiling point:

$$q = q_{FC} + q_B - q_{Bi} = q_{FC} + q_B \left[1 - \left(\frac{\Delta T_{sat,ib}}{\Delta T_{sat}} \right)^3 \right] \quad (1)$$

if $q \sim (\Delta T_{sat})^3$. This procedure is shown in Fig. 1. Here q_{Bi} is the magnitude of q_B at the incipient boiling wall superheat, $\Delta T_{sat,ib}$.

The term multiplying q_B accounts for the boiling suppression effect replacing the factor S of the Chen correlation.

Using the procedure of Bergles and Rohsenow [4], the wall superheat at the incipience of boiling is:

$$\Delta T_{sat,ib} = \frac{8 \sigma T_{sat} v_{fg} h_{FC}}{k_\ell h_{fg}} \quad (2)$$

51

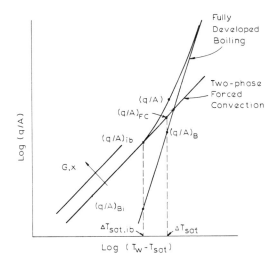

Fig. 1   Superposition Method

Traviss, et al. [5] developed a correlation for annular flow forced convection condensation inside tubes. The same analytical model applies to annular flow evaporation without nucleation. The recommended equation for q_{FC} in Eq. (1) is:

$$q_{FC} = \frac{Re_\ell^{0.9} Pr_\ell F(X_{tt}) k_\ell}{F_2} \frac{\Delta T_{sat}}{D}, \qquad (3)$$

where

$$F(X_{tt}) = 0.15 \left[\frac{1}{X_{tt}} + 2.0 \left(\frac{1}{X_{tt}} \right)^{0.32} \right], \qquad (4)$$

where

$$X_{tt} = \left(\frac{\rho_v}{\rho_\ell} \right)^{0.5} \left(\frac{\mu_\ell}{\mu} \right)^{0.1} \left(\frac{1-x}{x} \right)^{0.9}, \qquad (5)$$

$$Re_\ell = \frac{GD(1-x)}{\mu_\ell} \qquad (6)$$

and

$$\begin{cases} F_2 = 5Pr_\ell + 5\ln(1 + 5Pr_\ell) + 2.5\ln \qquad (7a) \\ \qquad (0.0031 Re_\ell^{0.812}) \, (Re_\ell > 1125) \\ \\ F_2 = 5Pr_\ell + 5\ln[1 + Pr_\ell(0.0964 Re_\ell^{0.585} - 1)] \\ \qquad (50 < Re_\ell < 1125) \qquad (7b) \\ \\ F_2 = 0.070 Pr_\ell Re_\ell^{0.5} \, (Re_\ell < 50) \qquad (7c) \end{cases}$$

The Mikic-Rohsenow correlation, used successfully for pool boiling data, was used to determine q_B in Eq. (1):

$$\frac{q_B}{\mu_\ell h_{fg}} \sqrt{\frac{g_o \sigma}{g(\rho_\ell - \rho_v)}} =$$

$$B_M \frac{k_\ell^{1/2} \rho_\ell^{17/8} C_{p\ell}^{19/8} \rho_v^{1/8}}{\mu_\ell^{7/8} h_{fg}^{9/8} (\rho_\ell - \rho_v)^{5/8} \sigma^{1/8} T_{sat}} (\Delta T_{sat})^3, \qquad (8)$$

where B_M is a dimensional constant which for pool boiling depends only on boiling surface properties, fluid properties, and gravity. This was determined for data to be $B_M = 1.89 \times 10^{-14}$ in SI units

(0.0000213 in engineering units) for forced convection boiling of water.

Subcooled and Low Quality Region

For these conditions, Bergles and Rohsenow [4] suggest the following superposition method:

$$q = [q_{FC}^2 + (q_B - q_{Bi})^2]^{1/2} = q = \left\{ q_{FC}^2 + q_B^2 \left[1 - \left(\frac{\Delta T_{sat,ib}}{\Delta T_{sat}} \right)^3 \right]^2 \right\}^{1/2} . \qquad (9)$$

The incipient boiling criterion for subcooled conditions is:

$$q_{ib} = \frac{h_{FC}}{(1-N)} \left(\frac{1}{4\Gamma N} + \Delta T_{sc} \right), \qquad (10)$$

where

$$\Gamma = \frac{k_\ell h_{fg}}{8\sigma T_{sat} v_{fg} h_{FC}}; N = \frac{h_{FC} r_{mx}}{k_\ell}$$

The recommended equation for the forced convection heat transfer coefficient, h_{FC}, is:

$$\left(\frac{h_{FC} D}{k_b} \right) = 0.023 \left(\frac{GD}{\mu_f} \right)^{0.8} \left(\frac{\mu_f C_{pb}}{\mu_f} \right)^{1/3} \qquad (11)$$

$$q_{FC} = h_{FC} \, (\Delta T_{sat} = \Delta T_{sc}) . \qquad (12)$$

The equation to be used to determine q_B and q_{Bi} is the Mikic-Rohsenow equation [8]. Good agreement with water data is found using the same value of B_M as is used in the higher quality region, 1.89×10^{-14} in SI units (0.0000213 in engineering units).

Comparison with much data [3] shows this procedure agrees with data with an average deviation of 15.0% compared with 17.4% for the Chen [2] correlation.

Future work is needed to refine this procedure for data other than water.

DISPERSED FLOW BOILING

Dispersed flow film boiling is a mode of flow boiling characterized by a dispersion of droplets entrained in the continuous vapor phase. Each droplet is cushioned from the heater surface by the layer of vapor generated by the droplet as it approaches the hot wall. One attribute of interest which results from this cushioned effect is the possibility of excessive heater material temperatures.

Most analyses of dispersed flow film boiling follow the superposition method originally proposed by Bennett [7]. The heat is assumed to be transferred from the wall to the flowing fluid by independent components: wall-to-vapor, vapor-to-drop, and wall-to-drop.

In his original work, Bennett [7] neglected the wall-to-vapor interaction. Later, researchers at

MIT (Laverty [8], Forslund [9]) included this interaction. Groeneveld [10] included all the cited
mechanisms and some of the secondary effects:
flashing, pressure drop and conversion of enthalpy
to potential and kinetic energy.

A study of the drop-wall interaction, performed by
Kendall [11], showed that for high enough wall
temperatures the droplet heat transfer effective-

$$\varepsilon = \frac{q_d}{m_d h_{fg}} \qquad (13)$$

ness is low, Figure 2. Since the void fractions
encountered in dispersed flow are high, and the
droplet effectiveness low, the drop-wall heat
transfer will be small compared to the heat input;
therefore only the wall-to-vapor heat transfer is
considered in determining wall temperature.

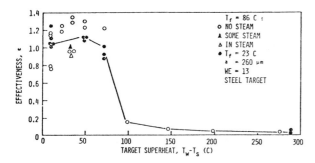

Fig. 2 Drop Heat Transfer Effectiveness

Recently, Yoder [12] used this approximation in
reducing the analysis to the solution of

$$K \frac{x^{3/4} x_{eq}}{(1 - x)^{7/12}} \frac{dx}{dx_{eq}} = (x_{eq} - x) \qquad (14)$$

where

$$K = \frac{2}{3} \frac{D_d}{D_T} \frac{1}{(1 - x_b)^{1/3}} \frac{q}{G h_{fg}} \Pr \frac{\rho_\ell}{\rho_v} \cdot$$

$$\frac{Re_d}{Nu_o x^{3/4} (1 - x)^{1/12}} \qquad (15)$$

is approximately constant with distance downstream.
Eq. (14) was integrated numerically for various
values of K, the results are plotted in Fig. 3.
With the value of K determined, these plots allow
the evaluation of the actual quality and hence
vapor temperature and thus all conditions at any
downstream point.

Then wall temperature is determined from

$$q = h (T_w - T_v) \qquad (16)$$

where

$$\frac{h D_T}{k_v} = 0.023 \left(\frac{G \times D_T}{\mu_v \alpha} \right)^{0.8} \Pr_v^{0.4}$$

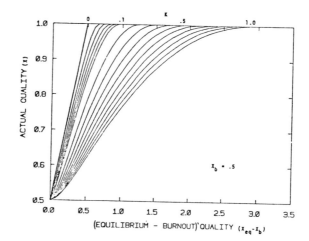

Fig. 3 Actual vs. Equilibrium Quality

Fig. 4 shows for one case the predicted and measured wall temperatures.

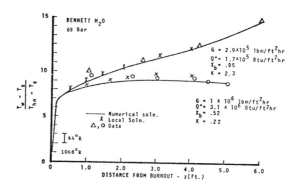

Fig. 4 Predicted and Actual Wall Temperatures in Film Boiling

An investigation of the region just beyond the dryout point is currently being performed at MIT by
Hull [13]. A simplified analysis shows that the
augmentation of the local two phases Nusselt number over the single phase case is indicated by

$$\frac{Nu}{Nu_{B=0}} = \left[\frac{\frac{B}{Re \, Pr} z/D_T}{1 - \exp\left(\frac{-B}{RePr}\right) \frac{z}{D_T}} \right]^{.204} \qquad (17)$$

where

$$B = n \pi D_d \, Nu_0 \, D_T^2$$

as can be seen in Fig. 5, the parameter B/Re Pr is
important in determining the heat transfer characteristics of dispersed flow in the region just
beyond dryout.

As can be seen by inspection of Eq. (1) the drop
diameter is an important quantity in the analysis
of dispersed flow. The drop diameter is difficult
to determine analytically: the drops may be formed
via many different mechanisms and may break up
after their initial formation. More detailed work

53

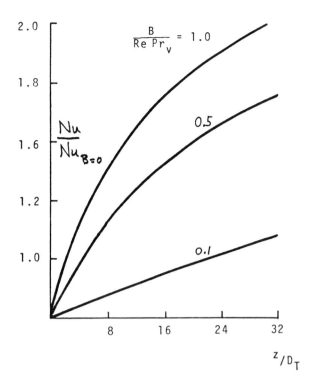

Fig. 5   Increase in Nu Downstream of Dryout

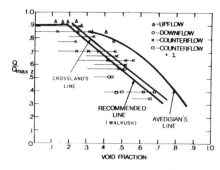

Fig. 6   CHF vs. Void Fraction [14]

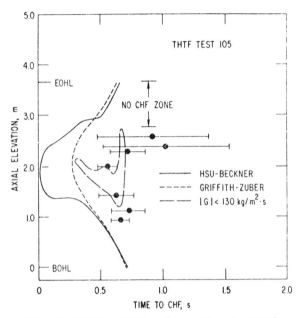

Fig. 7   CHF Prediction for a Flow Reversal

is needed in this area to improve the prediction methods. In particular, it would be helpful to have more information about: liquid entrainment flowrates and the corresponding drop sizes which occur in annular flow, the effect of the Weber number on droplet drag coefficients and droplet formation/breakup mechanisms.

In certain cases, it is important to know when the liquid has been completely evaporated. In this situation, the drop-wall interaction should be considered since this interaction may be of the same order of magnitude as the vapor-drop interaction. Experimental data and analytical analyses of this region are scarce.

Since the majority of the investigations have used circular tube geometry, further work using different geometrical arrangements may be fruitful for the predictions of the effects of grids, spacers, support structures which are present in many practical applications.

TRANSIENT CRITICAL HEAT FLUX

Prof. Peter Griffith investigates transient critical heat flux in up- and down-flow and found it could be related directly with local void fraction as shown in Fig. 6, where Q_{max} is the Zuber prediction in pool boiling [14]. This method was applied to various flow reversal tests at Oak Ridge (TH TF). During the transient the void fraction was predicted and from that the time to burnout at all locations was predicted from Fig. 6. Comparison of predicted vs. measured time to burnout is shown in Fig. 7. [15]. The agreement is quite good but not perfect. More work is needed to refine this prediction method.

BLOWDOWN HEAT TRANSFER

Prof. Griffith has also investigated heat transfer or wall temperatures during blowdown [16]. First, a set of q vs. ΔT curves for various p, G and x are calculated from previously established correlations. Then, for a set of blowdown conditions, p,x, and G and q are predicted vs. time. From the previously constructed graphs wall temperature vs. time is predicted. Fig. 8 shows the effect of quality on the boiling curves. Additional curves for the effects of pressure and mass velocity are also plotted. Fig. 9 shows the comparison of this prediction (BEEST) with data. Agreement is good, but further work is needed to refine this prediction procedure.

CRITICAL HEAT FLUX IN THIN FALLING FILMS

Professors J. Louis and M. El Masri are studying critical heat flux in flowing thin films as they might appear in gas turbine blade cooling passages. A film falling down and inclined plane simulates a film in the radial contrifugal field of a rotating blade cooling passage.

Using Dukler's [17] prediction of film thicknesses

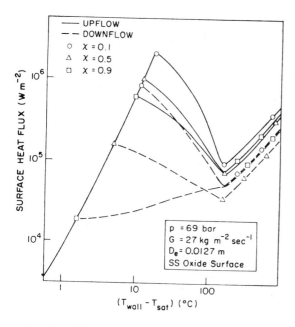

Fig. 8 Effect of Quality on Boiling Curves

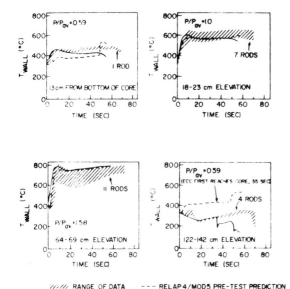

////// RANGE OF DATA   --- RELAP4/MOD5 PRE-TEST PREDICTION
--- BEEST PREDICTIONS (FLUID CONDITIONS FROM RELAP4/MOD5)
P/P$_{av}$ LOCAL PEAKING FACTOR (LOCAL TO AVERAGE POWER RATIO)
STAINLESS STEEL OXIDE SURFACE PROPERTIES USED

Fig. 9 Comparison of BEEST Prediction and
Data from Semiscale Test 5-04-5.

and Hsu's criterion for incipient boiling they
predict the heat flux for the onset of nucleate
boiling q_o. Kutateladze and Leontiev [18] postulated that CHF occurs due to hydrodynamic separation of the film due to transverse vapor flux at
the wall. Using this model they predict a CHF,
q_c.

At low pressure of 1 atm, they predict $q_c < q_o$
suggesting that burnout occurs at the onset of
nucleate boiling. This is verified by data of

General Electric Company as shown in Fig. 10, [19],
[20], [21].

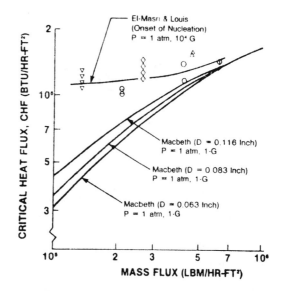

Fig. 10 CHF for Smooth and Crimped Passages

At higher pressure (10 atm) these predictions show
$q_c > q_o$, suggesting that burnout doesn't occur until well after nucleate boiling is established.
No data is available to verify this. Further work
is needed to understand this process before designing blade cooling passages with thin film evaporation.

DIRECT CONTACT HEAT TRANSFER BETWEEN IMMISSIBLE
FLUIDS

Professor M. Kazimi is studying the fuel-to-steel
heat transfer in a nuclear reactor core disruptive
accident.

A mathematical model for direct-contact heat
transfer between immiscible fluids was developed
and tested experimentally. The model describes
how heat is transferred from a hot fluid to an en-
semble of droplets of a cooler fluid that boils as
it passes through the hot fluid. The mathematical
model is based on single bubble correlations for
the heat transfer and a drift-flux model for the
fluid dynamics. The model yields a volumetric
heat transfer coefficient as a function of the
initial diameter, velocity and volume fraction of
the dispersed component. An experiment was constructed to boil cyclopentane droplets in water.
The mathematical and experimental results agreed
reasonably well [22].

Cyclopentane droplets emerge from holes in the bottom plate of a water column and evaporate as
shown in Fig. 11.

The results were applied to investigate the possibility of steel vaporization during a hypothetical core disruptive accident in a liquid metal
fast breeder reactor. The model predicts that substantial steel vaporization may occur in core disruptive accidents, if the steel reaches its saturation temperature rapidly enough. The potential
importance of steel vaporization is dependent on

the accident scenario.

Further work is need to predict both the interior and exterior heat transfer resistance at the droplet bubble interface.

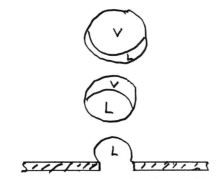

Fig. 11 Droplets Evaporating in Hot Liquid

CALCULATION CODES FOR NUCLEAR REACTORS

Many people are developing calculation procedures for predicting reactor performance. As an example, Prof. Kazimi has developed the THERMIT code which is a three-dimensional two-fluid code for light water reactor (LWR) transients. The formulation of the two-fluid model introduces interfacial exchange terms which have a controlling influence on the two-fluid equations. Therefore, the models which represent these exchange terms must be carefully defined and assessed.

In view of the importance of these interfacial exchange terms, a systematic evaluation of these models has been undertaken. This effort has been aimed at validating THERMIT for both subchannel and core-wide applications. The approach followed has been to evaluate THERMIT for simple cases first and then consider more complex flow conditions.

Fig. 12 is an example of the comparison of THERMIT prediction with data for axial wall temperature distribution for one case.

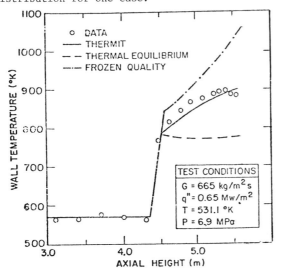

Fig. 12 Comparison of THERMIT Code Prediction with Data

As a result of these comparisons, the following conclusions can be made. First, it is found that THERMIT can adequately predict the void fraction for a wide range of flow conditions. This result implies that both the subcooled vapor generation model and the interfacial momentum exchange model are appropriate.

A second conclusion is that, while the heat transfer model is generally appropriate, specific parts of this model may need to be improved. For example, the pre-CHF wall temperatures are satisfactorily predicted and do not require improvement. The post-CHF temperatures are also adequately predicted and do not require improvement. The post-CHF temperatures are also adequately predicted even though there are some differences between the measured and predicted values. These differences are not uncommon for the post-CHF regime since the data base for the heat transfer correlations is limited. Nevertheless, some of these differences may be due to the method in which the heat transfer model is coupled to the fluid dynamic solution. Consequently, this coupling needs to be evaluated to insure that it is appropriate.

A third conclusion is that in order to accurately predict the flow and enthalpy distribution in subchannel geometry a turbulent mixing model must be added to THERMIT. Both single-phase and two-phase measurements illustrate this point. Without such a model, the mass flux and quality predictions are poorly predicted. Work to include the mixing effect is now in progress.

In addition to the above mentioned validation efforts, the core-wide and transient capabilities have also been assessed. From these comparisons it can be concluded that, on a core-wide basis, THERMIT can accurately predict the core exit temperature distribution. This conclusion is based on comparison with both measurements and COBRA-IV predictions.

It has also been found that THERMIT can accurately predict one-dimensional blowdown transients. For transients of this type, the wall friction and vapor generation rate have the greatest effect on the code predictions.

Finally, for multidimensional transients it can be concluded that the predictions of THERMIT appear to be qualitatively correct and, additionally, THERMIT is at least as computationally efficient as COBRA-IV (explicit). Differences between the predictions of the two codes may be anticipated in light of their respective two-phase flow models.

Research Needs

In April 1979 a Multiphase Flow Workshop was held in Florida. At that time an extensive discussion took place among a large number of research workers attempting to identify our research needs. The following is reproduced with permission of the Workshop Chairman, Dr. N. Veziroglu. It was prepared by the leaders of the discussion, Professors S.G. Bankoff and G.E. McCarthy.

The workshop discussion focused on the identification of several important physical mechanisms in boiling which are not currently understood. In parallel, specific system operation conditions were identified as potential research targets. The operating conditions, listed under the general heading of either Pool Boiling or Flow Boiling, represent practical applications considered important in the workshop discussion. To analyze a single set of operating conditions, it may be necessary to determine the nature and the role of several physical mechanisms. Therefore, the specific applications in pool or flow boiling would serve well not only as a target for boiling research, but also as a guide to selecting conditions for the study of individual boiling mechanisms.

Boiling Mechanisms

A. Nucleation in cavities after departure of previous bubbles, and the nature and distribution of nucleation sites, are areas for continued research. Rapid depressurization also raises questions of homogeneous as well as heterogeneous nucleation.

B. Bubble dynamics, including fluid temperature and velocity distributions, deserve further attention. The role of the microlayer is not well understood. Interactions between bubbles are important because nucleation sites are often closely spaced; but little work has been done in this area to date.

C. Thin film dryout mechanisms, including bubble break-through, surface evaporation, and rivulet formation, have yet to be assessed. Thin film flow itself is not well understood. The liquid surface tension, the liquid-solid interaction, the resulting contact angle, and their temperature dependence, are considered important. Thin film dryout is significant in the analysis of critical heat flux in slug flow, annular flow, and counter-current annular flow.

D. Drop impingement heat transfer depends very much on whether or not direct liquid-solid contact occurs. The simple models recently proposed are restricted to no-contact drop-surface interactions. The conditions of liquid-surface contact must be identified. Drop impingement heat transfer is important in dispersed flow in channels and on turbine blades.

E. The destabilization of a vapor film by pressure waves or other instabilities is important in understanding both dryout and rewet.

G. Explosive boiling (or vapor explosions) present a challenge to the experimentallist because of the small time scales involved.

H. Fluid motion in boiling systems can be significantly different not only from single phase systems, but in particular from unheated two-phase flows with the same velocities and void fraction. Therefore, fluid mechanisms in boiling systems deserves special attention. Examples include drop entrainment and deposition in dispersed or annular-dispersed flow, flow regime transitions, dryout.

I. Solid surface behavior in boiling systems is considered in terms of thermal, mechanical, and chemical response. Both short term response to

single drops and bubbles, and long term response to chemical deposit or mechanical deformation can be extremely important in determining heat transfer rates.

Pool Boiling

A. Enhancement of nucleate boiling heat transfer performance is the goal of special surface geometry and certain additives and coatings. An understanding of nucleation and bubble growth, as well as the surface response, is needed to select design parameters such as cavity size and coating materials.

B. Boiling with internal heat sources is an area which is just recently receiving some attention. There are problems with modeling a melted fuel pool (and boiling steel) using safe and simple techniques such as joule or microwave heating. Basic experimental results are needed.

C. Boiling in confined spaces, and boiling in porous media, present special geometric problems. Periodic flow and heat transfer have been observed. Problems include deposit buildup and thermal stress.

D. Boiling in immiscible liquid systems is considered important in high heat flux electronics and possibly desalination. The boiling surface has a deformable interface which makes this a unique problem.

Flow Boiling

A. Flow regime transitions have been mapped for certain conditions, but are not generally understood or easily predicted on the basis of physical mechanisms. The effect of flow history is difficult to assess. Fundamental small-scale tests are recommended.

B. The critical heat flux and dryout mechanisms in each flow regime are not understood. We have recently been introduced to the "Slow dryout".

C. Rewet of hot surfaces is a complicated process involving drop impingement and/or thin film-rivulet flow, and the thermal response of the surface. Fundamental experiments focusing on local parameters are recommended.

D. Transition boiling, with partial wet and dry surfaces, is different in every flow regime, none of which corresponds to pool boiling. The nature of the liquid-surface interaction, in terms of frequency, duration and energy transfer, is a subject for experimental and analytical work.

E. Subcooled boiling, including the onset of nucleation and the net vapor generation points, has received considerable attention in the past, but still presents some unanswered questions.

F. Heat transfer in the immediate vicinity of the dryout region has not been analyzed. Temperature profiles in the dryout region are difficult to predict. Conduction in the heater due to strong temperature gradients must be included with the analysis of the fluid flow and heat transfer in this drastic transition zone.

G. Blowdown heat transfer involves many flow regimes, including counter-current flow. Questions of transition in unsteady operation have to be answered.

H. Rapid depressurization magnifies the problem associated with blowdown heat transfer in that all transient terms become significant. As yet, it is difficult to describe even the most general parameters such as the flow direction, flow velocity, void fraction, and fluid enthalpy, anywhere in the system.

I. Fouling is extremely important in many flow boiling systems, and has a direct first-order effect on performance. Fouling should be studied from a mechanistic viewpoint.

General

Boiling of binary mixtures is important in many chemical engineering applications, as is boiling in porous media. Applications include most topics listed under both pool and flow boiling.

REFERENCES

1. Rohsenow, W.M., Transactions, ASME, 74, 969, (1952).
2. Chen, J.C., Industrial and Engineering Chemistry Process Design and Development, Vol. 5, 322 (1966).

3. Bjorge, R.W., Hall, G.R. and Rohsenow, W.M., submitted to Int. J. Heat Mass Transfer (1980).

4. Bergles, A.E. and Rohsenow, W.M., J. Heat Transfer, Transactions, ASME, (1964).

5. Traviss, D.P., Rohsenow, W.M. and Baron, A.B., Am. Society of Heating, Air Conditioning and Refrigeration Engineers, ASHRAE Reprint, 2272 RP-63 (1972).

6. Mikić, B.B. and Rohsenow, W.M., ASME, J. of Heat Transfer 91C, 2, (1969).

7. Bennett, A.W., Hewitt, G.G., Kearsey, H.A. and Keeys, R.K.F., AERE-R5372 (1967).

8. Laverty, W.F. and Rohsenow, W.M., J. of Heat Transfer, 89, 90-98 (1967).

9. Forslund, R.P., Ph.D. Thesis, MIT (1966).

10. Groeneveld, D.C., Ph.D. Thesis, U. of Western Ontario (1972).

11. McCarthy, G. and Rohsenow, W.M., MIT Report No. 58694-100 (1978).

12. Yoder, F., Ph.D. Thesis, MIT (1980).

13. Hull, L.M., S.M. Thesis (1980).

14. Smith, R.A., Price, F.A., and Griffith, P., ASME J. Heat Transfer, 98, 2 (1976).

15. Leung, JCM, ANL/RAS/LWR 80-2 (1980).

16. Bjornand, R.A. and Griffith, P., Light Water Reactor Symposium, Vol. I, ASME (1977).

17. Dukler, A.E., Chem. Eng. Prog. Symp. Ser., 56 30 (1960),

18. Kutateladze, S.S. and Leontiev, V.I., Proc. 3rd I.H.T.C., 3, Am. Inst. Chem. Eng. (1966).

19. El-Masri, M.A. and Louis, J.F., J. of Eng. for Power, 100, 586-591 (1978).

20. El-Masri, M.A., Ph. D. Thesis, MIT, 1979.

21. Sundell, R.E. et al., ASME 80-HT-14.

22. Smith, C.S., Ph.D. Thesis, Nucl. Engrg. Dept., MIT, (1980).

Toward the Systematic Understanding of CHF of Forced Convection Boiling: Case of Uniformly Heated Round Tubes

Y. KATTO

Department of Mechanical Engineering, University of Tokyo, Tokyo, Japan

ABSTRACT

The object of this paper is to study the possibility of finding empirically the underlying law which can predict the various trends of critical heat flux (CHF) of forced convection boiling in uniformly heated tubes. The author's generalized correlation of CHF, which classifies CHF into four characteristic regimes, is used for this purpose. The correlation equations concerned are presented in the tabular form for the reader's convenience, with information about the relations between CHF regimes and flow patterns. The propriety of the prediction for the effects of principal parameters (l, d, G and ΔH_i) on CHF is discussed by comparing the predicted CHF with the existing experimental results. The correspondence of the correlation equations with the scaling laws of fluid modeling is also discussed. Finally, future trends in the study of CHF are discussed with special reference to the role of the systematic understanding of CHF.

NOMENCLATURE

C = constant, Eq.(5)
d = I.D. of heated tube, m
F_G = mass velocity scaling factor
G = mass velocity, kg/m^{2}s
H_{fg} = latent heat of evaporation, J/kg
ΔH_i = enthalpy of inlet subcooling, J/kg
K = inlet subcooling coefficient, Eq.(10)
l = length of heated tube, m
p = absolute pressure, bar
q = heat flux, W/m^2
q_c = critical heat flux, W/m^2
q_{co} = q_c for $\Delta H_i = 0$
u_i = inlet velocity of liquid to the system, m/s
μ_l = viscosity of liquid, Pa.s
μ_v = viscosity of vapor, Pa.s
ρ_l = density of liquid, kg/m^3
ρ_v = density of vapor, kg/m^3
σ = surface tension, N/m
χ_{ex} = exit quality at CHF condition

1. INTRODUCTION

Critical heat flux (CHF) of forced convection boiling is a phenomenon which is important in energy problems, but is so complicated that we are apt to be as confused as the case of five blind men and the elephant. Systematic understanding of CHF is indispensable to establish proper relations between individual studies. Recently, excellent studies of theoretical analysis of CHF [1-6] have been made for annular steam-water flow with remarkable results for this purpose. In these studies, however, several assumptions or approximations must be made regarding two-phase flow and CHF condition, so that the limits of the proper application range of the analytical method are not necessarily

clear. At the present stage, therefore, it may be of use to establish empirical methods capable of predicting the whole trend of CHF systematically. In this paper, the discussion is restricted to the most common and simplest CHF which takes place under conditions where a subcooled liquid (or a saturated liquid as a limit) is fed to a uniformly heated tube with a sufficiently high mass velocity and the flow is stable with no oscillations. The generalized correlation recently derived by the author for this CHF [7-11] is used as a tool to test the possibility of systematic understanding.

2. GENERALIZED CORRELATION EQUATIONS OF CHF

2.1 Background of correlation

Fig.1 illustrates two typical boiling systems of saturated liquid flowing parallel to a flat heated surface, for which the author [12,13] has made experimental studies of CHF employing different fluids for the range of $l = 10 \sim 20$ mm, $\sigma\rho_l/G^2l = 2 \times 10^{-6} \sim 3 \times 10^{-3}$ (where $G = \rho_l u_i$), and $\rho_v/\rho_l = 0.000624 \sim 0.0103$ to give the following generalized correlation equations: for the system of Fig.1(a),

$$\frac{q_{co}}{GH_{fg}} = 0.0164 \left(\frac{\rho_v}{\rho_l}\right)^{0.133} \left(\frac{\sigma\rho_l}{G^2l}\right)^{1/3} \quad (1)$$

and for the system of Fig.1(b),

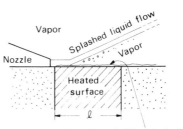

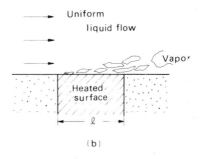

Fig.1 Two Typical Boiling Systems of Forced Convection Flowing Parallel to a Heated Flat Surface: (a) Plane Jet Flow, (b) Uniform Velocity Flow

$$\frac{q_{co}}{GH_{fg}} = 0.186 \left(\frac{\rho_v}{\rho_l}\right)^{0.559} \left(\frac{\sigma\rho_l}{G^2 l}\right)^{0.264} \qquad (2)$$

It is noted that either of Eqs.(1) and (2) is the function of the density ratio ρ_v/ρ_l and the Weber number $G^2 l/\sigma\rho_l$ composed of l, the heated surface length in direction of flow.

On the other hand, Fig.2 shows two limiting cases which may be postulated for forced convection boiling in tubes. The open flow system of Fig.1

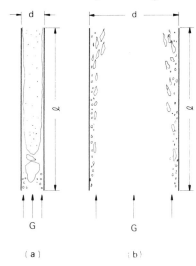

Fig.2 Two Limiting Cases Postulated for Forced Convection Boiling in Uniformly Heated Tubes: (a) Very High l/d or Very Low G, (b) Very Low l/d or Very High G

must be distinguished from the channel flow of Fig.2, but as far as flow patterns are concerned, there are similarities between Fig.1 and Fig.2, suggesting the possibility that the data of CHF accompanying the flow of Fig.2 may be correlated, in an approximate sense at least, by the following function:

$$\frac{q_{co}}{GH_{fg}} = f\left(\frac{\rho_v}{\rho_l}, \frac{\sigma\rho_l}{G^2 l}, \frac{l}{d}\right) \qquad (3)$$

Note: It seems desirable that the analysis of q_{co}/GH_{fg} in Chapter 2 of the author's paper [7] be replaced by the above-mentioned reasoning for the possibility of Eq.(3).

2.2 Correlation equations of CHF in tubes

Basic CHF: q_{co} Analyzing the existing data of 14 different fluids in the range of ρ_v/ρ_l = 0.000270 ~ 0.413, $\sigma\rho_l/G^2 l$ = 2.7 x 10^{-9} ~ 1.0 x 10^0, and l/d = 5.0 ~ 882, based on the assumption of Eq.(3), the author [7,8,10,11] derived the correlation equations for q_{co}, which are presented in Table 1 being classified into four characteristic regimes (L-, H-, N- and HP-regime). q_{co} is called 'basic critical heat flux' following Macbeth [14].

The numbers of the above equations are also tabulated in Table 2 with the naming of sub-regimes (L-1, L-2, H-1 and H-2); and Fig.3 (taken from [10]) shows a comparison with experimental results of various fluids for three nominal values of l/d, while Fig.4 (taken from [11]) represents the boundaries between characteristic regimes for three values of ρ_v/ρ_l. As is noticed in Fig.3, smooth continuous transition between adjoining regimes is neglected for the sake of simplicity.

Table 1. Prediction of q_{co}

L-regime :

$$\frac{q_{co}}{GH_{fg}} = 0.25 \frac{1}{l/d} \qquad (4)$$

$$\frac{q_{co}}{GH_{fg}} = C \left(\frac{\sigma\rho_l}{G^2 l}\right)^{0.043} \frac{1}{l/d} \qquad (5)$$

where C = 0.25 for l/d < 50, C = 0.34 for l/d > 150, and

$$C = 0.25 + \frac{(l/d) - 50}{150 - 50}(0.34 - 0.25) \text{ for } l/d = 50 - 150.$$

H- and N-regime :

$$\frac{q_{co}}{GH_{fg}} = 0.10 \left(\frac{\rho_v}{\rho_l}\right)^{0.133} \left(\frac{\sigma\rho_l}{G^2 l}\right)^{1/3} \frac{1}{1 + 0.0031 l/d} \qquad (6)$$

$$\frac{q_{co}}{GH_{fg}} = 0.098 \left(\frac{\rho_v}{\rho_l}\right)^{0.133} \left(\frac{\sigma\rho_l}{G^2 l}\right)^{0.433} \frac{(l/d)^{0.27}}{1 + 0.0031 l/d} \qquad (7)$$

Boundary dividing Eq.(7) into H- and N-regime is given by

$$\frac{\sigma\rho_l}{G^2 l} = \left(\frac{0.77}{l/d}\right)^{2.70} \qquad (8)$$

HP-regime :

$$\frac{q_{co}}{GH_{fg}} = 0.0384 \left(\frac{\rho_v}{\rho_l}\right)^{0.60} \left(\frac{\sigma\rho_l}{G^2 l}\right)^{0.173} \frac{1}{1 + 0.280\left(\frac{\sigma\rho_l}{G^2 l}\right)^{0.233}\frac{l}{d}} \qquad (9)$$

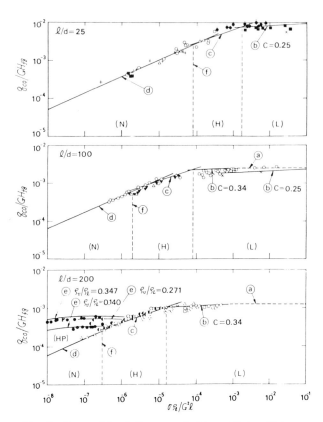

Fig.3 Generalized Representation of q_{co} Data of Various Fluids and Eqs.(4) to (9) for ρ_v/ρ_l = 0.048 in H- and N-regime (cf. Table 2)

Effect of inlet subcooling: K Fig.5 illustrates the relationship of q_c vs. ΔH_i observed generally for the existing data of CHF, that is, linear for L-, H- and HP-regime, while non-linear for N-regime in most cases [7]. In connection with this character, Table 3 may be given to compare

Table 2. Characteristic regimes of CHF

Regime	Sub-regime	Basic CHF q_{co}		Inlet subcooling	Occurrence conditions (see Figs.3 and 4)	Flow pattern at CHF condition (see Fig.6 and [19])	Common term for CHF
		Prediction of q_{co}	Sign in Fig.3	Prediction of K			
L	L-1	Eq.(4)	(a)	Eq.(11)	High l/d, or high $\sigma\rho_l/G^2 l$	(Annular)	Dryout
	L-2	Eq.(5)	(b)	Eq.(12)			
H	H-1	Eq.(6)	(c)	Eq.(13)	Intermediate	Spray annular	
	H-2	Eq.(7)*	(d)	Eq.(14)		(Wispy annular)	
N	—	Eq.(7)*		— **	Low l/d, or low $\sigma\rho_l/G^2 l$	Froth or bubbly	DNB
HP	—	Eq.(9)	(e)	Eq.(15)	High ρ_v/ρ_l, or high l/d with low $\sigma\rho_l/G^2 l$	———	———

* Boundary between H- and N-regime is given by Eq.(8) in Table 1.
** K is unobtainable because of non-linear q_c-ΔH_i relationship in N-regime. In Figs.6 and 10, however, Eqs.(10) and (14) are used tentatively for prediction of q_c in N-regime.

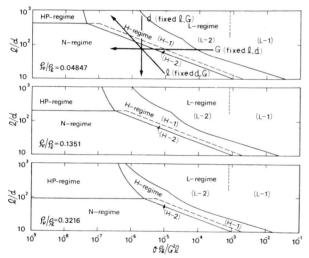

Fig.4 CHF-regime Map (ρ_v/ρ_l = 0.04847, 0.1351 and 0.3216 Correspond to p = 68, 137 and 196 bar Respectively in Case of Water)

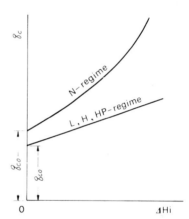

Fig.5 q_c vs. ΔH_i Relationship for Fixed l, d and G

Table 3. Comparison with two regions of Macbeth

G	— increases →			
Macbeth [14,15]	(Low mass velocity region)	(High mass velocity region)		—
The author	L-1 L-2 (L-regime)		H-1 H-2 (H-regime)	N-regime
$q_c - \Delta H_i$	linear			non-linear

Table 4. Prediction of K.

From Eq.(4), $K = 1$ (11)

From Eq.(5), $K = \dfrac{1.043}{4C(\sigma\rho_l/G^2 l)^{0.043}}$ (12)

From Eq.(6), $K = \dfrac{5}{6}\dfrac{0.0124 + d/l}{(\rho_v/\rho_l)^{0.133}(\sigma\rho_l/G^2 l)^{1/3}}$ (13)

From Eq.(7), $K = 0.416\dfrac{(0.0221 + d/l)(d/l)^{0.27}}{(\rho_v/\rho_l)^{0.133}(\sigma\rho_l/G^2 l)^{0.433}}$ (14)

From Eq.(9), $K = 1.12\dfrac{1.52(\sigma\rho_l/G^2 l)^{0.233} + d/l}{(\rho_v/\rho_l)^{0.60}(\sigma\rho_l/G^2 l)^{0.173}}$ (15)

the author's L-, H- and N-regimes with the well-known two regions of Macbeth [14,15] who assumed the linear q_c-ΔH_i relationship for both regions.
    From Fig.5, q_c is written as:

$$q_c = q_{co}\left(1 + K\frac{\Delta H_i}{H_{fg}}\right) \qquad (10)$$

where the dimensionless coefficient K is independent of ΔH_i for the regimes other than N-regime. In this linear case, if the boiling length concept is assumed to hold exactly near $\Delta H_i = 0$, then K can be derived theoretically from the respective equations of q_{co} in Table 1 (cf. [9],[11]). The results are shown in Eqs.(11) to (15) in Table 4, and the numbers of these equations are also tabulated in Table 2. The prediction of K by Eqs.(11) to (15) shows fairly good agreement with the experimental value of K in L-, H- and HP-regime.

2.3 Relation with flow pattern

    For uniformly heated channels, the onset of the CHF condition can be assumed to occur first at the

exit end of the channel (cf. Collier [16] for
example). Observations of the flow pattern near
the exit end of uniformly heated tubes fed with
subcooled water were made by Bergles and Suo [17],
and Bennett et al. [18]. Examples of the flow
pattern map thus obtained are shown in Fig.6
(taken from [19]) in terms of q/GH_{fg} and $\sigma\rho_l/G^2l$.
In Fig.6, the thick lines marked L, H and N represent the predictions of q_c made through Eq.(10)
and Eqs.(5) to (7) for the prescribed values of
ρ_v/ρ_l, l/d and $\Delta H_i/H_{fg}$, while the short vertical
broken line separating H- and N-regime is given
by Eq.(8). The results of Fig.6 suggest that the H-
regime accompanies spray annular flow, while the N-
regime accompanies froth or bubbly flow. This
relation of CHF regimes with flow patterns is
recorded in Table 2 along with the common terms
of dryout and DNB (departure from nucleate
boiling).

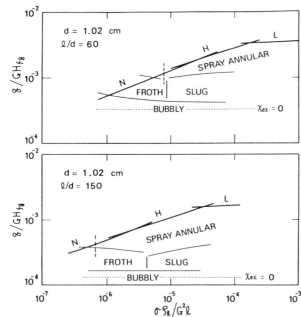

Fig.6 Flow Patterns Near the Exit End of Uni-
        formly Heated Tubes [17] and the Predic-
        tions of CHF: Water, p = 69 bar, $\Delta H_i/H_{fg}$
        = 0.0799

3. PARAMETRIC EFFECTS ON CHF

Fixing the fluid substance and the system
pressure, there are four principal parameters
affecting CHF, namely, l, d, G and ΔH_i. According to the author's correlation, l, d and G
participate in setting the regime of CHF within
the range of ΔH_i for which the q_c-ΔH_i relationship in Fig.5 holds; and three basic modes for
the change of l, d and G are illustrated in Fig.4
by three directed lines respectively.

3.1 Effect of G

For the effect of G on CHF in the case of
fixed l, d and ΔH_i, typical examples have already
been shown in Figs.3 and 6, so that any more
explanations will be omitted to save the space.
Next, Fig.7 shows the experimental data of
q_c vs. χ_{ex} obtained by Weatherhead [20] under
fixed l and d, changing ΔH_i alone at each indicated

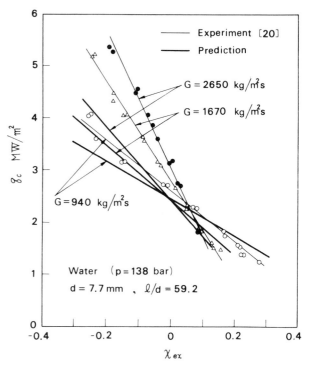

Fig.7 Cross-over of the Mass Velocity Effect
        on CHF ($\Delta H_i/H_{fg} \fallingdotseq 0 \sim 0.22$)

value of G. These data are quoted in leading text-
books (Collier [16], Rohsenow [21]) owing to the
interesting appearance of cross-over of the effect
of G on q_c. Now, the author's CHF regime corresponding to the above experimental condition can
be identified by means of Fig.4, and it reveals that
the cross-over character in Fig.7 is connected with
the experiments made in N-regime. Meantime, the
quantitative evaluation of q_c for $\Delta H_i > 0$ in N-
regime is beyond the limits of the author's
correlation (see Table 2). The predictions shown
in Fig.7 are those made tentatively through Eq.(7)
in Table 1, Eq.(10) and the well-known heat balance
equation for uniformly heated tubes:

$$\chi_{ex} = 4\,\frac{q_c}{GH_{fg}}\,\frac{l}{d}\, - \frac{\Delta H_i}{H_{fg}} \qquad (16)$$

where K is estimated by Eq.(13) in Table 4 because
Eq.(13) is found to be better than Eq.(14) when
the predictions are compared with the experimental
data of Fig.7.

3.2 Effect of d

Functions of l/d affecting Eqs.(4) to (7) in
Table 1 are represented in Fig.8, indicating the
trend that in L-regime, q_{co} is subject to the
direct effect of d; in H-regime, the effect of d
is not noticeable except the range of large l/d;
and in the N-regime, the effect of d is comparatively small.
Experimental results of Lee [22] shown in
Fig.9 are quoted by Lee [23,24] as examples sig-
nifying a somewhat strange effect of d on q_{co}.
The predictions of q_{co} in Fig.9, made by Eqs.(6)
and (7) in Table 1, accord with the above experimental results fairly well except minor deviations,
suggesting that the comparatively sharp rise of
q_{co} with d in the range of small d depends on the

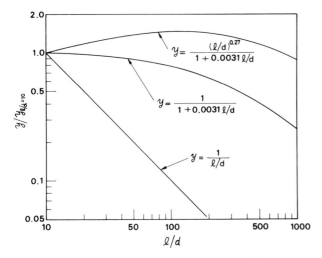

Fig.8 Functions of l/d affecting Eqs.(4) to (7)

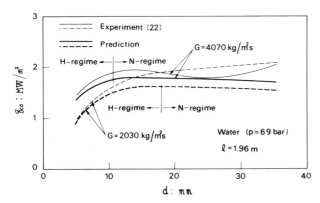

Fig.9 Effect of Diameter on CHF: Water, $p = 69$ bar, $l = 1.96$ m

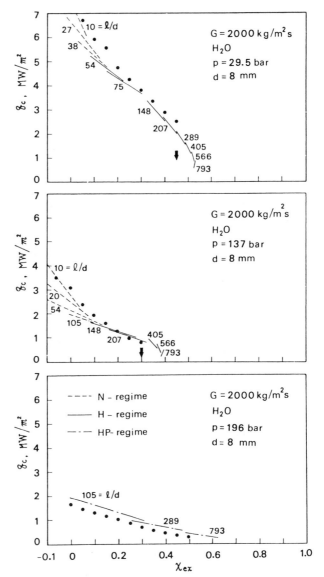

Fig.10 Relationship of q_c vs. χ_{ex} for Water Flowing Through Tubes of $d = 8$ mm With $G = 2000$ kg/m²s at $p = 29.5$, 137 and 196 bar (● and ↓ : USSR Standard Table [26])

experimental condition of H-regime with large l/d, while the dull effect of d on q_{co} in the range of large d is concerned with N-regime (see Fig.4 for the change of regime due to the variation of d).

3.3 Effect of l

If l varies under fixed d and G, the change of CHF regime takes place in the manner of the slanting directed line in Fig.4. As the system pressure or ρ_v/ρ_l increases, the value of $\sigma\rho_l$ decreases so that the directed line mentioned above moves to the left, making it easy to generate the HP-regime in the region of low $\sigma\rho_l/G^2 l$.

Fig.10 (taken from [25]) is concerned with the q_c-χ_{ex} relationship for water flowing through tubes of $d = 8$ mm with $G = 2000$ kg/m²s at $p = 29.5$, 137 and 196 bar. Each short segment in Fig.10 (--- signifies N-regime, —— H-regime, and —·— HP-regime) represents the prediction of q_c-χ_{ex} made by the author's correlation for the prescribed value of l, changing ΔH_i within the ordinary range of 0 to 400 kJ/kg (cf. the note ** of Table 2). Meanwhile, solid circles plotted in Fig.10 come from the standard table produced by the USSR Academy of Sciences [26] for the q_c-χ_{ex} relation of water in uniformly heated tubes of $d = 8$ mm. Then, the χ_{ex} indicated by the arrow pointing downward in Fig.10 is called 'the limiting

quality' in USSR, and it is said that the q_c vs. χ_{ex} relation has a flection point near this position [26]. It may be noted that there is a fairly good agreement between the author's prediction and the USSR standard data (apart from the different interpretations of the form of CHF curve [25]).

Next, Fig.11 represents the experimental data of q_c vs. χ_{ex} obtained by Stevens et al. [27] for R-12 fed to vertical tubes of $d = 16.1$ mm in the range of $\Delta H_i/H_{fg} \doteqdot 0 \sim 0.25$. It is noted that the groups of data for $G = 1021$, 510 and 203 kg/m²s are characteristic of a sharp bend, a linearity, and a curve being convex to the right respectively. Meanwhile, the prediction of q_c vs. χ_{ex} for the same parametric conditions as above are shown in Fig.12 along with the marks of CHF regimes. In Fig.12, Ⓗ ' and Ⓗ - Ⓛ indicate being very near the boundaries between H-1 and H-2, and H and L respectively, while Ⓛ " is fairly close to the state of Ⓗ - Ⓛ . For such conditions as above, the predic-

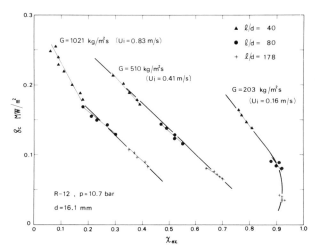

Fig.11 Experimental q_c [27] for R-12 in Tubes of d = 16.1 mm at 10.7 bar ($\Delta H_i/H_{fg} \fallingdotseq 0 \sim 0.25$)

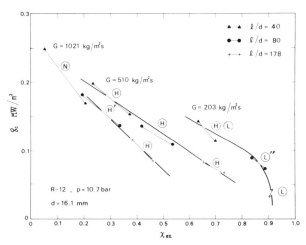

Fig.12 Predicted q_c for R-12 in Tubes of d = 16.1 mm at 10.7 bar ($\Delta H_i/H_{fg} = 0 \sim 0.25$)

tion of q_c for $\Delta H_i > 0$ is approximately given by averaging the predicted q_c's for two adjoining regimes. Thick lines in Fig.12 accord with those in Fig.11 in respect to the characteristics of curvature. As for the quantitative disagreements noticeable in the case of low G and small l/d, it must be noted that the velocity of liquid at the inlet of tube u_i is as low as 0.16 m/s for G = 203 kg/m²s, and 0.41 m/s for G = 510 kg/m²s. It has been mentioned in [7] that when u_i is very low and l/d is small for vertical tubes (VL-regime in [7]), gravity is apt to exert its effect on CHF and the author's correlation loses its accuracy.

4. FLUID MODELING OF CHF

The use of model fluids to substitute for water is of practical importance for the lower cost of CHF test, and there exist two approaches to this problem: an approach based on dimensional analysis, and another based on empirical parameter.

4.1 Dimensional analysis approach

The approach, initiated by Barnett [28,29],

has been extended skillfully by Ahmad [30], who suggests a scaling law corresponding to the idea that CHF is correlated by

$$\left.\begin{array}{l} \dfrac{q_c}{GH_{fg}} = f(\xi_{CHF}, \dfrac{\rho_v}{\rho_l}, \dfrac{l}{d}, \dfrac{\Delta H_i}{H_{fg}}) \\[2mm] \text{where } \xi_{CHF} = (\dfrac{\sigma\rho_l}{G^2 d})(\dfrac{Gd}{\mu_l})^{1/2}(\dfrac{\mu_l}{\mu_v})^{3/10} \end{array}\right\} \quad (17)$$

with empirically derived exponents of 1/2 and 3/10. Eq.(17) is clearly distinguished from the results of other studies [28, 29, 31 etc.] by the fact of being free from the dimensionless groups containing thermal transport properties of fluid.

Eq.(17) signifies that model fluid and water have the same value of q_c/GH_{fg} if the four dimensionless groups ξ_{CHF}, ρ_v/ρ_l, l/d and $\Delta H_i/H_{fg}$ are equal. Then the condition $(\xi_{CHF})_m = (\xi_{CHF})_w$ together with the assumption of equal size gives the mass velocity scaling factor F_G as follows:

$$F_G = \frac{(G)_w}{(G)_m} = \left[\frac{(\sigma\rho_l)_w}{(\sigma\rho_l)_m}\right]^{2/3}\left[\frac{(\mu_l)_m}{(\mu_l)_w}\right]^{2/15}\left[\frac{(\mu_v)_m}{(\mu_v)_w}\right]^{1/5} \quad (18)$$

Meanwhile, on the same principle as above, the author's CHF correlation:

$$\frac{q_c}{GH_{fg}} = f(\frac{\sigma\rho_l}{G^2 l}, \frac{\rho_v}{\rho_l}, \frac{l}{d}, \frac{\Delta H_i}{H_{fg}}) \quad (19)$$

gives F_G as follows:

$$F_G = \frac{(G)_w}{(G)m} = \left[\frac{(\sigma\rho_l)_w}{(\sigma\rho_l)_m}\right]^{1/2} \quad (20)$$

Eq.(18), containing μ_l and μ_v, seems to be of a different nature from Eq.(20). However, in Eq.(18), the effect of μ_l is not large because of the exponent as small as 2/15 (=0.133), and the effect of μ_v is small because the viscosity of saturated vapor does not differ so much between different substances. Therefore, it is not very strange that

Table 5. Comparison of F_G at the state of $\rho_l/\rho_v = 20$ (taken from [8]).

	Model fluid	R-12	R-22	R-113	CO₂
F_G	Empirical	1.40	1.40	1.46	1.13
	Ahmad [30]	1.38	1.38	1.34	1.2
	Eq.(20)	1.28	1.34	1.21	1.19

Eqs.(18) and (20) show almost equal values as shown in Table 5.

4.2 Empirical parameter approach

This approach was initiated by Stevens and Kirby [32], who found that the following relation holds between water and model fluid when ρ_v/ρ_l is equal:

$$\begin{aligned} \chi_{ex} &= f\left\{ (G)_w(d)_w^{1/4}(\frac{d}{l})_w^{0.59} \right\} \\ &= f\left\{ F_G \cdot (G)_m(d)_m^{1/4}(\frac{d}{l})_m^{0.59} \right\} \end{aligned} \quad (21)$$

where the dimensional parameter : $Gd^{1/4}(l/d)^{0.59}$ is empirically determined; and Staub [33] extended it to cover the case of $\Delta H_i/H_{fg} > 0$ by replacing l by the boiling length l_b.

Now, according to the author's correlation,

64

Eq.(21) is presumably the relation concerned with CHF in the H- and N- regime, because the following two expressions are obtained for $\Delta H_i/H_{fg} = 0$ from Eqs. (6) and (7) in Table 1 respectively through Eq.(16):

$$\chi_{ex} = f\left\{\frac{(\rho_l/\rho_v)^{0.2}}{(\sigma\rho_l)^{1/2}}Gd^{1/2}\frac{(1+0.0031\ l/d)^{3/2}}{l/d}\right\} \tag{22}$$

and

$$\chi_{ex} = f\left\{\frac{(\rho_l/\rho_v)^{0.15}}{(\sigma\rho_l)^{1/2}}Gd^{1/2}\right.$$
$$\left. \times\frac{(1+0.0031\ l/d)^{1.155}}{(l/d)^{0.965}}\right\} \tag{23}$$

Fig.13 shows that either function of l/d on the R.H.S. of Eqs.(22) and (23) may be replaced by $(d/l)^{0.59}$ approximately in the practical range of $l/d \doteqdot 40 \sim 350$. Therefore, Eqs.(22) and (23) may

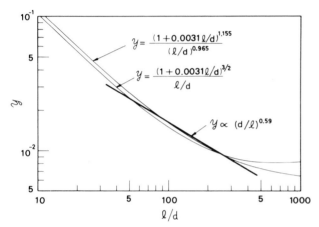

Fig.13   Functions of l/d Related to Eqs.(22) and (23)

be regarded as close to Eq.(21) except a little difference in the exponent of d. It is of interest to refer to Miles et al. [34] who showed that the exponent of d/l in Eq.(21) is diameter dependent.

5. FUTURE TRENDS

The results of this paper suggest that there is room for improvement of the author's CHF correlation equations. Nevertheless, it may be emphasized that it has been clarified that there are great possibilities for the systematic understanding of CHF of forced convection boiling in fundamental boiling systems.

When the systematic understanding of CHF is established for fundamental boiling systems, the current confusions such as deriving disparate conclusions to the same phenomenon may be avoided, and it enables us to make clearly-planned studies on the details of CHF including the elucidation of mechanims. Thus, in the future, CHF of forced convection boiling will become a fruitful area in the field of heat transfer, aiming to escape from the state of the art to that of the science.

The Weber number $G^2l/\sigma\rho_l$ composed with l instead of d plays important roles in the author's correlation, but it has never been taken into account in the studies made so far for CHF of forced convection boiling in uniformly heated channels [see Eq.(17) for example]. Studies are

needed on the validity or the physical meaning of the use of $G^2l/\sigma\rho_l$. In addition, the author's correlation will be improved by taking the effect of viscosity into account, and ξ_{CHF} in Eq.(17) is a suggestive result for this purpose.

REFERENCES

1. P.B.Whalley, P.Huctchinson and G.F.Hewitt, The calculation of critical heat flux in forced convection boiling, Proc. 5th International Heat Transfer Conference, Tokyo, Vol.IV, 290-294 (1974).
2. P.B.Whalley, P.Hutchinson and P.W.James, The calculation of critical heat flux in complex situations using an annular flow model, Proc. 6th International Heat Transfer Conference, Toronto, Vol.5, 65-70 (1978).
3. M.El-Shanawany, A.A.El-Shirbini and W. Murgatroyd, A model for predicting the dry-out position for annular flow in a uniformly heated vertical tube, Int.J.Heat Mass Transfer, 21, 529-536 (1978).
4. T.Saito, E.D.Hughes and M.W.Carbon, Multi-fluid modeling of annular two-phase flow, Nuclear Engineering and Design, 50, 225-271 (1978).
5. P.B.Whalley, The calculation of dryout in a rod bundle, Int.J.Multiphase Flow, 3, 501-515 (1977), and 4, 427-431 (1978).
6. J.Würtz, An experimental and theoretical investigation of annular steam-water flow in tubes and annuli at 30 to 90 bar, Risφ Report No.312 (1978).
7. Y.Katto, A generalized correlation of critical heat flux for the forced convection boiling in vertical unifromly heated round tubes, Int.J. Heat Mass Transfer, 21, 1527-1542 (1978).
8. Y.Katto, A generalized correlation of critical heat flux for the forced convection boiling in vertical uniformly heated round tubes——a supplementary report, Int.J.Heat Mass Transfer, 22, 783-794 (1979).
9. Y.Katto, An analysis of the effect of inlet subcooling on critical heat flux of forced convection boiling in vertical uniformly heated tubes, Int.J.Heat Mass Transfer, 22, 1567-1575 (1979).
10. Y.Katto, General features of CHF of forced convection boiling in uniformly heated vertical tubes with zero inlet subcooling, Int.J.Heat Mass Transfer, 23, 493-504 (1980).
11. Y.Katto, Critical heat flux of forced convection boiling in uniformly heated vertical tubes (Correlation of CHF in HP-regime and determination of CHF-regime map), Int.J.Heat Mass Transfer, 23, 1573-1580 (1980).
12. Y.Katto and K.Ishii, Burnout in a high heat flux boiling system with a forced supply of liquid through a plane jet, Proc. 6th International Heat Transfer Conference, Toronto, Vol. 1, 435-440 (1978).
13. Y.Katto and C.Kurata, Critical heat flux of saturated convective boiling on uniformply heated plates in a parallel flow, Int.J. Multiphase Flow, 6, 575-582 (1980).
14. R.V.Macbeth, Burn-out analysis. Part 4. Application of a local condition hypothesis to world data for uniformly heated round tubes and rectangular channels, AEEW-R 267 (1963).
15. B.Thompson and R.V.Macbeth, Boiling water heat transfer burnout in uniformly heated round tubes: a compilation of world data with

accurate correlations, AEEW-R 356 (1964).

16. J.C.Collier, Convective Boiling and Condensation, p.236, McGraw-Hill (1972).

17. A.E.Bergles and M.Suo, Investigation of boiling water flow regimes at high pressure, in Proc. 1966 Heat Transfer and Fluid Mechanics Institute, pp.79-99, Stanford Univ. Press (1966).

18. A.W.Bennett, G.F.Hewitt, H.A.Kearsey, R.K.F. Keeys and P.M.C.Lacey, Flow visualization studies of boiling at high pressure, Proc. Instn Mech.Engrs, 180-3C, 260-270 (1965-66).

19. Y.Katto, On the relation between critical heat flux and outlet flow pattern of forced convection boiling in uniformly heated vertical tubes, Int.J.Heat Mass Transfer, 24, 541-544 (1981).

20. R.J.Weatherhead, Nucleate boiling characteristics and the critical heat flux occurrence in sub-cooled axial flow water systems, ANL 6675 (1963).

21. W.M.Rohsenow, Boiling, in Handbook of Heat Transfer (Editors: W.M.Rohsenow and J.P. Hartnett),Section 13, McGraw-Hill (1973).

22. D.H.Lee, An experimental investigation of forced convection burn-out in high pressure water. Part III: Long tubes with uniform and non-uniform axial heating, AEEW-R 355 (1965).

23. D.H.Lee and J.D.Obertelli, An experimental investigation of burn-out with forced convection high-pressure water, Proc.Instn Mech. Engrs, 180-3C, 27-36 (1965-66).

24. D.H.Lee, Prediction of Burnout, in Two-Phase Flow and Heat Transfer (Editors: D.Butterworth and G.F.Hewitt), p.304, Oxford Univ.Press (1977).

25. Y.Katto, On the heat-flux/exit-quality type correlation of CHF of forced convection boiling in uniformly heated vertical tubes, Int.J.Heat Mass Transfer, 24, 533-539 (1981).

26. Scientific Council of USSR Academy of Sciences, Recommendations on calculating burnout when boiling water in uniformly heated round tubes, BTD OKB IVTAN, Moscow (1975). (in Russian).

27. G.F.Stevens, D.F.Elliott and R.W.Wood, An experimental investigation into forced convection burn-out in Freon, with reference to burn-out in water, uniformly heated round tubes with vertical up-flow, AEEW-R 321 (1964).

28. P.G.Barnett, Scaling of burn-out in forced convection boiling heat transfer, Proc.Instn Mech.Engrs, 180-3C, 16-26 (1965-66).

29. P.G.Barnett and R.W.Wood, An experimental investigation to determine the scaling laws of forced convection boiling heat transfer. Part 2: An examination of burnout data for water, Freon 12 and Freon 21 in uniformly heated round tubes, AEEW-R 443 (1965).

30. S.Y.Ahmad, Fluid to fluid modeling of critical heat flux: a compensated distortion model, Int.J.Heat Mass Transfer, 16, 641-662 (1973).

31. A.H.Mariy, A.A.El-Shirbini and W.Murgatroyd, Simulation of the region of annular flow boiling in high pressure steam generators by the use of refrigerants, in Two-Phase Flow and Heat Transfer (Editors: S.Kakaç and F.Mayinger), Vol.3, pp.1111-1132, Hemisphere Pub. Corp. (1977).

32. G.E.Stevens and G.J.Kirby, A quantitative comparison between burn-out data for water at 1000 lb/in^2 and Freon-12 at 155 lb/in^2 (abs) uniformly heated round tubes, vertical upflow, AEEW-R 327 (1964).

33. F.W.Staub, Two-phase fluid modeling——the

critical heat flux, Nuclear Science and Engineering, 35, 190-199 (1969).

34. D.N.Miles and K.R.Lawther, The predeiction of boiling crisis conditons using a modified empirical acaling technique, Proc. 5th International Heat Transfer Conference, Tokyo, Vol.IV, 255-259 (1974).

Critical Heat Flux Condition in High Quality Boiling Systems

TATSUHIRO UEDA

Faculty of Engineering, University of Tokyo, Bunkyo-ku,
Tokyo, Japan

ABSTRACT

The critical heat flux and the liquid film flow rate at the exit end of a heating section are measured with two different systems, one for falling film boiling and one for forced flow boiling. For forced flow boiling in uniformly heated tubes, the exit film flow rate at the critical condition is near zero in all cases of the exit qualities higher than 50%, however, the exit film flow rate increases as the heat flux is increased and the exit quality decreases less than 50%. A correlation is proposed for the critical condition in forced flow boiling with high heat fluxes, suggesting that liquid film separation by vapor generation on the heated surface is closely related to the occurrence of the critical condition.

NOMENCLATURE

D = tube diameter
G = mass velocity
g = acceleration of gravity
h_{fg} = heat of vaporization
L = length of heating section
L_s = boiling length
M_f = liquid film flow rate
p = pressure
q = heat flux
u = velocity
u_m = mean velocity of liquid film at exit end of heating section
x = quality
y = distance from wall surface
y_i = liquid film thickness
Γ = liquid film flow rate per unit periphery
μ_ℓ = liquid viscosity
ρ_ℓ = liquid density
σ = surface tension
τ_w = wall shear stress

Subscripts

c = critical heat flux condition
ex = exit of heating section
in = inlet of heating section

INTRODUCTION

The critical heat flux condition characterized by a sharp reduction in the local heat transfer coefficient gives a most important boundary in the performance of flow boiling systems. Although a great deal of work has been made on this problem, a complete understanding of the mechanisms of the critical heat flux condition is not yet obtained even in a simple geometry such as round tubes.

For uniformly heated tubes in which a liquid is vaporized to a relatively high quality of annular flow, it has been noted that the critical heat flux condition usually occurs first at the exit end of the tube. As the supplied heat flux is increased, the liquid film flow rate at the exit end decreases progressively, and then a point is reached where a sharp rise in wall temperature takes place at the exit end. This sharp rise in wall temperature is ascribed to be the desappearance of intimate contact of the heated surface with the liquid film. However, relatively little is known about the mechanisms for the occurrence of the critical condition.

The occurrence of the critical condition in the annular flow region is generally associated with dryout of the annular liquid film. By making measurements of the film flow rate of a climbing liquid film on a heated rod, Hewitt et al.[1] demonstrated that the critical condition occurred when the film flow rate decreased smoothly to zero. This concept was confirmed by the succeeding work of Hewitt et al.[2,3] for the usual forced flow boiling systems of a subcooled liquid at the tube inlet over a range of heat fluxes. Whalley et al. [4] proposed a method for calculating the critical heat flux taking into account droplet interchange between the liquid film and the vapor core along the tube. In this method, the critical condition was assumed to occur at a point where the liquid film flow rate at the exit end falls to zero. Ueda et al.[5] made measurements of R-113 high quality flow in a uniformly heated tube with relatively low heat fluxes, and showed that the position of the sharp rise in wall temperature coincided well with the position predicted the liquid film flow rate to be zero.

However, it may be possible to consider other mechanisms of the critical condition other than the liquid film dryout mentioned above, in the case of high heat fluxes where intensive nucleate boiling takes place under the liquid film. Todreas and Rohsenow [6] suggested the critical condition by nucleation-induced disruption of the annular film, in connection with the upstream critical condition for non-uniform axial heat flux distributions. Katto and Ishii [7] examined the critical phenomenon of a liquid plane jet flowing over a high heat flux surface. In this case, the main part of the liquid splashed away from the surface, and the sharp rise in surface temperature was accompanied with drying out the remaining thin liquid film on the surface by vigorous nucleate boiling.

Ueda and Kim [8] measured the critical heat flux and the liquid film flow rate at exit end of the heating section with R-113 upward flow in a uniformly heated tube of 2.45 m long. Fig.1 shows the results in which the solid plots on the abscissa represent the critical heat fluxes measured. The exit film flow rate at the critical condition is nearly equal to zero for low heat fluxes, however, it shows a trend to increase with increasing critical heat flux over a certain value. This

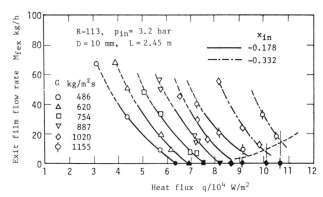

Fig.1 Variation of exit film flow
      rate with heat flux

trend has been found, although not as clear as Fig. 1, in the data of AERE Harwell [9] and also in the results of Staniforth and Stevens [10] obtained with R-12 upflow.

Therefore, the present experimental study was undertaken to examine the relationship between the critical heat flux and the exit film flow rate in a range of high heat fluxes. Two different apparatus were used. One was a system for boiling of a liquid film flowing downward on a heated surface. The other was for forced flow boiling in a uniformly heated tube in which R-113 was vaporized from subcooled liquid at the inlet to a relatively high quality at the exit.

APPARATUS AND PROCEDURE

A flow diagram of the apparatus for falling film boiling is shown in Fig.2(a). The test section was composed of an intake region made of a copper rod, a heating section and a copper tube silver-soldered to the lower end of the heating section. The heating section, a stainless-steel tube of 8.0

mm O.D. and 180 mm long, was heated uniformly by passing an alternative current through it. The liquid heated near the saturation temperature passed through a porous sintered tube and flowed down the outside surface of the intake region and the heating section as a film.

The liquid film flow rate decreased as the film flowed down the heating section, by vapor generation and droplet entrainment due to vapor bubble bursting through the film. Then, the film flow rate at the exit decreased with increasing power input to the heating section. The power input was increased step by step at small increments until the sharp rise in wall temperature was observed at the exit end of the heating section. At each power level, measurements were made of the wall temperature and the liquid film flow rate at the exit end of the heating section. The critical heat flux was determined as a heat flux just before the rise in wall temperature. The test liquid used in this study was water, R-113 and R-11 at near atmospheric pressure.

Figure 2(b) shows a flow diagram of the equipment used for forced flow boiling test. The test section arranged vertically is a stainless-steel tube of 10.0 mm I.D., 12 mm O.D., and the heating section is 1.50 m in length. R-113 liquid was supplied by a circulating pump to the test tube via a flow meter and a preheater, and flowed up inside the test tube. The test tube was heated uniformly by electrical resistance heating, and the axial temperature distribution was measured with thirty two thermocouples fitted along the tube length. At the exit of the test tube, a part of the liquid flows as an annular film on the wall surface and the remaining liquid is entrained in the vapor flow as droplets. For measuring the liquid film flow rate at the exit end, a method was used to extract the liquid film through a porous sintered tube of 10 mm I.D. provided in the upper tube.

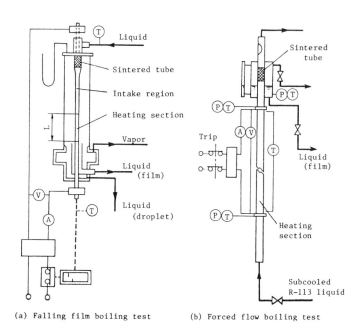

(a) Falling film boiling test   (b) Forced flow boiling test

Fig.2 Diagram of experimental apparatus

The power input to the test tube was increased in small steps keeping the inlet liquid subcool and the flow rate at a set of values. Measurements were repeated until the critical condition occurred at the exit end. The inlet pressure of the test tube was maintained at a value $p_{in} = 3.2 \times 10^5$ Pa through this experiment.

RESULTS OF FALLING FILM BOILING

Figure 3 shows the critical heat flux plotted against the liquid film flow rate at the inlet of the heating section

$$\Gamma_{in} = M_{fin}/\pi D.$$

The critical heat flux increases with increasing inlet film flow rate, showing a similar trend to that obtained for forced flow boiling systems. The dotted lines indicate, for reference, the heat fluxes to evaporate the liquid film completely on the heating section.

The sharp rise in wall temperature occurred at the exit end of the heating section. It will be, therefore, needed to consider the relation between the

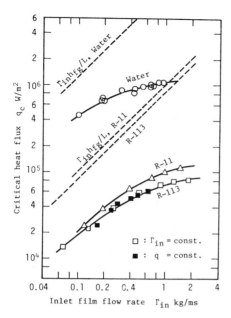

Fig.3 Critical heat flux data
for falling film boiling

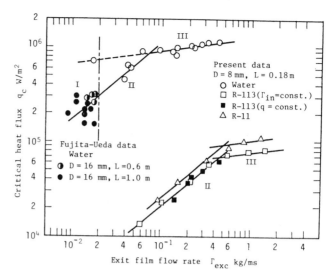

Fig.4 Relation between heat flux and exit
film flow rate at critical condition

critical heat flux and the exit film flow rate

$$\Gamma_{ex} = M_{fex}/\pi D.$$

Fig.4 shows the relationships obtained for the
falling film boiling. As is represented in this
figure, the critical heat flux shows three types
of characteristics in order of increasing
exit film flow rate.

In the region of type I, the critical condition
was caused by a reduction of the exit film flow
rate to some value near zero, irrespective of the
heat flux level. This situation was mainly observed in the experiment made by Fujita and Ueda
[11] with long heating sections. In their experiment, the sharp rise in wall temperature took
place by forming stable dry patches in the thin
film at the exit end of the heating section.
The critical phenomena of types II and III were
observed in the present experiment. In the region of type II, the liquid film involving vapor
bubbles was distorted around the tube periphery
with increasing heat flux, and a large stable dry
patch giving rise to a wall temperature excursion
was formed in a thinned film area. The film
distortion seemed to be caused by non-uniformity
of nucleation sites on the surface. In region III, there was a considerable amount of
liquid flowing at the exit end. The film flowed down covering the tube periphery almost entirely. However, the main part of the film ap-
peared to be separated from the heating surface
when the heat flux was increased to the critical
value. It seemed, from a photographic observation
of the film flow, that the sharp rise in wall
temperature occurred with local boiling up of a
thin film which was left on the surface by the
main film separation.

RESULTS OF FORCED FLOW BOILING

The liquid film flow rate was measured at the
exit of the test tube. Fig.5 shows an example
of the results obtained for an inlet quality x_{in}
= -0.178. The exit film flow rate decreased

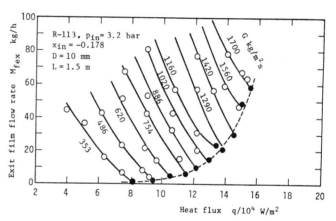

Fig.5 Variation of exit film flow rate
with heat flux

steadily as the heat flux was increased. A solid
symbol plotted at the end of each curve represents
the exit film flow rate at the critical condition.
In the same way as the results of Ueda and Kim
(Fig.1), the present results indicate clearly
that the exit film flow rate at the critical condition increases with increasing heat flux when
the heat flux is higher than about 10^5 W/m².

Figure 6 shows the exit film flow rate at the critical condition plotted against the exit quality
calculated by

$$x_{exc} = x_{in} + \frac{4 L q_c}{G D h_{fg}} . \qquad (1)$$

The open symbols in this figure represent data of
the present experiment and the solid symbols are
those obtained by Ueda and Kim. Both data indicate that the exit film flow rate at the critical condition is near to zero in all cases of
the exit qualities higher than 50 %, however, the
exit film flow rate increases steeply as the heat
flux is increased and the exit quality decreases
less than 50 %.

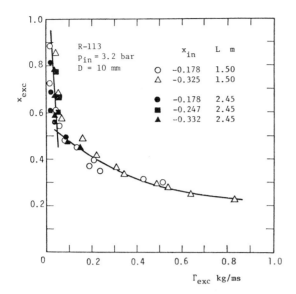

Fig.6 Exit film flow rate at the
critical condition

CORRELATION OF THE CRITICAL CONDITION IN HIGH HEAT FLUXES

For discussing the liquid film flow state at the critical condition, it is useful to introduce the velocity and thickness of the liquid film. The liquid film flow rate per unit periphery is expressed as

$$\Gamma = \rho_\ell \int_0^{y_i} u\,dy = \mu_\ell \int_0^{y_i^+} u^+ dy^+ \qquad (2)$$

where

$$u^+ = u / (\frac{\tau_w}{\rho_\ell})^{0.5},$$
$$y^+ = \frac{y\rho_\ell}{\mu_\ell}(\frac{\tau_w}{\rho_\ell})^{0.5}. \qquad (3)$$

Here, let us suppose for simplification that the liquid film involves no vapor bubble and the velocity distribution in the film can be approximated by the universal velocity profile. Then, the liquid film flow rate Γ can be expressed in terms of the non-dimensional film thickness

$$y_i^+ = \frac{y_i\rho_\ell}{\mu_\ell}(\frac{\tau_w}{\rho_\ell})^{0.5}. \qquad (4)$$

Therefore, the liquid film thickness y_i can be derived by applying the wall shear stress τ_w into Eq.(4), and the mean film velocity

$$u_m = \Gamma/\rho_\ell y_i \qquad (5)$$

is found from the measured film flow rate. In this calculation, the wall shear for the falling film boiling was determined by the relationship

$$\tau_w = \rho_\ell g y_i \qquad (6)$$

whereas, the wall shear for the forced flow boiling was derived by means of the Lockhart-Martinelli correlation [12,13]. Then, an attempt was made to correlate the critical condition with the mean

film velocity u_{mc} thus derived for the measured exit film flow rate at the critical condition Γ_{exc}.

Kutateladze and Leontev [14] gave the limiting condition for boundary layer separation from a permeable flat plate with fluid injection. On the basis of this concept, Tong [15] presented a semi-empirical correlation for the critical heat flux condition in forced flow boiling in the following form:

$$\frac{q_c/h_{fg}}{\rho_\ell U_0} = f(x)\,Re_\ell^{-0.6} \qquad (7)$$

where U_0 is the main stream velocity, $f(x)$ is a function of the bulk quality and Re_ℓ is the liquid Reynolds number. Although this correlation is the one proposed for the critical condition in subcooled and low-quality regions, the concept of the flow separation due to vapor generation on the wall surface seems to be applicable to the critical condition in film flow boiling. Katto and Ishii [7] showed that their critical heat flux data obtained with a saturated liquid jet flowing over the heated surfaces of L = 0.01 – 0.02 m were well correlated, irrespective of the liquid jet thickness, by the following equation:

$$\frac{q_c/h_{fg}}{\rho_\ell u_e} = 0.0164 (\frac{\rho_g}{\rho_\ell})^{0.133}$$
$$\times (\frac{\rho_\ell u_e^2 L}{\sigma})^{-0.33} \qquad (8)$$

where u_e denotes the liquid jet velocity.

Taking into account these results mentioned above, the present data of the type III in falling film boiling were plotted on the coordinates shown in Fig.7. The abscissa represents the Weber number of the exit film flow defined for the heating length

$$We = \rho_\ell u_{mc}^2 L/\sigma. \qquad (9)$$

Fig.7 shows that the critical condition of the type III can be expressed as

$$\frac{q_c/h_{fg}}{\rho_\ell u_{mc}} = 0.0135 (\frac{\rho_g}{\rho_\ell})^{0.08}$$
$$\times (\frac{\rho_\ell u_{mc}^2 L}{\sigma})^{-0.33}. \qquad (10)$$

This correlation agrees well with Eq.(8) in spite of a great difference in heating length used in the tests. This fact along with the visual observation of the film flow state suggests that the critical condition of the type III is closely related to the liquid film separation resulting from high vapor generation.

As shown in Figs.5 and 6, the exit film flow rate at the critical condition in forced flow boiling increased when the heat flux was raised over a value. Then, the critical heat flux data of the exit film flow rate Γ_{exc} greater than 0.05 kg/ms (Fig.6) were plotted, as shown in Fig.8, against the Weber number

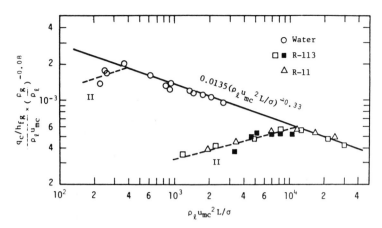

Fig.7 Correlation of the critical heat flux of type III
in falling film boiling

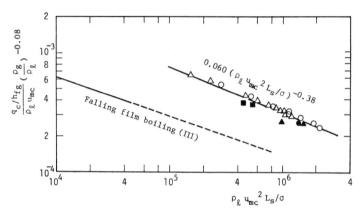

Fig.8 Correlation of the critical heat flux
in forced flow boiling

$$We = \rho_\ell u_{mc}^2 L_s/\sigma \qquad (11)$$

where L_s is the boiling length, i.e., the tube length from the location of $x = 0$ to the exit end of the heating section.   Since the inlet liquid was subcooled in the forced flow boilng systems, the boiling length was substituted in Eq.(11) instead of the heating length L.

Figure 8 represents that the critical condition data in forced flow boiling with high heat fluxes are well correlated by the following equation:

$$\frac{q_c/h_{fg}}{\rho_\ell u_{mc}} = 0.060 \left(\frac{\rho_g}{\rho_\ell}\right)^{0.08}$$

$$\times \left(\frac{\rho_\ell u_{mc}^2 L_s}{\sigma}\right)^{-0.38}. \qquad (12)$$

Eq.(12) gives a value about 2.5 times as high as Eq.(10), however, the gradient of both equations for the Weber number is in fairly good agreement. In the forced flow boiling, it has been noted that an amount of heat is transferred by convection in the annular liquid film and subsequent evaporation at the interface, and then the vapor generation by bubble nucleation on the heated surface is relatively limited.   The liquid film in forced flow boiling was thin enough, and its velocity u_{mc} was

about five times as high as that in falling film boiling. The discrepency in the quantity of $q_c/(h_{fg} \rho_\ell u_{mc})$ between Eq.(12) and Eq.(10) may be ascribed to some extent to the relatively low bubble generation rate in forced flow boiling. Therefore, it seems to be that the liquid film separation by bubble nucleation on the heated surface takes an important part of the critical phenomenon in forced flow boiling with high heat fluxes.

CONCLUSIONS

The conclusions derived by the present investigations on the critical heat flux condition are as follows:

1) The characteristics of the critical heat flux in falling film boiling can be divided into three types according to the exit film flow rate.   In a type of high exit film flow rates, the onset of the critical condition is closely related to liquid film separation from the heated surface by vapor generation.

2) For forced flow boiling in uniformly heated tubes, the exit film flow rate at the critical condition is near to zero in all cases of the exit qualities higher than 50 %, however, the exit film flow rate increases steeply as the heat flux is increased and the exit quality decreases less than 50 %.

3) The critical condition data in forced flow boiling with high heat fluxes are correlated by equation (12). Comparing the characteristics with those for falling film boiling, it seems to suggest that the liquid film separation by bubble nucleation on the heated surface is connected with the occurrence of the critical condition in forced flow boiling with high heat fluxes.

REFERENCES

1. Hewitt,G.F., Kearsey,H.A., Lacey,P.M.C. and Pulling,D.J., Int. J. Heat Mass Transfer, 8, 793-814 (1965).
2. Hewitt,G.F., Kearsey,H.A., Lacey,P.M.C. and Pulling,D.J., Proc. Instn. Mech. Engrs., 1965-66, 180, Pt.3C, 206-215.
3. Bennett,A.W., Hewitt,G.F., Kearsey,H.A., Keeys, R.K.F. and Pulling,D.J., Trans. Instn. Chem. Engrs., 45, T319-333 (1967).
4. Whalley,P.B., Hutchinson,P. and Hewitt,G.F., Proc. Fifth Int. Heat Transfer Conf., Tokyo, 4, 290-294 (1974).
5. Ueda,T., Tanaka,H. and Koizumi,Y., Proc. Sixth Int. Heat Transfer Conf., Toronto, 1, 423-428 (1978).
6. Todreas,N.E. and Rohsenow,W.M., Proc. Third Int. Heat Transfer Conf., Chicago, 3, 78-85 (1966).
7. Katto,Y. and Ishii,K., Proc. Sixth Int. Heat Transfer Conf., Toronto, 1, 435-440 (1978).
8. Ueda,T. and Kim,K.K., Preprint of 16th Japan Heat Transfer Symposium, 211-213 (1979) (in Japanese).
9. Hewitt,G.F. and Hall Taylor,N.S., Annular Two-Phase Flow, Pergamon Press(1970), 227.
10. Staniforth,R. and Stevens,G.F., Proc. Instn. Mech. Engrs., 1965-66, 180, Pt3C, 216-225.
11. Fujita,T. and Ueda,T., Int. J. Heat Mass Transfer, 21, 109-118 (1978).
12. Lockhart,R.W. and Martinelli,R.C., Chem. Eng. Prog., 45, 39-48 (1949).
13. Soliman,M., Schuster,J.R. and Berenson,P.J., Trans. ASME, Ser.C, 90, 267-276 (1968).
14. Kutateladze,S.S. and Leonte'v,A.I., Proc. Third Int. Heat Transfer Conf., Chicago, 5, 1-6 (1966).
15. Tong,L.S., Int. J. Heat Mass Transfer, 11, 1208-1211 (1968).

Heat Transfer Characteristics of Evaporation of a Liquid Droplet on Heated Surfaces

I. MICHIYOSHI

Department of Nuclear Engineering, Kyoto University, Kyoto, Japan

K. MAKINO

Department of Mechanical Engineering, Maizuru Technical College, Maizuru, Kyoto, Japan

ABSTRACT

This paper presents heat-transfer characteristics of evaporation of a single droplet of pure water placed on smooth surfaces of copper, brass, carbon steel and stainless steel at temperature ranging from 80 to 450°C, droplet-dia. 2.54–4.50mm. From time-dependent measurements of surface temperature just below the droplet, and those of change in the geometrical shape of the droplet, we can obtain the time-averaged heat flux q and the surface temperature T_{ws}. The heat transfer characteristics are analyzed by correlating the heat flux q with $\Delta T_{sat} = T_{ws} - T_{sat}$. The heat from this boiling curve, it is found that the present data in the nucleate boiling region come on a certain line parallel to the nucleate pool boiling curve for thin water film, no matter what materials are used for a heated plate, while in the film boiling region, the theoretical curve derived by Baumeister et al. does not always agree with the present data. And new formulae are presented for predicting the film boiling curve better. The transition boiling is also discussed.

NOMENCLATURE

A = projected base area of droplet, m^2
$\bar{A}$ = time-averaged projected base area of droplet, m^2
D = diameter of droplet at any time, m
D_o = initial diameter of droplet, m
L = latent heat of vaporization, J/kg
L_o = enthalpy difference between initial droplet and of saturated water, J/kg
L^* = reduced latent heat of vaporization, J/kg
　　$L^* = L[1 + (7/20) c_{pv}(\Delta T_{sat}/L)]^{-3}$
T_B = temperature T_{wo} at the beginning of bubble formation, °C
T_F = temperature T_{wo} having the maximum evaporation time, i.e. Leidenfrost temperature, °C
T_M = temperature T_{wo} having the minimum evaporation time, °C
T_{sat} = saturation temperature, °C
T_w = surface temperature just below droplet at any time, °C
T_{wo} = initial surface temperature, °C
T_{ws} = time-averaged surface temperature, °C
T_L = liquid temperature of droplet, °C
ΔT = $T_w - T_L$, K
ΔT_{drop} = maximum temperature difference between T_{wo} and T_w, K
ΔT_{sat} = $T_{ws} - T_{sat}$, K
Q = heat required to evaporate completely one droplet, J
V = volume of droplet at any time, m^3
V_o = initial volume of droplet, m^3
a = function of T_{wo}, in equation (5)
c_{pv} = specific isobaric heat capacity of steam, J/(kg·K)

Experimental data were originally published in Int. J. Heat and Mass Transfer, Vol.21, pp.605–613(1978) and Vol.22, pp.979–981(1979).

g = acceleration of gravity, m/s^2
n = function of T_{wo}, in equation (5)
q = time-averaged heat flux, W/m^2
t = time, s
t^* = reversed time, its origin is the time when the droplet just disappears, s
α = heat-transfer coefficient between surface and droplet at any time, $W/(m^2 \cdot K)$
$\bar{\alpha}$ = time-averaged α, $W/(m^2 \cdot K)$
δ = thickness of thin water film, m
κ = thermal diffusivity of surface material, m^2/s
λ = thermal conductivity of surface material, $W/(m \cdot K)$
λ_v = thermal conductivity of steam, $W/(m \cdot K)$
μ_v = viscosity of steam, Pa·s
ρ = density of pure water, kg/m^3
ρ_v = density of steam, kg/m^3
τ = total evaporation time, s.

INTRODUCTION

The information on the heat-transfer characteristics of a liquid droplet on a heated surface is currently important in safety considerations of various types of nuclear reactors. At the present time, there are many works related to the evaporation of a single droplet. But most works are concerned with the "Leidenfrost point" and/or with the temperature range higher than that point. In this temperature range, many analytical models from a small droplet to a larger one are presented with various assumptions. They derive theoretical or semi-empirical equations to calculate the heat-transfer characteristics of the droplet, such as evaporation times and coefficients of heat-transfer and so on [1-3,12]. Those calculated values do not always agree with experimental data, as will be shown later.

Baumeister et al.[4] presented the prediction technique of "Leidenfrost point" by considering effects of the critical temperature of the liquid, thermal properties of the solid, surface energy of the liquid, and surface energy of the solid. But considerable differences exist between those predicted "Leidenfrost point" and the experimental data obtained by many research workers. And other investigators [5,6] discussed about the dynamic instability of droplet. Recently Nishio et al.[16-18] presented the dynamic behaviour of contact of a droplet on a heated surface by visual study, and the systematic measurements of the Leidenfrost temperature T_F for wide range of variables. They proposed a theoretical model, which predicts the T_F point, based upon those observations. Yokoya et al.[20] investigated the mode of contact between a water droplet and a heated surface in the higher temperature range than T_F by solving the differential equations of heat conduction. On the other hand, Styricovich et al. [14] studied the mechanism of heat and mass transfer between a water droplet, about 2 mm in dia, and a heated plate of nickel-chrome foil, 0.1 mm in thick-

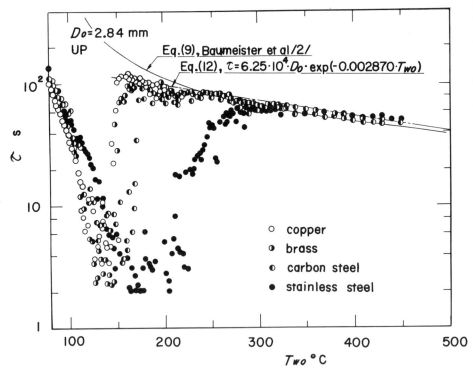

Figure 1   Total Evaporation Time τ vs T_{wo}

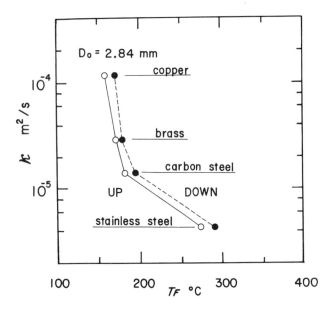

Figure 2   Thermal Diffusivity of Plate vs Leidenfrost Temperature

ness, for the impact velocity ranging slower than 1.5 m/s.

Emmerson[7] conducted an experimental study, dealing with effects of pressure and surface material on the "Leidenfrost temperature, T_F," and the temperature range over it. It is full of interest that he paid as much attention to the wettability and thermal diffusivity of surfaces as to the heat-transfer characteristics of a droplet.

On the other hand, there are few works in the temperature range lower than T_F point. In this temperature range Hiroyasu et al.[3] investigated the evaporation of a droplet of various liquids.

To find out experimentally the heat-transfer characteristics of a droplet, we must conduct time-dependent measurements of the projected base area A of a droplet and of the surface temperature T_w just below it and the liquid temperature T_L within it. The projected base area A and volume of droplet were measured in a temperature range higher than T_F by Gottfried et al.[1] and Wachters et al.[12], respectively. Measurements of the surface temperature T_w were conducted [8,9], using the surface made of stainless steel together with various organic liquids as well as water. They used the chromel-alumel thermocouple of 50 μm dia in [8] and the resistance thermometer of thin film nickel in [9]. No measurement of liquid temperature T_L of a droplet has ever been made.

Rhodes et al.[15] studied experimentally and analytically the Leidenfrost phenomenon for Freon R 114 droplet up to its critical pressure and developed an analytical model, based upon the Gottfried-Lee-Bell model[1]. Ueda et al.[21] and Narazaki et al.[22] investigated experimentally the heat-transfer characteristics of the impinging drops of water by using the cooling curve of heated surface. In the latter one of their purposes is to get the effect of the thickness of a coated layer on the surface.

Recently Akiyama [26] studied the Leidenfrost temperature of a liquid droplet which contains some impurities. For a fuel droplet, Mizomoto et al.[23-25] investigated its evaporation and ignition processes, and also the effect of initial diameter of the droplet on them.

When a single pure droplet having the same size is settled on a heated surface, the total evaporation time τ of a droplet, arranging it with the initial surface temperature T_{wo}, varies mainly with the materials and conditions of the surfaces. In the case of different surface materials and the same surface treatment, one example of the present experimental results of total evaporation time τ is shown in Fig.1. It shows that the evaporation time τ is considerably affected by the surface materials in the range lower than T_F, while it is not different in the range higher than T_F temperature, which has relations to the thermal diffusivities of each surface material as shown in Fig.2. In Fig.1, the theoretical curve of Baumeister et al.[2], which can be calculated taking into account the change in volume of a droplet, is also drawn. It can be seen that the theory does not always agree with the experimental data, as already mentioned.

This paper presents the heat-transfer characteristics

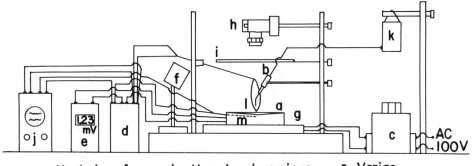

a. Heated surface   b. Hypodermic syringe   c. Variac

d. Cold junction   e. Digital-volt-meter   f. Strobo-scope

g. Heating unit   h. Strobo-streak camera   i. Glass
                     or 8 mm cinecamera

j. Dual-beam   k. Water tank   l. 25 μm diameter
   synchroscope   thermocouple

m. 0.3 mm diameter
   thermocouple

Figure 3   Schematic Diagram of Apparatus

of evaporation of a small droplet of pure water settled on the various heated plates (materials: copper, brass, carbon steel and stainless steel) at the initial surface temperature T_{wo} ranging from 80-450°C. The initial temperature of a droplet is about 20°C. Two thermocouples are made use of for the time-dependent measurements of surface temperature T_w just below the droplet and of liquid temperature T_L within the droplet. Change in geometrical shape of a droplet on the heated plate is recorded by a strobo-streak camera or 8 mm cine-camera. From these measurements we can calculate the time-averaged heat flux q and the time-averaged surface temperature T_{ws}, and then we discuss the so-called boiling curve of a single droplet on the heated plate. To the authors' knowledge, such a boiling curve has not yet been obtained.

EXPERIMENTAL APPARATUS AND PROCEDURE

In the range of temperature T_{wo} less than T_F, we use droplets of water of 2.54, 2.86, 3.29 and 4.50 mm dia and four plates of various materials (copper, brass, carbon steel and stainless steel). These plates, having the same size of 50 mm dia and 20 mm thickness, are engraved with a very small groove on the surface from the edge to the center. In this groove a 0.3 mm dia chromel-alumel sheath type thermocouple is buried with solder. The surface soldered along the groove is plain and finished until the sheath of the thermocouple is almost bared. At room temperature the whole surface of the plate is polished with chrome oxide to a mirror finish and then carefully cleaned by benzene. The surface thus treated is used for heat-transfer experiments at any temperature T_{wo}, which are conducted stepwise from room temperature to higher temperature, which is called UP, and thereafter are conducted again stepwise from that high temperature to room temperature, which is called DOWN. In such a way, the surface is rusted slowly, and the wettability increases.

In the range of temperature T_{wo} higher than T_F, in order to study the effects of the initial droplet size D_o on the heat-transfer characteristics, we use four initial droplet sizes, 2.54, 2.78, 2.86 and 3.29 mm dia, and two plates made of copper and brass, be-

cause in this temperature range the evaporation time is independent of the surface materials as shown in Fig.1. The two plates have the same size, 150 mm in dia and 20 mm in thickness, and those surfaces are slightly concaved so that a droplet may not spill off.

The experimental apparatus is shown schematically in Fig.3. A droplet is deposited softly from the hypodermic syringe (b) to the heated plate (a). The time duration until the droplet disappears after it touched the plate, i.e. the evaporation time τ of droplet, is measured by a stop watch. At the same time, the evaporating droplet is photographed by the strobo-streak camera (h) under the light of strobo-scope (f) or by 8 mm cine-camera (h). From these photographs we can obtain the change in the projected base area A of droplet. Of course, the droplet adheres to or floats over the surface according to the surface temperature T_{wo}. In order to measure the surface temperature T_w just below the evaporating droplet, we use the 0.3 mm dia chromel-alumel sheath type thermocouple (m), as mentioned before. The liquid temperature T_L of a droplet is measured with a 25 μm dia chromel-alumel bare type thermocouple (l), fastened at the tip of a bar, the hot junction of which is set up about 0.1 mm over that for T_w-measurement. The droplet falls so as to pass through these two hot junctions. The signals from these two thermocouples, T_w and T_L, are led to the dual-beam synchroscope (j). To measure the temperature T_{wo}, we use two additional thermocouples, and we adopt the mean temperature of the two as T_{wo}. The temperature T_{wo} ranges from 80-450°C in the present experiment.

RESULTS AND DISCUSSION

The heat required to evaporate one droplet, Q, is given by equation (1),

$$Q = (\pi/6) D_o^3 \rho (L + L_o) \qquad (1)$$

If the initial diameter D_o and temperature of the droplet are fixed, Q is kept constant. On the other hand, Q is expressed by equation (2),

$$Q = \int_0^\tau \alpha \, \Delta T \, A \, dt \qquad (2)$$

The time-averaged heat flux q of the evaporating droplet may be expressed by equation (3),

$$q = Q/(\int_0^\tau A \, dt) \qquad (3)$$

where

$$\int_0^\tau A \, dt$$

is the time-integration of the projected base area A of the droplet during its life-time, and it can be obtained experimentally as mentioned before.

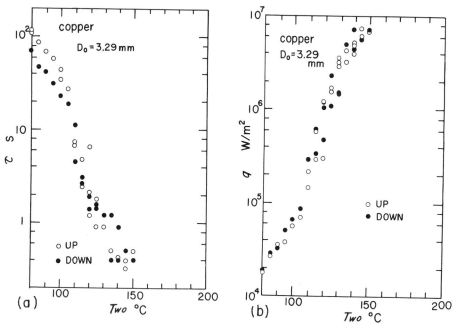

Figure 4   Effects of the Surface Condition of Copper
Plate: (a) Total Evaporation Time; (b)
Time-averaged Heat Flux.

The following expression can be written,

$$\int_0^\tau A \, dt = \bar{A} \tau \qquad (4)$$

In the range of $T_{wo} \leq T_F$

Effects of surface conditions.   Figure 4(a) shows
the evaporation curve, τ vs T_{wo} in the case of 3.29
mm dia, for the copper plate and Fig.5(a) shows
that for the carbon steel plate.   These figures
show that in the case of UP the evaporation time τ
is longer than in the case of DOWN at fixed value
of T_{wo}.   These differences of τ may be due to the
wettability of oxidized film on the surface, which

governs the time-averaged
projected base area $\bar{A}$.
Hence, if we correlate the
time-averaged heat flux q
expressed by equations (3)
and (4) with the surface
temperature T_{wo}, the two
curves for UP and DOWN agree
as shown in Figs. 4(b) and
5(b).   Since this fact is
recognized for each of the
four materials of the plates
the diagrams of q vs T_{wo}
are shown in Fig.6 without
distinction of UP and DOWN.
The appearances of the
evaporating droplet are
illustrated also in this
figure, from which it can
be seen that the same ap-
pearances take place in
the range of the same time-
averaged heat flux q for
any plate materials.   In
other initial diameters'
cases these tendencies are
about the same as 3.29 mm.

Changes in T_W and T_L with
time.   Figure 7 shows
examples of changes in T_W and T_L with time obtained
by using the dual-beam synchroscope for $D_0 = 3.29$mm,
in which(a) is the case of the brass plate, UP, T_{wo}
= 155°C, and (b) is the carbon steel plate, DOWN,
T_{wo} = 185°C.   The surface temperature T_W just below
the droplet falls considerably immediately after
the droplet touched the surface, and then it is kept
at almost constant temperature until the droplet
disappears on the surface, regardless of the material of the plate.   On the other hand, the water
temperature T_L of droplet rises, and then it is
kept at almost constant temperature, which is nearly equal to the saturation temperature T_{sat}.

The more the thermal dif-
fusivity of the plate ma-
terial is, the less the
temperature drop, ΔT_{drop}.
When T_{wo} is lower than T_B,
ΔT_{drop} is nearly constant
regardless of T_{wo}.   This
corresponds to the non-
boiling region.   When T_{wo}
is between T_B and T_M,
ΔT_{drop} increases with increasing T_{wo}.   This corresponds to the nucleate
boiling region.   If T_{wo} is
higher than T_M, ΔT_{drop}
takes nearly the same value
as the maximum value in
the former temperature
range.   This corresponds
to the transition and film
boiling regions.

Such changes of T_W are summarized in Figs 8(a)-(c).
If we subtract the value,
that is calculated by di-
viding the hatched area in
Fig.8 by the evaporation time
τ, from T_{wo}, we can obtain
the surface temperature

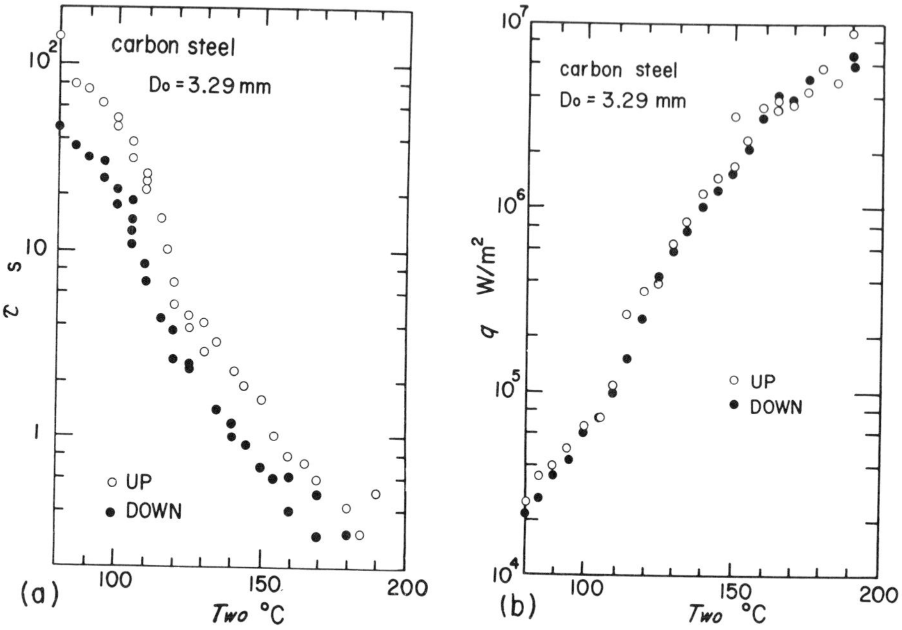

Figure 5   Effects of the Surface Condition of Carbon
Steel Plate: (a) Total Evaporation Time;
(b) Time-averaged. Heat Flux.

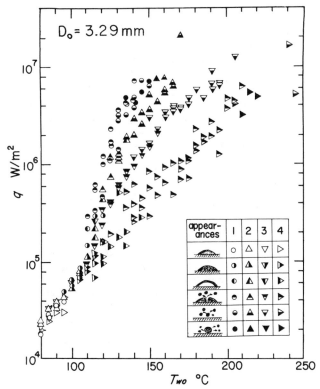

Figure 6   Time-averaged Heat Flux vs Initial Surface Temperature for Four Materials: (1) Copper, (2) Brass; (3) Carbon Steel; (4) Stainless Steel.

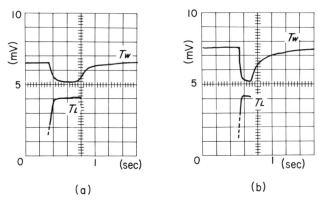

(a)   (b)

Figure 7   Changes in T_w and T_L with Time obtained by Dual-beam Synchroscope, $D_o = 3.29$ mm: (a) Brass Plate, UP, $T_{wo} = 155^{\circ}$C; (b) Carbon Steel, DOWN, $T_{wo} = 185^{\circ}$C.

T_{ws}. Correlating the time-averaged heat flux q with the surface temperature T_{ws} in place of the initial surface temperature T_{wo} in Fig.6 for $D_o = 3.29$ mm, we can obtain Fig.9. This shows that the whole data for the various plate materials coincide very well. Consequently, Fig.9 presents the heat-transfer characteristics of a single water droplet of 3.29 mm dia, independently of the surface conditions and surface materials.

Figures 10(a) and (b) show the effects of the initial droplet diameter D_o (q vs ΔT_{sat} diagram) in the nucleate and transition boiling regions. Although there is a little scatter of experimental data, the effect of D_o on the boiling curve can not be recognized.

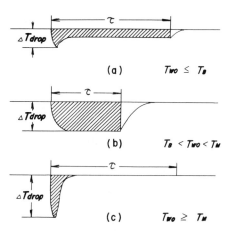

Figure 8   Schematic Diagram of Change in T_w with Time: (a) $T_{wo} \leqq T_B$; (b) $T_B < T_{wo} < T_M$; (c) $T_M \leqq T_{wo}$.

In the range of $T_{wo} > T_F$

The instantaneous diameters of droplet taken from the photographs are presented for two cases, i.e. T_{wo} is 450 and 200°C, in Fig.11. In this figure the time t* is reversed to the normal one, i.e. its origin is the time when the droplet just disappeared. Changes in diameter are almost independent of the initial droplet diameter, but they are influenced mainly by T_{wo}. Then we may write the relations between the diameter of droplet D at any time and the reversed time t* such as equation (5),

$$D = a\ t*^{1/n} \qquad (5)$$

where a and n are functions of T_{wo}. Substituting equation (5) into equation (4), we can obtain the time-averaged projected base area $\bar{A}$ by the calculation,

$$\bar{A} = \int_0^\tau A\ dt*/\tau = (\pi/4)\int_0^\tau D^2\ dt*/\tau$$
$$= (\pi/4)\ [n/(n+2)]\ D_o^2 \qquad (6)$$

where $\bar{A}/D_o^2 = (\pi/4)[n/(n+2)]$ is independent of the initial droplet diameter D_o.

The relation between $\bar{A}/D_o^2$ obtained from the data and $\Delta T_{sat}(= T_{wo}^\dagger - T_{sat})$ is shown in Fig.12, where the calculated values for n = 2 and n = 1 in equation(6) are also shown. It can be seen that n exists between 1 and 2, and n is close to 1 for low ΔT_{sat} but it approaches 2 for higher ΔT_{sat}.

On the other hand, the time-averaged heat flux q obtained from the data with equations (3) and (4) is correlated with ΔT_{sat} in Fig.13, which intimates that the relation between q and ΔT_{sat} is almost independent of the initial droplet diameter D_o. But the data of q for $D_o= 2.54$ mm are somewhat high. This cause is under investigation.

Boiling curve

According to the correlation technique mentioned before, we can get the heat-transfer characteristics of a single water droplet of 2.54-4.50 mm dia for various surface materials of heated plates in the whole temperature range in the present experimental

---

† In this temperature range, since T_w changes as shown in Fig.8(c), T_{ws} is nearly equal to T_{wo}.

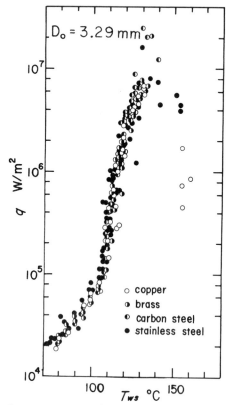

Figure 9 Time-averaged Heat Flux q vs T_{ws} for Four
Plate Materials.

study. The results are shown in Fig.14. This is
the so-called boiling curve.

In the nucleate boiling region the present data provide a straight line parallel to the nucleate pool
boiling curve for thin water film of 1 mm thickness
on a heated plate of wide area [11, 13], no matter

what materials are used for heated plate and no
matter which sizes of droplet are used for initial
diameter as well as no matter how the surface
conditions are made. Our present data are more
close to Shibayama et al.'s recent experimental
data [19], which were obtained by using thin water
film of 2 mm thickness on the horizontal heating
surface of 22 mm dia, than other investigators'
data [11,13]. This might be due to the fact that
Shibayama et al. conducted the experiment of nucleate boiling on the heating surface of small area as
compared with other investigators'. The reason why
the present data are located somewhat above the
nucleate pool boiling curve for thin water film may
be due to the violent nucleate boiling in a very
small domain of which appearances have already been
illustrated in Fig.6.

In the transition boiling region, the data may be
correlated with the same technique as other boiling
regions. The curve for the heated plate of lower
thermal diffusivity (stainless steel) comes to a higher ΔT_{sat} than for higher thermal diffusivity (copper)
independently of D_O. This corresponds to the fact
that the higher the thermal diffusivity of the plate,
the lower the Leidenfrost temperature T_F (see,Fig.2).

In the film boiling region, the relations of q vs
ΔT_{sat} are the same as those presented in Fig.13.
But as shown in Fig.14, ΔT_{sat} at the T_F point is
dependent on the surface materials. The higher the
thermal diffusivity of the plate, the lower the ΔT_{sat}
at the T_F point, as already shown in Fig.2.
Therefore, we may get   four different curves for
four surface materials of the plate in the transition
boiling region, but only one curve in the film boiling region regardless of surface materials and
surface conditions by virtue of floating of droplet
over the surface.

According to the theory of Baumeister et al.[2], the
heat transfer coefficient α at any time defined by
the projected base area A of droplet in the film

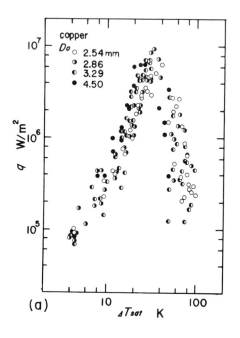

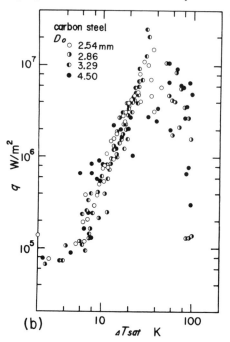

Figure 10 Effects of Initial Droplet Diameter D_O on
the Boiling Curve: (a) Copper, (b) Carbon
Steel.

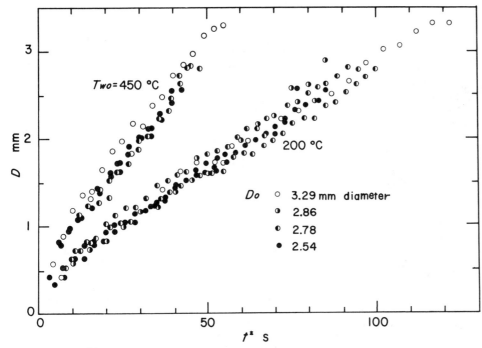

Figure 11 Instantaneous Diameter of Droplet obtained from Photographs vs Reversed Time.

The time-averaged heat-transfer coefficient $\bar{\alpha}$ can be calculated by

$$\bar{\alpha} = \frac{1}{\tau} \int_o^\tau \alpha \ dt$$

$$= \frac{1}{\tau} \int_{V_o}^o \alpha(V) \ \frac{dt}{dV} \ dV \qquad (10)$$

together with equations (7) and (8), and hence it gives the time-averaged heat flux q,

$$q = \bar{\alpha} \ \Delta T_{sat}$$

$$= \frac{1.5}{\tau} \times \frac{\rho \ L}{1.813} \times 3 \ V_o^{1/3} \qquad (11)$$

where τ is given by equation (9).

The relationship of q vs ΔT_{sat} thus obtained is also shown in Fig.14. The present experimental data do not always agree with the theory in the film boiling region. This corresponds to the differences between

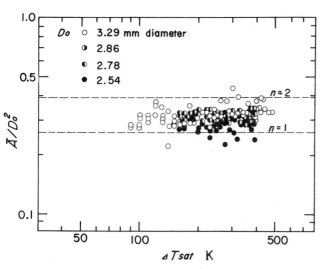

Figure 12 Relations between $\bar{A}/D_o^2$ obtained from the Data and ΔT_{sat}.

boiling region is given by

$$\alpha = 1.1 \times 1.5 \ (\frac{\lambda_v^3 \ \rho \ \rho_v \ L^* \ g}{\mu_v \ V^{1/3} \ \Delta T_{sat}})^{1/4} \qquad (7)$$

On the other hand, the following expression can be derived from the heat balance,

$$- \rho \ L \ \frac{dV}{dt} = \alpha(V) \ A(V) \ \Delta T_{sat} \qquad (8)$$

where V is the volume of droplet at any time. Integration of this equation gives the total evaporation time τ as follows:

$$\tau = \frac{\rho \ L}{1.1 \times 1.813} \ \{\frac{\mu_v}{\lambda_v^3 \ \rho \ \rho_v \ L^* \ g}\}^{1/4} \times \Delta T_{sat}^{-3/4}$$
$$\times \frac{12}{5} \ V_o^{5/12} \qquad (9)$$

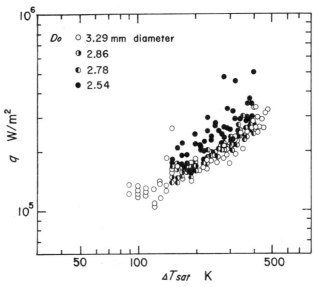

Figure 13 Time-averaged Heat Flux q vs ΔT_{sat} for Four Different Initial Diameters D_o.

equation (9) and the present data which are shown in the diagram of τ vs T_{wo} of Fig.1.

On the other hand, the time-averaged projected base area $\bar{A}$ is given by equation (6), and the total evaporation time τ can be expressed, according to the present experimental data, in the film boiling region by,

$$\tau = 6.25 \times 10^4 \times D_o \ \exp(-0.002870 \cdot T_{wo}) \qquad (12)$$

which is shown in Fig.1. Since $q = Q/(\bar{A} \ \tau)$, we can get the relationship between q and ΔT_{sat}. Figure 14 also indicates this relationship which satisfies the present data in the film boiling region if the value of parameter n is 1-2, as already pointed out

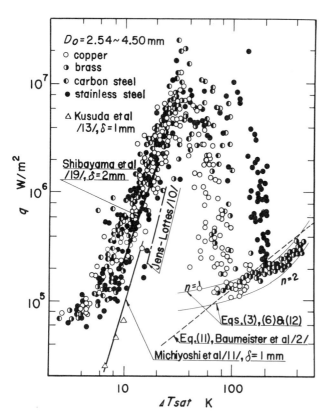

Figure 14 Boiling Curve for Various Heated Plates,
D_O = 2.54-4.50 mm.

in Fig.12.

CONCLUSION

The heat-transfer characteristics of a single droplet of pure water which is placed on smooth surfaces
of four plate materials at temperatures ranging from
80-450°C are summarized by analyzing the so-called
boiling curve (q vs ΔT_{sat} diagram, Fig.14) as
follows:

(1) In the nucleate boiling region, the present
data provide a straight line parallel to the nucleate pool boiling curve for thin water film, no
matter what material is used for a heated plate and
no matter how surface conditions are made as well
as no matter which sizes of droplet are used for
the initial diameter.

(2) In the transition boiling region, the curve
for the stainless steel plate goes to a higher
ΔT_{sat} than for the copper plate, corresponding to
the fact that the higher the thermal diffusivity
of the plate, the lower the Leidenfrost point T_F,
as shown in Fig.2.

(3) In the film boiling region, the present data
provide a certain curve which does not always
agree with the theoretical curve derived by
Baumeister et al. On the other hand, the total
evaporation time τ and the time-averaged projected
base area of droplet $\bar{A}$ can be expressed by equations
(12) and (6), respectively. The time-averaged heat
flux q derived from these equations satisfies quite
well the present film boiling curve in the whole
range of temperature.

REFERENCES

1. Gottfried,B.S., Lee,C.J. and Bell,K.J., Int. J.
    Heat Mass Transfer, 9, 1167-1187(1966).
2. Baumeister,K.J., Hamill,T.D. and Schoessow,G.J.,
    in proceedings of the 3rd International Heat
    Transfer Conference, Vol.4, 66, New York(1966).
3. Hiroyasu,H., Kadota,T. and Senda,T., Trans.
    Japan Soc. Mech. Engrs, 39(328), 3779-3787
    (1973).
4. Baumeister,K.J. and Simon,F.F., J. Heat Transfer,
    95C, 166-173(1973).
5. Hall,W.B., in proceedings of the 5th Inter-
    national Heat Transfer Conference, Vol.4, 125-
    129, Tokyo(1974).
6. Kaji,S., Trans. Japan Soc. Mech. Engrs, 39(328),
    3771-3778(1973).
7. Emmerson,G.S., Int. J. Heat Mass Transfer, 18,
    381-386(1975).
8. Kotake,S., Trans. Japan Soc. Mech. Engrs, 36
    (287), 1146-1152(1970).
9. Seki,M., Kawamura,H. and Sanokawa,K., J. Heat
    Transf., 100C, 167-169(1978).
10. Jens,W.H. and Lottes,P.A., ANL-4627(1951).
11. Michiyoshi,I. and Ueno,N., in proceedings of the
    9th Japan Heat Transfer Symposium, 185-188,Heat
    Transfer Soc. Japan, Tokyo(1972).
12. Wachters,L.H.J., Bonne,H. and van Nouhuis,H.J.,
    Chem. Engng Sci., 21, 923-936(1966).
13. Kusuda,H. and Nishikawa,K., Trans. Japan Soc.
    Mech. Engrs, 34(261), 935-943(1968).
14. Styricovich,M.A., Baryshev,Yu.V., Tsiklauri,G.V.
    and Grigorieva,M.E., in proceedings of the 6th
    International Heat Transfer Conference, Vol.1,
    239-243, Toronto(1978).
15. Rhodes,T.R. and Bell,K.J., in proceedings of
    the 6th International Heat Transfer Conference,
    Vol.1, 251-255, Toronto(1978).
16. Nishio,S. and Hirata,M., in proceedings of the
    6th International Heat Transfer Conference, Vol.
    1, 245-250, Toronto(1978).
17. Nishio,S. and Hirata,M., Trans. Japan Soc. Mech.
    Engrs, 43(374), 3856-3867(1977).
18. Nishio,S. and Hirata,M., Trans. Japan Soc. Mech.
    Engrs, 44(380), 1335-1346(1978).
19. Shibayama,S., Katsuta,M., Suzuki,K., Kurose,T.
    and Hatano,Y., Trans. Japan Soc. Mech. Engrs,
    44(384), 2429-2438(1978).
20. Yokoya, S. and Katto,Y., in proceedings of the
    15th Japan Heat Transfer Symposium, 181-183,
    Heat Transfer Soc. Japan, Tokyo(1978).
21. Ueda,T., Enomoto,T. and Kanetsuki,M., in pro-
    ceedings of the 15th Japan Heat Transfer
    Symposium, 283-285, Heat Transfer Soc. Japan,
    Tokyo(1978).
22. Narazaki,M., Fuchizawa,S., Bannai,F. and Takeda,
    N., in proceedings of the 17th Japan Heat
    Transfer Symposium, 547-549, Heat Transfer Soc.
    Japan, Tokyo(1980).
23. Mizomoto,M., Hayano,H. and Ikai,S., Trans.
    Japan Soc. Mech. Engrs, 44(380), 1366-1373
    (1978).
24. Mizomoto,M. and Ikai,S., Trans. Japan Soc. Mech.
    Engrs, 44(380), 1374-1382(1978).
25. Mizomoto,M., Morita,A. and Ikai,S., Trans.
    Japan Soc. Mech. Engrs, 45(389), 136-146(1979).
26. Akiyama,M., in proceedings of the 17th Japan
    Heat Transfer Symposium, 217-219, Heat Transfer
    Soc. Japan, Tokyo(1980).

Condensation

J. W. WESTWATER
Department of Chemical Engineering, University of Illinois,
Urbana, Illinois, U.S.A.

ABSTRACT

This writing is concerned with the subject of condensation. It is intended to be a review of the present state of knowledge and to indicate what future work needs to be done. It is written from the viewpoint of one who directs research on heat transfer in a university setting.

INTRODUCTION

In this paper it is convenient to discuss filmwise condensation and dropwise condensation separately. The phenomena are the same in a very gross way -- that is in both cases a vapor changes into a liquid. But the details are very different, and no equation has ever been presented which is claimed to describe both forms of condensation.

FILMWISE CONDENSATION: PRESENT STATUS

Filmwise condensation was first treated analytically by Nusselt in 1916 [1]. His paper is one of the classic publications in heat transfer, and it seems unlikely that any subsequent paper on filmwise condensation will ever be of equal importance.

Nusselt showed a clear understanding of the physics of filmwise condensation. He realized that vapor condenses as fast as it can. What prevents the rate from being faster is the resistance of the liquid film itself. Nusselt set up a force balance and a heat balance and produced his famous equation (1) for the heat transfer coefficient on a vertical, solid, cold surface.

$$h = \frac{4}{3} \left[\frac{k_L^3 \rho_L (\rho_L - \rho_V) g \lambda}{4 \mu_L \Delta T} \right]^{\frac{1}{4}} \qquad (1)$$

Slight modifications of this equation also were worked out by Nusselt for filmwise condensation on inclined plates, outside horizontal tubes, and inside horizontal tubes.

Nusselt made a very large number of assumptions in his derivation. It is amazing that his equations are useful in real life. His simple equations are adequate for a great number of applications. Collier [2] has an excellent discussion of Nusselt's derivation and suggests that the theoretical heat transfer coefficient be multiplied by 1.2 to allow for discrepancies between theory and fact for ordinary applications. There are few other design equations for heat transfer for which a safety factor of 20% is large enough.

No one has disputed Nusselt's general description of the physics of filmwise condensation. It is obvious that he had a clear understanding of the phenomenon. However, the many assumptions which he made have attracted much expert attention by scientists desiring to reduce the number of assumptions.

Included in the original assumptions are the following: the condensate flow is streamline; no ripples, waves, or vibrations exist; the liquid is not subjected to electric fields or magnetic fields; gravity is the only force causing the liquid to flow downward; acceleration is negligible; viscosity is the only force restraining the flow of liquid; heat transfer across the liquid film is by conduction only and is unaffected by buoyancy or by surface tension; the vapor is of uniform temperature; the vapor is a pure single component; the vapor exerts no drag on the liquid film; the cold wall is isothermal; and subcooling of the condensate is negligible compared to latent heat effects. Until large digital computers became available, many of Nusselt's assumption could be evaluated only by calculations requiring extraordinary effort. But during our modern age of computers, nearly all the assumptions have been tested in detail by computer calculations.

A brief survey of contributors to improvements in Nusselt's equations follows. Subcooling of the condensate was studied by Drew [3], Bromley [4], Rohsenow [5], and Minkowycz and Sparrow [6]. Turbulent flow of the condensate has been studied by Kirkbride [7], Colburn [8], Kutateladze [9], Labuntsov [10], and Seban [11]. The effect of vapor shear was studied by Carpenter and Colburn [12], Chen [13], Koh, Sparrow, and Hartnett [14], Berenson et al [15], Berman and Tumanov [16], Shekriladze and Gomelauri [17], Rohsenow, Webber, and Ling [18], Jakobs [19], Dukler [20], and Kunz and Yerazunis [21]. The effect of acceleration of the condensate film was studied by Sparrow and Gregg [22]. The effect of non condensible gases was studied by Othmer [23], Colburn and Hougen [24], Akers, Davis, and Crawford [25], Sparrow and Eckert [26], Minkowycz and Sparrow [6], and Sparrow, Minkowycz, and Saddy [27]. The effect of variable physical properties has been studied by Voskresenskiy [28], and by Labuntsov [10].

FILMWISE CONDENSATION: NEEDED WORK

The preceding section seems to imply that the subject of filmwise condensation has been exhausted. But this is not so. Cases still in need of study involve special fluids and special geometries.

Liquid Metals. Nusselt's equation does not work well for liquid metals. Fig. 1 is a graph by Sukhatme and Rohsenow [29] as shown by Collier [2]. The uppermost line is the prediction by Nusselt's theory. The three lines immediately thereunder employ corrections by Sparrow et al [14] and Chen [13]. The theoretical lines give very poor fits to the observed data from three experimenters.

Two reasons have been suggested to account for the disagreement in Fig. 1. One reason is theoretical, namely that one must deal with the heat transfer by considering two phenomena in series. One phenomena is the flux of vapor molecules from the bulk vapor toward the vapor-liquid interface as derived from the kinetic theory of gases. The second phenomena is conduction of the heat released by condensation at the interface across the liquid film to the metal well. For ordinary vapors, the temperature drop from the bulk vapor to the liquid-

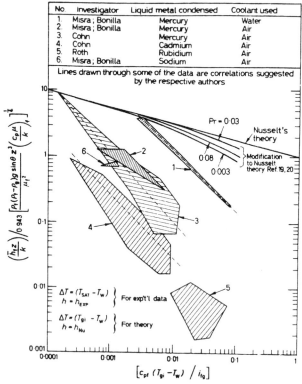

No.	Investigator	Liquid metal condensed	Coolant used
1.	Misra ; Bonilla	Mercury	Water
2.	Misra ; Bonilla	Mercury	Air
3.	Cohn	Mercury	Air
4.	Cohn	Cadmium	Air
5.	Roth	Rubidium	Air
6.	Misra ; Bonilla	Sodium	Air

Lines drawn through some of the data are correlations suggested by the respective authors

Figure 1 Condensation Curves for Four Liquid Metals. Nusselt's Theory is included for comparison (Sukhatme and Rohsenow).

clear how to calculate this case. Suggested correlations in the literature vary from having the heat duty be a weak function to a strong function of the number of rows. The tube clearance, the tube pitch, and vapor velocity obviously are factors which are important here. More work is needed.

Condensation With Two-Phase Flow. When a vapor enters a tube at low velocity, condensation occurs with a distinct segregation of the liquid phase onto the tube wall. This is the model treated by Nusselt. However, if the vapor velocity is high, this simple separation of phases may not occur. Instead, any of several regimes of two-phase flow may exist: slug flow, bubbly flow, mist flow, etc. In fact, all these can exist in one long tube at various positions along the length.

The reverse of this process is boiling during two-phase flow. That subject has been the object of massive research programs by industry, universities, and government laboratories. These have resulted in excellent correlations for design purposes for flow boiling in selected geometries. Generalization has not been achieved. Flow condensation may be analogous to flow boiling, but the amount of information available on flow condensation is meager by comparison.

The so-called Baker [32] flow chart is a useful means for describing two-phase flow in a tube. It was established for air and water flowing simultaneously inside a single horizontal tube, with no heat transfer. This flow map, and modifications of it, are used to attack problems of boiling during

vapor interface is trivial, but for liquid metals it is significant. Knowledge of the accommodation coefficient (condensation coefficient) is needed for the kinetic theory calculation. Fig. 2 shows that most writers get values less than unity. Wilcox and Rohsenow [30] present an error analysis and conclude that the values less than unity are based on temperature measurements beyond the limits of accuracy. They state that a value of unity should be used for the coefficient in the kinetic theory calculation.

The second reason for the poor agreement in Fig. 1 is experimental. Elimination of non-condensible gases from a system operating at liquid metal temperatures is much more difficult than for ordinary systems. Rohsenow [31] concludes that noncondensibles are present in all condensing tests. The Nusselt equation assumes that none are present. Work is needed to arrive at a satisfactory method of predicting the heat flux during the condensation of metal vapors.

Horizontal Tubes in Banks. When vapor condenses outside horizontal tubes in a bundle, condensate falls off the bottom of every tube. Nusselt's equation serves to describe the heat duty for the top tube in each vertical row. The lower tubes have an increased liquid flow, because of the drainage from above. If the condensate flow is assumed to be ideal as shown in Fig. 3a, it is possible to calculate the heat duty for every tube. Unfortunately, much splashing and agitation occurs, as indicated in Fig. 3b. It is un-

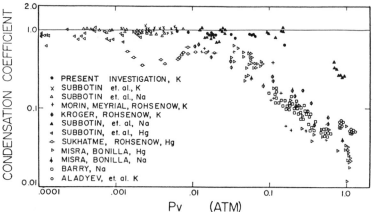

- • PRESENT INVESTIGATION, K
- x SUBBOTIN et. al., K
- ▲ SUBBOTIN et. al., Na
- + MORIN, MEYRIAL, ROHSENOW, K
- ◆ KROGER, ROHSENOW, K
- ▲ SUBBOTIN, et. al., Na
- ◄ SUBBOTIN, et. al., Hg
- -◇- SUKHATME, ROHSENOW, Hg
- ● MISRA, BONILLA, Hg
- ▲ MISRA, BONILLA, Na
- □ BARRY, Na
- ◇ ALADYEV, et. al. K

Figure 2 Condensation Coefficient for Liquid Metals (Wilcox and Rohsenow).

two-phase flow. Bell, Taborek, and Fenoglio [33] show that the idea is useful also for condensation during two-phase flow. Further study is needed on this topic. In particular we need verification whether flow regimes are the same for zero heat transfer as for large heat transfer.

In 1976, Taitel and Dukler [34] used theoretical arguments to obtain an improved flow chart. Fig. 4 shows how flow parameters can change along the flow path. Breber, Palen, and Taborek [35] tested the Taitel-Dukler flow map against existing data for single pure components and concluded that the agreement is very encouraging. They suggested the flow map containing transition

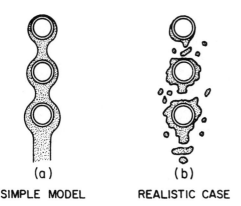

SIMPLE MODEL   REALISTIC CASE

Figure 3 Condensation on a Vertical Row Horizontal Tubes.

zones shown in Fig. 5. This graph is very useful, but admittedly it is oversimplified. More work is needed here.

For condensing mixed vapors, or a vapor plus non-condensible gas, inside tubes at very high flow rates, very little is known. These topics are wide open for study.

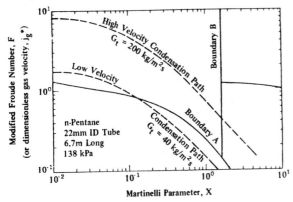

Figure 4 Example Condensation Paths on Taitel-Duckler Flow Map (Breber, Palen, and Taborek).

Direct Contact Condensation. The literature on direct contact condensation is very small compared to that for surface condensers. When a vapor is injected directly into a colder liquid, oscillations (chugging) frequently appear. These are objectionable from the structural standpoint and also constitute a noise nuisance. Marks and Andeen [36] propose a flow map including a region of oscillations; verification and generalization of this map is needed. Oscillations were studied also by Sargis, et al [37]. These authors concluded that the oscillations occur with a random frequency.

An unusual technique for carrying out direct contact condensation is by use of a packed bed. Jacobs, et al [38] studied this method and recommended it for conditions of very small ΔT. Other investigations who have used packed beds for direct contact condensation include Wilke, et al [39], Rai and Pinder [40], Mao and Hickman [43], Finkelstein and Tamir [44], Murty and Sastri [45], Jacobs, Fannar, and Beggs [46], and Jacobs and Cook [47]. Much more work is needed.

Condensation on Fins or Fluted Surfaces. Although it is well established that fins are sometimes very desirable for boiling heat transfer, it does not follow that fins are also good for condensation. Experimental and theoretical studies are

needed to explore the analogy between boiling on a fin and condensing on a fin.

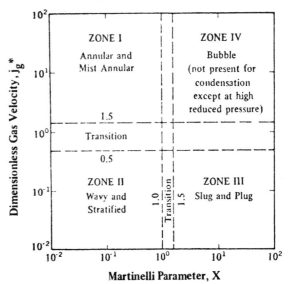

Figure 5 Simplified Criteria for Horizontal, Tubeside Condensation Flow Regimes, Based on Available Data (Breber, Palen, and Taborek).

Tests have been done with condensation on surfaces having low fins, on surfaces with wires attached, and on grooved surfaces. Contributors have included Gregorig [48], Thomas [49], Mehta and Rao [50], Marto, et al [51], Yamamoto and Ishibachi [52], Fujii and Honda [53], Panchal and Bell [54], Mori, Hijikata, Hirasawa, and Nakayama [55], Nakayama, Daikoku, Kuwahara, and Kakizaki [56], Hirasawa, Hijikata, Mori, and Nakayama [57], Hirasawa, Hijikata, Mori, and Nakayama [58], Edwards, et al [59], and Webb [60]. The usual result has been a significant increase in the heat duty compared to a plain surface at the same ΔT. The common explanation is that the fins or grooves cause the condensate flow to be non-uniform. Well-defined rivulets occur for the liquid drainage. Between the rivulets the condensate film is unusually thin, and the heat transfer rate in those regions is unusually great. A similar improvement of condensate drainage and increase in heat duty was achieved by Mikic, et al [61] who put strips of Teflon tape on a condenser tube at intermittent positions. At present it is not possible to do reliable design work for condensation on finned or fluted surfaces. Much additional work is needed. The subject looks very promising.

DROPWISE CONDENSATION: PRESENT STATUS

The first study of dropwise condensation was by E. Schmidt and coworkers [62] in 1930. Progress was extremely slow for 20 years until the subject was taken up by Hampson and coworkers [63-69]. In the span of 10 years Hampson showed clearly that the heat transfer coefficient for dropwise condensation is superior to that for filmwise condensation; that a variety of compounds can serve as promoters; that non condensible gases are deleterious; that the condenser geometry is important (particularly from the standpoint of the vent location); that some promoters have a lifetime of just a few hours, but others are effective for more than 3530 hours; that fouling will ruin promoter activity; that increasing vapor velocity increases the heat transfer coefficient; that varying the metal seems to affect the coefficient; that the angle of inclination is of some importance; and that the so-called spreading coefficient is of value in predicting whether dropwise condensation is apt to occur. He used photography to measure drop sizes and their distribution. He developed an apparatus for measuring the non-condensibles in steam, which is still used in various laboratories today. If Nusselt is the father of filmwise condensation because of his theoretical contributions, then Hampson is the father of dropwise condensation because of his experimental contributions.

In the 20 years subsequent to Hampson's work, a great many researchers have studied dropwise condensation. Most use techniques which are more sophisticated and accurate than those available to Hampson. Thus our knowledge of the subject becomes broader and more detailed. Most modern findings substantiate Hampson's general conclusions. In fact, the only major statement by him which is controversial and unresolved is his statement that all clean metals produce filmwise condensation.

There is agreement that for ordinary (non-noble) metals, a promoter is needed to produce dropwise condensation. Many promoters are high-molecular weight waxes, oils, greases, or soaps. Mercaptans, silicone resins, and several polymers have been used on occasion. Systematic evaluations of various promoters have been carried out by Bromley et al [70,71], Drew et al [72], Osment et al [73], Tanner et al [74,75], and Vylkov [76]. Most promoters wash off in a matter of hours, days or weeks. This handicap has led to widespread tests with Teflon coatings on metal [70,76-89]. Unfortunately if the Teflon is thick enough to be durable, it has a significant resistance to conduction and thus the dropwise condensation is of no advantage. If the Teflon is thin enough to be a heat-transfer advantage, it eventually deteriorates according to Bromley [70]. However, Butcher and Honour [79] report that it endures for at least 2500 hours.

Dropwise condensation on noble metals is of great interest because no promoter need be added. Gold surfaces have been used by many [90-105]. Erb reports a life of at least 4.7 years for dropwise condensation continuously on such a surface [98]. Other noble metals which have been used with some success include silver [98], rhodium [98], palladium [98], and chromium [62,72,76,98,106-109, 132]. Some workers do use promoters with these noble metals, whereas others use no promoter and report excellent results. Chromium may not belong on this list, for two references report filmwise condensation unless an organic promoter is added [72,109].

Liquids which have been condensed in the dropwise mode include steam, ethylene glycol [77,89, 114,115] nitrobenzene [77,89], aniline [77,89], ethanol [109], carbon disulfide [76,110] isooctane [110], mercury [111,112] water-methanol mixture [113], water-methanol-acetone mixture [113], phenol [76], pyridene [76], bromoform [76], ethylene diamine [76], furfural [76], dibromomethane [76], and glycerine [76,115]. Vylkov [76] concludes that dropwise condensation cannot occur unless the surface tension is greater than 18 dynes/cm. Kosky [82] states a cut-off value of 18.5 dynes/cm. If this is the correct limit, most organic liquids will show dropwise condensation only at low temperature (vacuum) operation. In addition, at a sufficiently

elevated temperature (high pressure) approaching its thermodynamic critical point, no liquid will condense in the dropwise mode.

O'Bara et al [116] condensed steam on copper with cupric oleate promoter. At 125 psig, dropwise condensation was in action. At 200 psig, this changed to filmwise condensation. The change in surface tension between these two conditions is from 45 dyne/cm to 41 dyne/cm. This hardly seems enough to cause the observed change in the mode of condensation. Thus in this case we must suspect degradation of the promoter at the higher temperature (197°C) but not at the lower 178°C. All organic promoters must degrade at some sufficiently high temperature, the values of which are unknown at present.

Most observers who have studied the effect of pressure on dropwise condensation have covered small pressure ranges insufficient to cause much change in surface tension. Usually they note that increasing pressure causes an increase in the heat transfer coefficient [78,88,114,118,119]. The relationship between surface tension and temperature however is such that for any given solid-to-vapor ΔT there must exist one pressure which gives a maximum heat transfer coefficient. One observer describes this [117].

A variety of non-noble metals have been used successfully as surfaces for dropwise condensation with promoters. Nearly every investigator has used copper at times. Other metals tested have been copper alloys [76,78,79,94,120,121,132], stainless steel [75,76,110,111,132], aluminum [86,121,132], nickel [76], and iron [132]. Attempts to get dropwise condensation on titanium have failed [132]. Most observers conclude that metals with high thermal conductivity give higher heat transfer coefficients than metals of low conductivity. Some controversy exists, as will be discussed later. Venkatram and Kuloor [122] reported dropwise condensation of steam on stainless steel without a promoter. They used rubber gaskets and rubber connections which could serve as a source of organic contaminants. No one else reports success with stainless steel in the absence of promoters. Drew and coworkers reported dropwise condensation on unpromoted chromium [106,107] when they used rubber seals. Later when cotton seals were used in place of rubber, only filmwise condensation occurred on the unpromoted chromium [72]. It was also noted that different results were obtained for steam from a glass boiler compared to steam from a power plant [72]. Very little promoter is needed to cause dropwise condensation. Tanner et al [75,109,123] state that 30% of a monolayer is adequate. This amount is in sharp contrast to the minimum thickness needed for a noble metal plated on a base metal to obtain dropwise condensation For this, one needs about 500 atomic layers [105].

The geometry chosen for dropwise condensation tests is most often a vertical plate. Tests with flat plates at various angles [104,124,125] show that the angle is only a weak variable except for positions very close to horizontal face up or horizontal face down. A few tests have been made with dropwise condensation on the outside of horizontal tubes of rather small diameter and rather short lengths. Tests on big commercial-size exchangers would be welcome. The largest equipment described for dropwise condensation with an organic promoter is a double effect evaporator with each unit having 45 sq. ft. [131]. The largest gold-plated condenser reported is 11.5 sq. ft.[98]. No data are available for dropwise condensation inside tubes.

Two reports are concerned with the effect of dropwise condensation on horizontal tubes in a vertical row. Erb et al [98] show an improvement of about 20% for the lower tubes compared to a single tube. Birt et al [131] state that flooding of lower tubes does not decrease the heat transfer to the lower tubes.

The effect of vapor velocity is well established [68,125,126,130]. As would be expected, increasing the velocity causes an increase in the heat transfer coefficient for dropwise condensation. Velocities used have been as high as 80 ft/sec [126].

The effect of noncondensible gases also is well established [63,90,118,126-129]. As would be expected, increasing the amount of noncondensibles causes a decrease in the heat transfer coefficient. The amounts of noncondensibles in the test vapors have been as high as 15,000 p.p.m. [118]. Studies of the location of the vent in the condenser show that this is of great importance [127].

The relationship between the heat transfer coefficient, h, for dropwise condensation and the vapor-to-solid surface temperature difference, ΔT, is the concern of all. If dropwise condensation is a nucleation phenomenon, many h vs ΔT curves are possible for a vapor. McCormick and Westwater [133] give evidence that nucleation sites are microscopic pits, scratches, and specks of impurities on the surface. They show that a different miminum ΔT is required for each site to become active. The number of active sites always increases at the ΔT increases. Fig. 6 is a graph relating heat flux, ΔT, and the number of sites taken with numerous separate tests of steam at low pressure condensing on a horizontal copper plate promoted with benzyl mercaptan. For a single, uninterrupted test with a gradual increase in ΔT, the heat flux would rise from curve to curve. The exact shape of the curve would be fixed by the precise number and size distribution of the nucleation sites on the surface. That of course is determined by the surface preparation. The heat flux vs ΔT line may be straight, of almost any slope, or curved. As a result, h vs ΔT can be a curve of many arbitrary shapes. Most experiments give h increasing as ΔT increases. Rose et al argue that h must increase with ΔT changes [92,101,111,120,131]. Others obtain h decreasing as ΔT is increased [92,101,113,116,134, 136]. All these functions are possible.

Diligent efforts have made to obtain a theory of heat transfer during dropwise condensation. To be successful such a theory would result in a design method for commerical equipment. No such satisfactory design scheme is yet at hand. The theory would have to account for the arbitrary number and size distribution of the nucleation sites, the growth rate of every drop, the coalescences which occur, and the roll-off of large drops. So far the first item (surface texture) has been insurmountable. It has been necessary to guess at the size distribution of nucleation sites smaller than about 10 micron. The number of these sites is prodigious, so the guess cannot be trivial. For larger drops, the distribution can be measured. Once the drops start to grow, a conduction model can be formulated to give the growth rate [100,112, 125,135-137,139-148,150-157,159-162]. Conduction through the metal under the drops should be accounted for, and some researchers do consider it [94,137,138]. Most neglect it [100,136,145,146,148, 152,159,162]. Coalescences of drops occur whenever two drops touch. To handle the enormous numbers of drops and coalescences (for example:

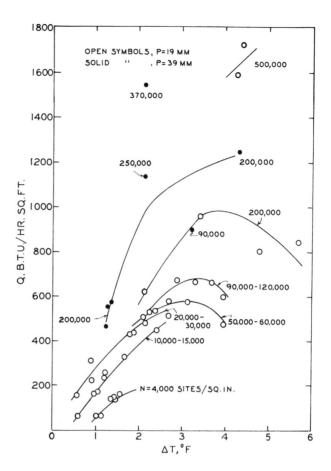

Figure 6   Heat Flux During Dropwise Condensation of Steam at Low Pressure on Promoted Horizontal Copper Plate. The parameter on the Curves is the Population of Active Nucleation Sites Per sq cm. (McCormick and Westwater).

400,000 coalescences [114] to produce one final drop) computer simulation is used [141,142, 144,147,156,160] Glicksman and Hunt [142] used six successive stages of growth and coalescence to handle 3.24×10^8 sites/sq. cm. Computer simulations lead to universal drop size distributions which describe all events averaged over space and time [108,125,143,153,156-158]. The authors are careful to point out that the "universal" distributions probably are not valid for the very small drops with diameters below 10 microns. Fig. 7 shows the Graham-Griffith universal distribution.

With the various equations suggested for dropwise condensation heat transfer, usually the fit to data is very limited. The broadest claim is made by Tanaka [157] whose expression fits the data for steam from three laboratories, mercury from two, and ethylene glycol from one. The Tanaka method requires knowledge of (or a prediction of) the size of departing drops as well as several other parameters which are ordinarily not given. Photographic techniques have been used by various workers to measure the size of departing drops. An equation for predicting the size is available [159].

The majority of people who do heat transfer research are convinced that dropwise condensation is a nucleation phenomenon. Visual evidence by high speed motion picture photography is most easily explained by that argument. The strongest

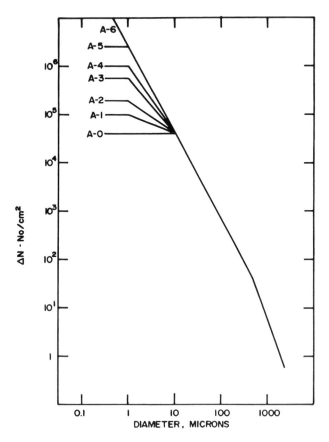

Figure 7 Universal Drop-Size Distribution for Steam
on Promoted Copper. For sizes smaller
than 10 microns, 7 guesses are shown,
and A-3 was selected (Graham and Griffith).

evidence was published in 1965 by Umur and Griffith
[100]. They used elliptically polarized light to
examine the surface between drops. The reflected
optical signal indicated that no layer of liquid
thicker than a monolayer exists between drops.

Prior to that paper, the film rupture model
was supported by certain writers [115,163-166].
Two others continue to support the film rupture
model. According to this model, first proposed
by Jakob [164], vapor condenses on drop-free areas
in the form of a film. The film grows to a critical thickness and then ruptures into tiny drops.
The critical thickness for water on promoted copper
was estimated to be about 0.5 micron [115]. In
1966, Sugawara and Katsuta [167] used interference
photography through a microscope to examine the
area between drops of water condensing on promoted
copper. Interference stripes were obtained, and
that signal indicated the existence of a liquid
layer of about 0.63 micron thickness. In 1974,
Tanasawa [168] wrote that the interferometry evidence was strong, and he stated that a thin liquid
film may exist on the surface between drops.

DROPWISE CONDENSATION: NEEDED WORK

The topics on dropwise condensation which
follow appear to be in serious need of further
work. The interested reader may also consult
earlier review articles by others [169-172].
Critical Heat Flux. During dropwise condensation, vapor moves toward the cold surface, and
liquid drops roll off. There must be a maximum
rate at which these two phases can pass one another.

It may be a hydrodynamic crisis, similar to that
which leads to the maximum heat flux during nucleate boiling. The condensation case seems more
complicated [173]. No theory of a maximum heat
transfer during condensation has been proposed
other than that given by the kinetic theory of
gases.

The maximum heat flux (first crisis) for condensing steam has been determined experimentally
by Takeyama and Shimizu [174] who obtained about
5.5×10^6 kcal/m^2 hr at a ΔT of about 60 K.
Tanasawa and Utaka [175] obtained a flux of about
twice that magnitude at a ΔT of about 60K for a
steam velocity of 12 m/sec. Wilmshurst and Rose
[89] obtained maxima for aniline, glycol, and
nitrobenzene. Much more needs to be done to verify the existence of the maximum heat flux and to
show how it may be predicted.
The Condensation Curve. Fig. 8 shows the
Takeyama-Shimizu results. The authors noted that
for pool boiling a single curve describes nucleate
boiling, transition boiling, and film boiling.
They deduced that a similar curve exists for condensation. Their results show a condensation
curve with a maximum and a minimum and three
branches corresponding to dropwise, transition,
and filmwise condensation. Wilmshurst and Rose
[89] and Tanasawa and Utaka [175] present similar
curves. Obviously the subject of the complete
condensation curve has hardly been touched, and
this should prove to be an exciting field for
future work. Note that for water the maximum flux
of reference [174] for condensation is about four
times the maximum flux for boiling. Shea and Krase
[130] suggested in 1940 that a maximum flux for
condensation must exist, but they did not envision
the actual curve with two crises.

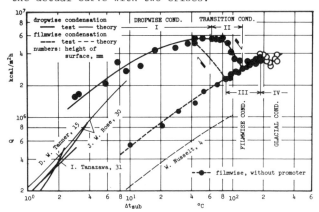

Figure 8 Complete Condensation Curve for Steam on
Promoted Copper. Other results are shown
for comparison (Takeyama and Shimizu).

Noble Metals. The wettability of noble metals
has been a question of dispute for years. Tennyson
Smith [176] shows that since 1934 there have been
at least 8 studies, led by Zisman and his coworkers,
which prove that gold is wettable by water. Simultaneously there were 18 studies, led by Zettlemoyer
and his coworkers, which prove that gold is not
wettable. Smith carried out a first study to get
direct evidence of the surface cleanliness by Auger
electron spectroscopy. He concludes that clean
gold is wet by water. He shows that less than one
monomolecular layer of organic contamination will
cause non-wetting. He shows that serious amounts
of carbon compounds and oxygen are adsorbed from
air in 10 minutes. Bromley states that water condenses in the filmwise fashion on pure clean gold

[93]. Woodruff [103] got the same result. Erb [97,99] says that pure steam condenses as drops on pure gold. Umur and Griffith [100] state that gold gives dropwise condensation (without a promoter). Considering Smith's results it is clear that in a real heat exchanger gold surfaces are contaminated. Studies are needed to determine what contaminants cause dropwise condensation on noble metals and what contaminants will ruin it.

Effect of Metal Conductivity. Mikic and coworkers [102,177-179] write of the "constriction effect". Drops on a metal surface act as resistance to heat transfer and alter the flux lines in the metal. The authors conclude that the product of the physical properties, $k \rho c_p$, for the solid is an important parameter during dropwise condensation. Others agree [94,137]. Fig. 9 is an illustration of the effect of thermal conductivity. It predicts that the heat transfer coefficient decreases toward zero as the metal conductivity decreases toward zero. Also in Fig. 9 are data by

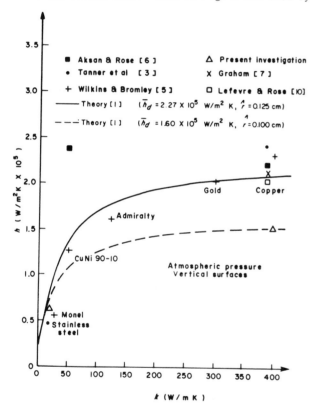

Figure 9 Effect of Metal Thermal Conductivity on Heat Transfer Coefficient for Drop-wise Condensation of Steam (Hannemann and Mikic).

Rose which disagree. Rose [121,180,181] and his coworkers argue that the bulk metal conductivity is not important, but that the surface conditions are important. This matter must be resolved.

Deficiencies with Theory. The outstanding difficulty is that of describing quantitatively the surface texture. This is needed so that the size distribution of the nucleation sites may be used as a starting point for theories. The sites are much smaller for condensation than for boiling, thus this problem is extremely onerous. Can one make artifical sites of the right size and distribution? This problem is worthy of study.

Nearly all theoretical attacks on the heat transfer use an accommodation coefficient (condensation coefficient) in conjunction with the kinetic

theory of gases. Values used for this have included 0.04 [126], 0.1 [118], and 1.0 [144,182]. This is a serious discrepancy, and more work is needed. It is not at all clear as to what goes on between drops on a surface. The surface is colder than the vapor, thus the second law of thermodynamics demands some energy be transferred there. Various writers assume zero heat flux at the area between drops [147,156,183] or state that no condensation occurs between drops [100]. Proof is needed that we can (or cannot) ignore the so-called bare areas.

The temperature under a nucleation site has been shown to rise and fall for each drop growth and departure there [184]. Thus transverse temperature gradients surely exist on the bare areas and these may constitute a driving force for movement of adsorbed molecules. We need to know more about adsorption and surface diffusion.

As to whether drops may have internal circulation because of surface tension driving forces, Lorenz and Mikic [185] conclude that the Marangoni effect is insignificant during dropwise condensation.

The problem of noncondensible gases is really a problem of geometry. Apparently the only case which can be handled with some confidence is that for filmwise condensation of vapor plus noncondensibles inside a straight round vertical tube with gentle down flow. It seems wise to try this geometry with dropwise condensation.

Big Equipment. Knowledge is sadly lacking concerning dropwise condensation in big equipment. Nothing as large as 100 sq. ft. has been tested. We have no idea whether fouling, cleaning, and maintenance will be problems of ordinary routine or whether they will be insurmountable. Work is needed on big exchangers.

ACKNOWLEDGMENT

Support for the preparation of this writing was furnished by the U.S. Department of Energy, Grant DE-AC02-80ER10596.

REFERENCES

1. Nusselt, W., "Die Oberflachenkondensation des Wasserdampfes", Zeitschr. Ver. Deutsch. Ing., 60, 541 (1916).
2. Collier, J.G., Convective Boiling and Condensation, McGraw-Hill Co., New York, 301-346 (1972).
3. Drew, T.B., described in McAdams, W. H., Heat Transmission, 3rd ed., McGraw-Hill Co., New York (1954).
4. Bromley, L. A., "Effect of Heat Capacity of Condensate", Ind. Eng. Chem., 44, 2966 (1952).
5. Rohsenow, W.M., "Heat Transfer and Temperature Distribution in Laminar Film Condensation", Trans. ASME, 78, 1645-48 (1956).
6. Minkowycz, W.J. and Sparrow, E.M.. Condensation Heat Transfer in the Presence of Non-Condensibles, Interfacial Resistance, Superheating, Variable Properties, and Diffusion", Int. J. Heat Mass Transfer, 9, 1125-44 (1966).
7. Kirkbride, C.G., "Heat Transfer by Condensing Vapor on Vertical Tubes", Trans. A.I.Ch.E., 30, 170 (1933-34).
8. Colburn, A.P., "Calculation of Condensation Where a Portion of the Condensate Layer is in Turbulent Motion", Trans. A.I.Ch.E., 30, 187 (1933-34).
9. Kutateladze, S.S., Fundamentals of Heat Trans-

fer, Academic Press, Inc., New York, 309 (1963).

10. Labuntsov, D.A., "Heat Transfer in Film Condensation of Pure Vapors on Vertical Surfaces and Horizontal Tubes", Teploenergetika, No. 5, (1960).

11. Seban, R.A., "Remarks on Film Condensation With Turbulent Flow", Trans. ASME, 76, 299 (1954).

12. Carpenter, F.G. and Colburn, A.P., "Effect of Vapor Velocity on Condensation Inside Tubes", Proc. of General Discussion on Heat Transfer, Inst. Mech. Engr. and ASME, 20-26 (1951).

13. Chen, M.M., "An Analytical Study of Laminar Film Condensation", J. Heat Transfer, 83, 48-55 (1961).

14. Koh, J.C.Y., Sparrow, E.M., and Hartnett, J.P., "Two-Phase Boundary Layer in Laminar Film Condensation", Int. J. Heat Mass Transfer, 2, 69-82 (1961).

15. Soliman, M., Schuster, J.R., and Berenson,P.J., "A General Heat Transfer Correlation for Annular Flow Condensation", J. Heat Transfer, 90, 267-276 (1968).

16. Berman, L.D. and Tumanov, Yu A., "Condensation Heat Transfer in Vapor Flow Over a Horizontal Tube", Teploenergetika, 10, 77-84 (1962).

17. Shekriladze, I.G. and Gomelauri, V.I., "Theoretical Study of Laminar Film Condensation of Flowing Vapor", Int. J. Heat Mass Transfer, 9, 581-91 (1966).

18. Rohsenow, W.M., Webber, J.H., and Ling, A.T., "Effect of Vapor Velocity on Laminar and Turbulent Film Condensation", Trans. ASME 78, 1637-43 (1956).

19. Jacobs, H.R., "An Integral Treatment of Combined Body Force and Forced Convection in Laminar Film Condensation", Int. J. Heat Mass Transfer, 9, 637-648 (1966).

20. Duckler, A.E., "Fluid Mechanics and Heat Transfer in Vertical Falling Film Systems", Chem. Eng. Prog. Symposium Series, No. 30, 56, 1 (1960).

21. Kunz, H.R. and Yerazunis, S., "Analysis of Film Condensation, Film Evaporation, and Single Phase Heat Transfer for Liquid Prandtl Numbers from 10^{-3} to 10^4", J. Heat Transfer, 91, 413-420 (1969).

22. Sparrow, E.M. and Gregg, J.L., "A Boundary Layer Treatment of Laminar Film Condensation", J. Heat Transfer, 81, 13 (1959).

23. Othmer, D.F., "Condensation of Steam", Ind. Eng. Chem., 21, 576-583 (1929).

24. Colburn, A.P. and Hougen, O.A., "Design of Cooler Condensers for Mixtures of Vapors with Non-Condensing Gases", Ind. Eng. Chem., 26, 1178-82 (1934).

25. Akers, W.W., Davis, S.H., and Crawford, J.E., "Condensation of a Vapor in the Presence of a Non-Condensing Gas", Chem. Eng. Prog. Symposium Series, No. 30, 56, 139-144 (1960).

26. Sparrow, E.M. and Eckert, E.R.G., "Effects of Superheated Vapor and Non-Condensible Gases on Laminar Film Condensation", A.I.Ch.E.J., 7, 473-77 (1961).

27. Sparrow, E.M., Minkowycz, W.J., and Saddy, M., "Forced Convection Condensation in the Presence of Non-Condensibles and Interfacial Resistance", Int. J. Heat Mass Transfer, 10, 1829-45 (1967).

28. Voskresenskiy, K.D., "Calculation of Heat Transfer in Film Condensation Allowing for the Temperature Dependence on the Physical Properties of the Condensate", USSR Acad. Sci., OTK, (1948).

29. Sukhatme, S.P. and Rohsenow, W.M., "Heat Transfer During Film Condensation of a Liquid Metal Vapor", Report MIT-2995-1, Mass. Inst. of Tech., Cambridge, April (1964).

30. Wilcox, S.J. and Rohsenow, W. M., "Film Condensation of Potassium Using Copper Condensing Block for Precise Wall-Temperature Measurement", J. Heat Transfer, 92, 359-371 (1970).

31. Rohsenow, W.M., "Film Condensation of Liquid Metals", Canadian Soc. Mech. Engineering Paper 71, Annual Meeting (1971).

32. Baker, O., "Design of Pipe Lines for Simultaneous Flow of Oil and Gas", Oil and Gas J., 26, July (1954).

33. Bell, K.J., Taborek, J., and Fenoglio, F., "Interpretation of Horizontal In-Tube Condensation Heat Transfer Correlations Using a Two-Phase Flow Regime Map", Chem. Eng. Prog. Symposium Series, No. 102, 66, 150-63 (1970).

34. Taitel, Y. and Dukler, A.E., "A Model for Predicting Flow Regime Transitions in Horizontal and Near Horizontal Gas-Liquid Flow". A.I.Ch.E.J. 22, 47-55 (1976).

35. Breber, G., Palen, J.W., and Taborek, J., "Prediction of Horizontal Tubeside Condensation of Pure Components Using Flow Regime Criteria", Condensation Heat Transfer, ASME, New York, 1-8 (1979).

36. Marks, J.S. and Andeen, G.B., "Chugging and Condensation Oscillation", Condensation Heat Transfer, ASME, New York, 93-102 (1979).

37. Sargis, D.A., Masiello, P.J., and Stuhmiller, J.H., "A Probabilistic Model for Predicting Steam Chugging Phenomena", Condensation Heat Transfer, ASME, New York, 85-91 (1979).

38. Jacobs, H.R., Thomas, K.D., and Boehm, R.F., "Direct Contact Condensation of Immiscible Fluids in Packed Beds", Condensation Heat Transfer, ASME, New York, 103-110 (1979).

39. Wilke, C.R., Cheng, C.T., Ledesma, V.L., and Porter, J.W., "Direct Contact Heat Transfer for Sea Water Evaporation", Chem. Eng. Progr., 59, No. 12, 69-75 (1963).

40. Rai, V.C. and Pinder, K.L., "Direct Contact Condensation of Steam in a Packed Column with Immiscible Heat Transfer Agents", Canadian J. of Chem. Eng., 45, 170-74 (1967).

41. Harriott, P. and Wiegandt, H., "Countercurrent Heat Exchanger with Vaporizing Immiscible Transfer Agent", A.I.Ch.E.J., 10, 755-58 (1968).

42. Tamir, A. and Rachmilew, I., "Direct Contact Condensation of an Immiscible Vapor on a Thin Film of Water", Int. J. Heat Mass Transfer, 17, 1241-51 (1974).

43. Mao, J.R. and Hickman, K., "Direct Contact Condensation of Steam on a Modified Oil Coolant", Desalination, 10, 95-111 (1972).

44. Finkelstein, Y. and Tamir, A., "Interfacial Heat Transfer Coefficients of Various Vapors in Direct Contact Condensation" Chem. Eng. J., 12, 199-209 (1976).

45. Murty, N.S. and Sastri, V.M.K., "Condensation of a Falling Laminar Liquid Film", Heat Transfer 1974, 5th International Heat Transfer Conference, Tokyo, 3, 231-35 (1974).

46. Jacobs, H.R., Fannar, H., and Beggs, G.C., "Collapse of a Bubble of Vapor in an Immiscible Liquid", Proceedings of the 6th International Heat Transfer Conference. Toronto, 2, 383-88 (1978).

47. Jacobs, H.R. and Cook, D.S., "Direct Contact Condensation on a Non-Circulating Drop", Proceedings of the 6th International Heat Transfer Conference, Toronto, 2, 389-93 (1978).

48. Gregorig, R., "Hautkondensation an Feingewell-ten Oberflachen bei Berucksichtigung der Ober-flachenspannungen", Zeitschrift fur Angew. Math. und Physik, 5, 36-49 (1954).

49. Thomas, D.G., "Enhancement of Film Condensa-tion Heat Transfer Rates on Vertical Tubes by Vertical Wires", I & EC Fundamentals, 6, No. 1, 97-103 (1967).

50. Mehta, M.H. and Rao, M.R., "Heat Transfer and Frictional Characteristics of Spirally En-hanced Tubes for Horizontal Condensers", Advan-ces in Enhanced Heat Transfer, ASME, New York, 11-21 (1979).

51. Marto, P.J., Reilly, D.J., and Fenner, J.H., "An Experimental Comparison of Enchanced Heat Transfer Condenser Tubing", Advances in En-hanced Heat Transfer, ASME, New York, 1-9 (1979).

52. Yamamoto, H. and Ishibachi, T., "Calculation of Condensation Heat Transfer Coefficients of Fluted Tubes", Heat Transfer Jap. Res., 6, 61-68 (1977).

53. Fujii, T. and Honda, H., "Laminar Filmwise Condensation on a Vertical Single Fluted Plate", Proceedings of the 6th International Heat Transfer Conference, Toronto, 2, 419-24 (1978).

54. Panchal, C.B. and Bell, K.J., "Analysis of Nusselt-Type Condensation on a Vertical Fluted Surface", Condensation Heat Transfer, ASME, New York, 45-53 (1979).

55. Mori, Y., Hijikata, K., Hirasawa, S., and Nakayama, W., "Optimized Performance of Con-densers with Outside Condensing Surface", Condensation Heat Transfer, ASME, New York, 55-62 (1979).

56. Nakayama, W., Daikoku, T., Kuwahara, H., and Kakizaki, K., "High-Flux Heat Transfer Surface THERMOEXCELL", Hitachi Review, 24, No. 8, 329-334 (1975).

57. Hirasawa, S., Hijikata, K., Mori, Y., and Nakayama, W., "Effect of Surface TEnsion on Laminar Film Condensation Along a Vertical Plate with a Small Leading Radius", Proceedings of the 6th International Heat Transfer Con-ference, Toronto, 2, 413-18 (1978).

58. Hirasawa, S., Hijikata, K., Mori, Y., and Nakayama, W., "Effect of Surface Tension on Laminar Film Condensation: Study of Condensate Film in a Small Groove", Transactions JSME, 44, 2041-48 (1978).

59. Edwards, D.K., Gier, K.D., Ayyaswamy, P.S., and Cotton, I., "Evaporation and Condensation in Circumferential Grooves on Horizontal Tubes", ASME Paper 73-HT-251, (1973).

60. Webb, R. L., "A Generalized Procedure for the Design and Optimization of Fluted Gregorig Condensing Surfaces", J. Heat Transfer, 101, 335-39 (1979).

61. Glicksman, L.R., Mikic, B.B., and Snow, D.F. "Augmentation of Film Condensation on the Out-side of Horizontal Tubes", A.I.Ch.E. J., 19, 636-7 (1973).

62. Schmidt, E., Schurig, W., and Sellschopp, W., "Versuche uber die Kondensation von Wasser-dampf in Film und Tropfenform", Tech. Mech. Thermodynamik, 1, 53-63 (1930).

63. Hampson, H., "The Condensation of Steam on a Metal Surface", Proc. of the General Discussion on Heat Transfer, London, 58-61 (1951).

64. Hampson, H. and Ozisik, N., "An Investigation Into the Condensation of Steam", Proc. Inst. Mech. Engrs., 1B, 282-93 (1952).

65. Hampson, H., "Dropwise Condensation on a Metal Surface", Engineering, 179, 464-9 (1955).

66. Hampson, H., "Modes of Condensation and Their Effect on Heat Transfer", British Chem. Eng., 532-35, October (1957).

67. Blackman, L.C.E., Dewar, M.J.S., and Hampson, H., "An Investigation of Compounds Promoting the Dropwise Condensation of Steam", J. Applied Chem., 7, 160-71 (1957).

68. Furman, T. and Hampson, H., "Experimental In-vestigation Into the Effects of Cross Flow-with Condensation of Steam and Steam-Gas Mix-tures on a Vertical Tube", Proc. Inst. Mech. Engrs., 173, 147-69 (1959).

69. Hampson, H., "Condensation of Steam on a Tube with Filmwise or Dropwise Condensation and in the Presence of a Non-Condensible Gas", Intern. Developments in Heat Transfer, 310-18 (1961).

70. Bromley, L.A., Porter, J.W., and Read, S.M. "Promotion of Drop-by-Drop Condensation of Steam from Seawater on a Vertical Copper Tube", A.I.Ch.E.J. 14, 245-50 (1968).

71. Bromley, L. A. and Read, S.M., "Dropwise Con-densation", 21, 391-92 (1975).

72. Drew, T.B., Nagle, W.M., and Smith, W.Q., "The Conditions for Dropwise Condensation of Steam", Trans. A.I.Ch.E., 31, 605-21 (1935).

73. Osment, B.D.J., Tudor, D., Speirs, R.M.M., and Rugman, W., "Promoters for the Dropwise Condensation of Steam", Trans. Inst. Chem. Engrs., 40, 152-60 (1962).

74. Tanner, D.W., Poll, A., Potter, J., Pope, D., and West, D., "Promotion of Dropwise Conden-sation by Montan Wax. II. Composition of Mon-tan Wax and the Mechanism of Promotion", J. Applied Chem., 12, 547-52 (1962).

75. Tanner, D.W., Pope, D., Potter, C.J., and West, D., "Heat Transfer in Dropwise Condensation Part II", Intern. J. Heat Mass Transfer, 8, 427-36 (1965).

76. Vylkov, V.T., "Heat Transfer Intensification During Dropwise Condensation of Organic Liq-uids", Heat and Mass Transfer Sourcebook: Fifth All-Union Conference, Minsk, John Wiley and Sons, 134-41 (1977).

77. Topper, L. and Baer, E., "Dropwise Condensation of Vapors and Heat Transfer Rates", J. Colloid Sci., 10, 225-26 (1955).

78. Brown, A.R. and Thomas, M.A., "Filmwise and Dropwise Condensation of Steam at Low Pressure", Proc. of Third Intern. Heat Transfer Conf., 2, 300-5 (1966).

79. Butcher, D.W. and Honour, C.W., "Tetrafluoro-ethylene Coatings on Condenser Tubes", Intern. J. Heat Mass Transfer, 9, 835 (1966).

80. Edwards, J.A. and Doolittle, J.S., "Tetra-fluoroethylene Promoted Dropwise Condensation", Intern. J. Heat Mass Transfer 8, 663-66 (1965).

31. Eibling, J. A. and Hyatt, D.L., "Methods of Improving Heat Transfer in Evaporators of Small Thermocompression Sea-Water Stills", Second Natl. Heat Transfer Conf., Chicago, A.I.Ch.E. Preprint 28, 38 pages (1958).

82. Kosky, P.G., "Tetrafluorethylene Coatings on Condenser Tubes", Intern. J. Heat Mass Trans-fer, 11, 374-75 (1968).

83. Mizushina, T., Kamimura, H., and Kuriwaki, Y., "Tetrafluoroethylene Coatings on Condenser Tubes", Intern. J. Heat Mass Transfer, 10, 1015-16 (1967).

84. Depew, C.A. and Reisbig, R.L., "Vapor Conden-sation on a Horizontal Tube Using Teflon to Promote Dropwise Condensation", Ind. Eng. Chem. Process Design Develop., 3, 365-69 (1964).

85. Davies, G.A. and Ponter, A.B., "Prediction of the Mechanism of Condensation on Condenser

Tubes Coated with Tetrafluoroethylene", Intern. J. Heat Mass Transfer, 11, 375-77 (1968).

86. DePew, C.A. and Reisbig, R.L., "Vapor Condensation on a Horizontal Tube Using Teflon to Promote Dropwise Condensation", I.&EC Process Design and Develop., 3, 365-69 (1964).

87. Reisbig, R.L., "Macroscopic Growth Mechanisms in Dropwise Condensation", Proc. of 5th Intern. Heat Transfer Conf., Paper Cs 2, 3, 255-58 (1974).

88. Wilmshurst, R. and Rose, J.W., "Dropwise Condensation: Further Heat Transfer Measurements", Proc. Fourth Intern. Heat Transfer Conf., Paper Cs 1.4, 6, 11 pages (1970).

89. Wilmshurst, R. and Rose, J.W., "Dropwise and Filmwise Condensation of Aniline, Ethanediol and Nitrobenzene", Proc. of Fifth Intern. Heat Transfer Conf., Paper Cs 2.4, 3, 269-73 (1974).

90. Abdul-Hadi, M.I., "Dropwise Condensation of Different Steam-Air Mixtures on Various Substrate Materials", Canadian J. Chem. Eng., 57, 451-59 (1979).

91. Nijaguna, B.T. and Abdelmessih, A.H., "Precoalescence Drop Growth Rates in Dropwise Condensation", Proc. Fifth Intern. Heat Transfer Conf., 3, 264-68 (1974).

92. Abdelmessih, A.H., Neumann, A.W., and Yang, S.W., "Effect of Surface Characteristics on Dropwise Condensation", Letters in Heat Mass Transfer, 2, 285-92 (1975).

93. Wilkins, D.G., Bromley, L.A., and Read, S.M., "Dropwise and Filmwise Condensation of Water Vapor on Gold", A.I.Ch.E.J., 19, 119-23 (1973).

94. Wilkins, D.G. and Bromley, L.A., "Dropwise Condensation Phenomena", A.I.Ch.E.J., 19, 839-45 (1973).

95. Detz, C.M. and Vermesh, R.J., "Nucleation Effects in the Dropwise Condensation of Steam on Electroplated Gold Surfaces", A.I.Ch.E.J., 22, 87-93 (1976).

96. Erb, R.A. and Thelen, E., "Dropwise Condensation", First Intern. Symposium on Water Desalination, Washington, D.C., 13 pages (1965).

97. Erb, R.A., "The Wettability of Gold", J. Phys. Chem., 72, 2412-17 (1968).

98. Erb, R.A., Haigh, T.I., and Downing, T.M., "Permanent Systems for Dropwise Condensation for Distillation Plants", Symposium on Enhanced Tubes for Desalination Plants, U.S. Dept. of Interior, Washington, D.C., 177-201 July (1970).

99. Erb, R.A., "Dropwise Condensation on Gold", Gold Bulletin, 6, No. 1, 2-6 (1973).

100. Umur, A. and Griffith, P., "Mechanism of Dropwise Condensation", J. Heat Transfer, 87, 275-82 (1965).

101. Griffith, P. and Lee, M.S., "Effect of Surface Thermal Properties and Finish on Dropwise Condensation", Intern. J. Heat Mass Transfer, 10, 697-707 (1967).

102. Hannemann, R.J. and Mikic, B.B., "An Experimental Investigation into the Effect of Surface Thermal Conductivity on the Rate of Heat Transfer in Dropwise Condensation", Intern. J. Heat Mass Transfer, 19, 1309-17 (1976).

103. Woodruff, D.W., "Gold Surfaces for Dropwise Condensation", Ph.D. Thesis, Chemical Engineering Department, University of Illinois, Urbana, 1980.

104. Tower, R.E. and Westwater, J.W., "Effect of Plate Inclination on Heat Transfer During Dropwise Condensation", Chem. Eng. Progress Symposium Series, 66, No. 102, 21-25 (1970).

105. Woodruff, D.W. and Westwater, J.W., "Steam Condensation on Electroplated Gold: Effect of Plating Thickness", Intern. J. Heat Mass Transfer, 22, 629-32 (1979).

106. Nagle, W.M. and Drew, T.B., "The Dropwise Condensation of Steam", Trans. A.I.Ch.E., 30, 217-55 (1934).

107. Nagle, W.M., Bays, G.S., Blenderman, L.M., and Drew, T.B., "Heat Transfer Coefficients During Dropwise Condensation of Steam", Trans. A.I.Ch.E., 31, 593-600 (1935).

108. Tanaka, H., "Measurements of Drop-Size Distributions During Transient Dropwise Condensation", 97, 341-46 (1975).

109. Tanner, D.W., Pope, D., Potter, C.J., and West, D. "Promotion of Dropwise Condensation by Monolayers of Radioactive Fatty Acids. II. Chromium Surfaces", J. Appl. Chem. 14, 439-44 (1964).

110. Bobco, R.P. and Gosman, A.L., "Promotion of Dropwise Condensation of Several Pure Organic Vapors", ASME Paper 57-S-2, 6 pages (1957).

111. Necmi, S. and Rose, J.W., "Heat Transfer Measurements During Dropwise Condensation of Mercury", Intern. J. Heat Mass Transfer, 20, 877-81 (1977).

112. Rose, J.W., "Dropwise Condensation of Mercury", Intern. J. Heat Mass Transfer, 15, 1431-34 (1972).

113. Tamir, A., "Mixed Pattern Condensation of Multi-component Mixtures", Chem. Eng. J., 17, 141-56 (1979).

114. Peterson, A.C. and Westwater, J.W., "Dropwise Condensation of Ethylene Glycol", Chem. Eng. Progress Symposium Series, 62, No. 64, 135-42, (1966).

115. Welch, J.F. and Westwater, J.W., "Microscopic Study of Dropwise Condensation", Proc. of Second Intern. Heat Transfer Conf., 302-9 (1961-62).

116. O'Bara, J.T., Killian, E.S. and Roblee, L.H.S., "Dropwise Condensation of Steam at Atmospheric and Above Atmospheric Pressures", Chem. Eng. Sci., 22, 1305-14 (1967).

117. Dolloff, J.B., Metzger, N.H., and Roblee, L.H.S., "Dropwise Condensation of Steam at Elevated Pressures", Chem. Eng. Sci., 24, 571-83 (1969).

118. Tanner, D.W., Pope, D., Potter, C.J., and West, D., "Heat Transfer in Dropwise Condensation at Low Steam Pressures in the Absence and Presence of Non-Condensable Gas", Intern. J. Heat Mass Transfer, 11, 181-90 (1968).

119. McCormick, J.L. and Westwater, J.W., "Drop Dynamics and Heat Transfer During Dropwise Condensation of Water Vapor on a Horizontal Surface", Chem. Eng. Progress Symposium Series, 62, No. 64, 120-34 (1966).

120. Fitzpatrick, J.P., Baum, S., and McAdams, W.H. "Dropwise Condensation of Steam on Vertical Tubes", Trans. A.I.Ch.E., 35, 97-107 (1939).

121. Rose, J.W., "Effect of Condenser Tube Material on Heat Transfer During Dropwise Condensation of Steam", Intern. J. Heat Mass Transfer, 21, 835-40 (1978).

122. Venkatram, T. and Kuloor, N.R., "Condensation of Steam on Stainless Steel", Indian J. Tech. 2, 73-6, March (1964).

123. Tanner, D.W., Pope, D., Potter, C.J., and West, D., "Promotion of Dropwise Condensation by Monolayers of Radioactive Fatty Acids. I. Stearic Acid on Copper Surfaces", J. App. Chem. 14, 361-69 (1964).

124. Citakoglu, E. and Rose, J.W., "Dropwise Condensation: The Effect of Surface Inclination", Intern. J. Heat Mass Transfer, 12, 645-51 (1969)

125. Tanasawa, I., Tachibana, F., and Ochiai, J.,

"Dropwise Condensation (I)", Report of the Institute of Industrial Science, Univ. of Tokyo, 23, No. 2, 40 pages, August (1973).

126. Tanner, D.W., Potter, C.J., Pope, D., and West, D., "Heat Transfer in Dropwise Condensation, Part I", Intern. J. Heat Mass Transfer, 8, 419-26 (1965).

127. Citakoglu, E. and Rose, J.W., "Dropwise Condensation: Some Factors Influencing the Validity of Heat Transfer Measurements", Intern. J. Heat Mass Transfer, 11, 523-37 (1968).

128. LeFevre, E.J. and Rose, J.W., "Heat Transfer Measurements During Dropwise Condensation of Steam", Intern. J. Heat Mass Transfer, 7, 272-73 (1964).

129. LeFevre, E. J. and Rose, J.W., "An Experimental Study of Heat Transfer by Dropwise Condensation", Intern. J. Heat Mass Transfer, 8, 1117-33 (1965).

130. Shea, F.L. and Krase, N.W., "Dropwise and Film Condensation of Steam", Trans. A.I.Ch.E., 36, 463-90 (1940).

131. Birt, D.C.P., Brunt, J.J., Shelton, J.T., and Watson, R.G.H., "Methods of Improving Heat Transfer from Condensing Steam and Their Application to Condensers and Evaporators", Trans. Instn. Chem. Engrs. 37, 289-94 (1959).

132. Watson, R.G.H., Birt, D.C.P., Honour, C.W., and Ash, B.W., "Promotion of Dropwise Condensation by Montan Wax. I. Heat Transfer Measurements", J. Appl. Chem., 12, 539-46 (1962).

133. McCormick, J.L. and Westwater, J.W., "Nucleation Sites for Dropwise Condensation", Chem. Eng. Sci., 20, 1021-36 (1965).

134. Tanasawa, I. and Shibata, Y., "Dropwise Condensation at Low Heat Flux and Small Surface Subcooling", Report of the Inst. of Ind. Sci., Univ. of Tokyo, 30, No. 10, 25-8 (1978).

135. LeFevre, E.J. and Rose, J.W., "A Theory of Heat Transfer by Dropwise Condensation", Proc. of 3rd Intern. Heat Transfer Conf., 2, 362-75 (1966).

136. Reisbig, R.L. and Lay, J.E., "A Nucleation Theory and Experimental Study of Dropwise Condensation", Proc. 4th Intern. Heat Transfer Conf., 6, Paper Cs 1.2, 11 pages (1970).

137. Nijaguna, B.T. and Abdelmessih, A.H., "Precoalescence Drop Growth Model for Dropwise Condensation", A.S.M.E. Paper 71-WA/HT-47, 7 pages (1971).

138. Neumann, A.W., Abdelmessih, A.H., and Hameed, A., "Role of Contact Angels and Contact Angle Hysteresis in Dropwise Condensation Heat Transfer", Intern. J. Heat Mass Transfer, 21, 947-53 (1978).

139. McCormick, J.L. and Baer, E., "On the Mechanism of Heat Transfer in Dropwise Condensation", J. Colloid Sci., 18, 208-216 (1963).

140. McCormick, J.L. and Baer, E., "Dropwise Condensation on Horizontal Surfaces", Developments in Mechanics, Pergamon Press, N.Y., 749-75 (1965).

141. Gose, E.E., Mucciardi, A.N., and Baer, E., "Model for Dropwise Condensation on Randomly Distributed Sites", Intern. J. Heat Mass Transfer, 10, 15-22 (1967).

142. Glicksman, L.R. and Hunt, A.W., "Numerical Simulation of Dropwise Condensation", Intern. J. Heat Mass Transfer, 15, 2251-69 (1972).

143. Rose, J.W. and Glicksman, L.R., "Dropwise Condensation-The Distribution of Drop Sizes", Intern. J. Heat Mass Transfer, 16, 411-25 (1973).

144. Graham, C. and Griffith, P., "Drop Size Distributions and Heat Transfer in Dropwise Condensation", Intern. J. Heat Mass Transfer, 16, 337-45 (1973).

145. Fatica, N. and Katz, D.L., "Dropwise Condensation", Chem. Eng. Progress, 45, 661-74 (1949).

146. Isachenko, V.P., Solodov, A.P. and Mal'Tsev, A.P., "Moving Picture Study of Dropwise Condensation of Steam", Heat Transfer Soviet Res., 11, 146-48 (1979).

147. Hurst, C.J. and Olson, D.R., "Conduction Through Drops During Dropwise Condensation", J. Heat Transfer, 95, 12-20 (1973).

148. Sadhal, S.S. and Martin, W.W., "Heat Transfer Through Drop Condensate Using Differential Inequalities", Intern. J. Heat Mass Transfer, 20, 1401-7 (1977).

149. Sadhal, S.S. and Plesset, M.S., "Effect of Solid Properties and Contact Angle in Dropwise Condensation and Evaporation", J. Heat Transfer, 101, 48-54 (1979).

150. Plesset, M.S., "Note on the Flow of Vapor Between Liquid Surfaces", J. Chem. Physics, 20, 790-93 (1952).

151. Plesset, M.S. and Prosperetti, A., "Flow of Vapor in a Liquid Enclosure", J. Fluid Mech., 78, 433-444 (1976).

152. Rose, J.W., "On the Mechanism of Dropwise Condensation", Intern. J. Heat Mass Transfer, 10, 755-62 (1967).

153. Rose, J.W. and Glicksman, L.R., "Dropwise Condensation: The Distribution of Drop Sizes", Intern. J. Heat Mass Transfer, 16, 411-425 (1973).

154. LeFevre, E.J. and Rose, J.W., "A Theory of Heat Transfer by Dropwise Condensation", Proc. Third Intern. Heat Transfer Conf., 2, 362-75 (1966).

155. Rose, J.W., "Dropwise Condensation Theory", Intern. J. Heat Mass Transfer, 24, 191-194 (1981).

156. Tanaka, H., "A Theoretical Study of Dropwise Condensation", J. Heat Transfer, 97, 72-8 (1975).

157. Tanaka, H., "Further Developments of Dropwise Condensation Theory", J. Heat Transfer, 101, 603-611 (1979).

158. Sadhal, S.S., "Further Developments of Dropwise Condensation Theory", J. Heat Transfer, 102, 394 (1980).

159. Sugawara, S. and Michiyoshi, I., Dropwise Condensation", Memoirs Faculty of Engineering, Kyoto University, 18, 84-111 (1956).

160. Tanasawa, I. and Tachibana, F., "A Synthesis of the Total Process of Dropwise Condensation Using the Method of Computer Simulation", Proc. Fourth Intern. Heat Transfer Conf., 6, Paper Cs 1.3, 11 pages (1970).

161. Wenzel, H., "Der Warmeubergang Bei der Tropfenkondensation", Linde-Ber. Tech. Wiss., 18, 44-56 (1964).

162. Wenzel, H., "Extended Theory of Heat Transfer in Dropwise Condensation", Warme und Stoffubertragung, 2, 6-18 (1969).

163. Baer, E. and McKelvey, J.M., "Heat Transfer in Dropwise Condensation", Amer. Chem. Soc. Delaware Sci. Symposium, Univ. of Del, Newark, (1958).

164. Jakob, M., "Heat Transfer in Evaporation and Condensation-II", Mech. Eng. 58, 729-39 (1936).

165. Ruckenstein, E., "About Drop-Wise or Film Condensation on a Solid Surface", Revue De Physique (Rumania), 5, 405-14 (1960).

166. Ruckenstein, E. and Metiu, H., "On Dropwise Condensation on a Solid Surface", Chem. Eng. Sci., 20, 173-80 (1965).

167. Sugawara, S. and Katsuta, K., "Fundamental Study on Dropwise Condensation", Proc. Third Intern. Heat Transfer Conf., 2, 354-61 (1966).

168. Tanasawa, I., "What We Don't Know About the Mechanism of Dropwise Condensation", Proc. of Fifth Intern. Heat Transfer Conf., 7, 186-91 (1974).

169. Westwater, J.W., "Dropwise Condensation", Advanced Heat Transfer, University of Illinois Press, Urbana, 233-62 (1969).

170. Griffith, P., "Dropwise Condensation", Handbook of Heat Transfer, ed. by Rohsenow, W.M. and Hartnett, J.P., McGraw-Hill Book Co., 12-34 to 12-47 (1973).

171. Merte, H., "Condensation Heat Transfer", Advances in Heat Transfer, 9, Academic Press, 181-272 (1973).

172. Tanasawa, I., "Dropwise Condensation-The Way to to Practical Applications", Proc. of Sixth Intern. Heat Transfer Conf., 6, 393-405 (1978).

173. Labuntsov, D.A. and Kryukov, A.P., "Analysis of Intensive Evaporation and Condensation", Intern. J. Heat Mass Transfer, 22, 989-1002 (1979).

174. Takeyama, T. and Shimizu, S., "On the Transition of Dropwise Condensation", Proc. of Fifth Intern. Heat Transfer Conf., 3, 274-78 (1974).

175. Tanasawa, I. and Utaka, Y., "Measurement of Condensation Curve for Dropwise Condensation Heat Transfer", Condensation Heat Transfer, Amer. Soc. Mech. Engineers, New York, N.Y., 63-68 (1978).

176. Smith, T., "The Hydrophilic Nature of a Clean Gold Surface", J. Colloid Interface Sci., 75, No. 1, 51-55 (1980).

177. Mikic, B.B., "On Mechanism of Dropwise Condensation", Intern. J. Heat Mass Transfer, 12, 1311-23 (1969).

178. Horowitz, J.S. and Mikic, B.B., "Effect of Surface Thermal Properties on Dropwise Condensation", Proc. of Fifth Intern. Heat Transfer Conf., Paper Cs 2.2, 259-63 (1974).

179. Hannemann, R.J. and Mikic, B.B., "An Analysis of the Effect of Surface Thermal Conductivity on the Rate of Heat Transfer in Dropwise Condensation", Intern. J. Heat Mass Transfer, 19, 1299-1307 (1976).

180. Aksan, S.N. and Rose, J.W., "Dropwise Condensation: The Effect of Thermal Properties of the Condenser Material", Intern. J. Heat Mass Transfer, 16, 461-67 (1973).

181. Rose, J.W., "Effect of Surface Thermal Conductivity on Dropwise Condensation Heat Transfer", Intern. J. Heat Mass Transfer, 21, 80-81 (1978).

182. Wenzel, H., "On the Condensation Coefficient of Water Estimated from Heat Transfer Measurements During Dropwise Condensation", Intern. J. Heat Mass Transfer, 12, 125-26 (1969).

183. Ivanovskiy, M.N., Milovanov, Yu, V., Sorokin, V. P., Subbotin, V.I., and Chulkov, B.A., "Heat Transfer in Film and Dropwise Condensation of Liquid Metal Vapors", Heat Transfer Soviet Research, 1, 96-102 (1969).

184. Ohtani, S., Chiba, Y., and Ohwaki, M., "Heat Transfer in Dropwise Condensation of Steam: Correspondence of Drop Behavior and Surface Temperature Fluctuation", Kagaku Kogagu, 36, No. 4, 412-18 (1972).

185. Lorenz, J.J. and Mikic, B.B., "Effect of Thermocapillary Flow on Heat Transfer in Dropwise Condensation", J. Heat Transfer, 92, 46-52 (1970).

A Study of Metal Vapour Condensation

R. ISHIGURO, K. SUGIYAMA, and T. HISAMATSU
Hokkaido University, Sapporo, Japan

ABSTRACT

Studies of metal vapour condensation are briefly reviewed including recent results of intensive mass transport conditions. Information on an experiment presently being conducted in the author's laboratory with potassium vapour is also included. The experiment covers vapour temperatures from 580 to 670 K, and vapour pressures from 53 to 530 Pa. Scrupulous cautions were taken in the design and operation of the experimental apparatus for correct measurement of condensation surface temperature and for minimizing noncondensable gas in the vapour. The condensation coefficients obtained by the measurement fell between 0.9 and 1.0 except in the data taken under conditions of intensive condensation. For the intensive cases, the data was compared with Necmi and Rose's results, which was obtained by using mercury.

NOMENCLATURE

h_{fg} = latent heat of vapourization;

M_a = Mach number;

m = net mass flux;

P_s = saturation pressure corresponding to condensate surface temperature, T_s;

P_v = bulk saturation pressure of the vapour;

Q = heat flux in test condenser block;

R = specific ideal-gas constant;

T = temperature;

T_s = condensate surface temperature;

T_v = bulk staturation temperature of the vapour;

T_v' = vapour temperature;

ΔT = $T_v - T_s$;

v_g = specific volume of vapour.

Greek symbols

κ = ratio of specific heat;

σ = condensation coefficient.

INTRODUCTION

Metal vapour condensation has been actively studied in the past thirty years since metal vapours or liquid metals were recognized as excellent working substances in nuclear or space power plants. Although there have been many analytical and experimental examinations made on condensation in recent years, many have exhibited different features from nonmetalic ones and the intrinsic nature of condensation has not been completely understood.

From the viewpoint of analytical treatment, the uncertainty exists in a nonequilibrium thermodynamical condition at the liquid-vapour interphase. Some theoretical methods involving the condition have been tried to estimate the condensation rates, but it seems that none of them has been entirely successful. Therefore, analytical attempts to clarify the phenomenon are still in order.

On the other hand, many experimental investigations have been performed with alkali metals or mercury which have moderate boiling temperature levels. However, the former are active materials which produce chemical reactions, and the latter vapour is poisonous. For this reason many restrictions must be placed on the design of the test apparatus, thereby making precise measurements difficult to obtain. Thus, more accurate experimental evaluations are continuously needed.

The effective use of thermal energy is a high priority goal imposed on modern engineering. The development of working substances such as metal vapours and liquid metals which are sustainable under high temperatures is also an important need. The study of the condensation phenomenon of metal vapours is of significance then to these problems.

PREVIOUS WORKS

A remarkable difference between the condensation of metal and nonmetal vapour is the prominent effect of thermal resistance at the vapour-liquid interface. It is common knowledge that heat transfer associated with film condensation of nonmetallic vapour can be calculated well by Nusselt's film resistance theory [1] or its modifications [2]. However, thermal conductivity of liquid metals is so great that liquid film may not cause the biggest resistance, as the thermodynamical nonequilibrium at the liquid-vapour interface often makes a prominent temperature jump. Schrage [3] calculated the magnitudes of the temperature jump by a theory of gas kinetics using the Maxwellian distribution of molecules assumption and obtained the following relation for the mass exchange process:

$$m = \left(\frac{2\sigma}{2 - \sigma}\right)\frac{(P_v - P_s)}{\sqrt{2\pi RT_v}} \quad . \qquad (1)$$

The σ in Eq. (1) is called the condensation coefficient and refers to the fraction captured by the liquid surface in all molecules impinging on the surface.

Some modifications of Schrage's theory have been attempted. For example, Labuntsov (4) gives the following equation for small Knudsen number regions using the "double-flow" distribution function:

$$m = \left(\frac{2\sigma}{2 - 0.798\sigma}\right)\frac{(P_v - P_s)}{\sqrt{2\pi RT_v}} \quad . \qquad (2)$$

Huang et al. (5) have suggested that certain molecules at low grazing angles to the surface are not captured, and their analysis led to the following result:

$$m = \frac{\pi}{2}\frac{1}{\sqrt{2\pi RT_v}}(P_v - P_s) \quad . \qquad (3)$$

This equation shows that $2\sigma/(2-\sigma)$ in Eq. (1) and $2\sigma/(2-0.798\sigma)$ in Eq. (2) are replaced by $\pi/2$.

Much experimental research (6 ∿ 15) has been conducted using mercury, sodium and potassium. The obtained σ values are scattered from 0.02 to 1.0, though corresponding values to vapour pressures less than 0.01 atm are always close to unity.

Wilcox & Rohsenow (14) performed an experiment of potassium vapour condensation employing strict precautions in order to obtain precise temperature measurements and they analyzed the magnitude of systematic errors of the measurement. They proposed that the σ values are always unity. The proposition seems to be natural according to their discussion and it has the possibility of becoming an established opinion. The development of this research up to this stage is well reviewed by Rohsenow (16).

On the other hand, Necmi and Rose (17) recently published a paper about their experimental research of mercury vapour condensation. Their data covers a rather intensive condensation region. They indicated the mass flux by following equation:

$$m = \xi(P_v - P_s)/\sqrt{RT_v'} \qquad (4)$$

and proposed that the ξ can be expressed by a function of $Qv_g/h_{fg}\sqrt{RT_v'}$. In comparing Eq. (4) with Eq. (1) or Eq. (2), the relations between σ and ξ are as follows:

$$\xi = \left(\frac{2\sigma}{2 - \sigma}\right)/\sqrt{2\pi}$$

by Schrage's Equation,

$$\xi = \left(\frac{2\sigma}{2 - 0.798\sigma}\right)/\sqrt{2\pi}$$

by Labuntsov's Equation. Since $Q = mh_{fg}$, $Qv_g/h_{fg}\sqrt{RT_v'}$ is approximately equal to $\kappa^{0.5}\cdot M_a$. As a result, they concluded that ξ is a function of M_a, where M_a is a Mach number based on the velocity of vapour molecules impinging on the condensation surface.

Quite recently, Labuntsov and Kryukov (18) published an analytical treatment on one-dimensional problems to clarify the basic relationship which governs the processes of intensive evaporation and condensation. They divided the field into two regions: a Knudsen layer, which was several lengths of the free paths of molecules, and an outer region where ordinary macroscopic equations were applied. The Knudsen layer was analyzed from the viewpoint of the molecular-kinetic theory and matched with the continuous medium region. They also made a comparison of their analytical results with these of Necmi and Rose on intensive mercury condensation, which yielded a rather high agreement.

TEST APPARATUS

Fig. 1 shows our test apparatus. It consists of a metal vapour loop and a vacuum pumping system. The loop is made of stainless steel (SUS 316) and includes a metal boiler, a condensing test surface and an auxiliary condenser. The heater capacity of the boiler is 20 kW. The test condensing surface is 9 cm² and the auxiliary condenser has a heat transfer area of 600 cm². The inventory of potassium is 3 kg. Metal vapours generated in the boiler are passed through the test section into the auxiliary condenser, where all metal vapours are condensed before returning to the boiler by gravitational force. Three horizontal thermo-wells made of thin stainless steel pipes are positioned in front of the test surface to measure the vapour temperature. The loop is connected to an oil diffusion vacuum pump of 6" bore, via a liquid nitrogen cold trap. A vacuum valve made of metal is used to shut off the vacuum pumping system from the loop.

The test condensing surface is arranged on one end of a copper block. The other end of the block prepared variable numbers of fins of the screw in-bolt type is cooled by air. The temperature of the condensing surface and the heat flux are estimated by a temperature gradient in the block, which is measured by twelve sheathed thermo-couples of 0.25 mm (outside diameter) placed in six holes, each two of them is set in one hole from opposite openings. The holes are arranged perpendicularly to the block axis. Precision of measurement largely depends upon the smallness of the hole diameters. To minimize errors due to heat conduction along the thermocouples,

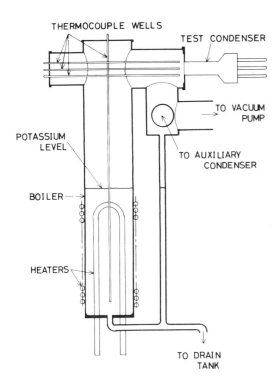

Fig. 1 Test Apparatus

Photo 1 Condensing Copper Block

the holes should be located along the isotherms. To prepare the hole with a small enough diameter and sufficient depth, the copper block was cut in half along the axis, and six grooves of 0.35 mm width were machined on one side of the mating surfaces. The other mating surface was silver plated. Both halves were then fastened together with screw bolts, and the block was reassembled by diffusion welding in a vacuum furnace of 1070 K. A stainless steel flange of 1.2 mm thickness was soldered to the reassembled copper block with silver-copper eutectic bounding.

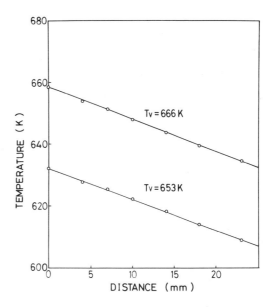

Fig. 2 Temperature Distributions in Copper Block

The soldering process was also done in the vacuum furnace to prevent chemical reactions. Photo 1 shows the condensing block. The distance from each thermocouple hole to the condensing surface is 4, 7, 10, 14, 18, and 23 mm. Lastly, the condensing surface was plated with nickel to prevent the solder from melting into the potassium. The stainless steel flange was welded by the method of TIG to the loop for the final set up.

Wilcox and Rohsenow proposed a method to estimate the systematic error of temperature measurement caused by the size of the thermocouple wells. Applying this method to our system, the uncertainty of the estimated surface temperature was 0.04 K for the case of 157 kW/m^2 heat flux. This magnitude of error was considerably smaller than that obtained in systems used by other researchers. For example, in the Wilcox and Rohsenow's system the uncertainty was estimated as 0.13 K, and in the Bakulin et al. system, it was 0.11 K for the same heat flux.

Fig. 2 shows an example of temperature distribution on the axis of the copper block. The measured value lies almost on a straight line, and the magnitude of scatterings is less than 0.35 K. This magnitude is within the uncertainty of the thermocouples calibrations. All thermocouples except the ones used for subsidiary purposes were inconel sheathed chromel-alumel thermocouples of 0.25 mm diameter taken from one reel. The e.m.f.'s were measured by a digital volt meter, and the minimum reading was 1 μ Volt. Calibration was performed at the melting points of Zn, Pb and Sn, and the e.m.f.'s of the three melting points were fit on a quadratic curve for each thermocouple. To confirm each calibration curve, we compared the e.m.f.'s evaluated from the curves with those of all the thermocouples set in a furnace under uniform temperatures. Thermal conductivity of the condensing copper block, which was made by 99.99 % pure copper, was measured by a method of stepwise heating (19) and the value coincided well with the TPRC table (20). Thermal resistance of the plated nickel to the heat fluxes on the condensing surface was not evaluated for these measurements.

Many earlier reports have stressed that the mixing of a very small amount of noncondensable gas in the metal vapour has a significant effect on the condensation performance. With this in mind, our apparatus was arranged with maximum caution to maintain vacuum tightness. The vessel is made of stainless steel with all seams welded by TIG except one, where the copper condensing block is connected to the stainless steel frange. The joint to the vacuum pumping system is a frange coupling with a copper gasket. A helium leak detector with a sensitivity of 10^{-11} Pa m^3/s was used to check the tightness of all parts of the apparatus. Generally if a vacuum vessel with no leakage is not baked well, its pressure returns to a level 1 ∿ 0.1 Pa soon after evacuation due to evolved gases from the wall materials. We checked the pressure condition of our vessel. The loop was well evacuated without baking and left closed. A pressure of 0.4 Pa was kept over an interval of one week. We then baked the vessel at 520 ∿ 620 K for about 10 hours a day and this procedure was repeated for several days. A total baking time of 50 hours was needed to bring the pressure to 5×10^{-7} Pa at room temperature. Finally the loop was isolated. After 30 days the pressure was found still to be less than 0.1 Pa.

The potassium charged in the vessel contained less than 20 ppm oxygen and 1 ppm hydrogen. The purity of argon gas to cover the potassium was better than 99.9995 %.

EXPERIMENTAL PROCEDURE

It is desirable that noncondensable gas in the potassium loop be kept as low as possible. If the loop is held in a vacuum with less than 0.1 Pa, its gas particles are reduced to one-millionth numbers of atmospheric conditions at most. This rate is lower than the impurity of the argon gas used in this experiment. In addition, if argon gas enters the loop to drain the potassium, the gas will be absorbed into the inner wall of the loop. For these reasons and due to the vacuum holding performance of the loop described in the preceding section, we maintained the potassium in the boiler without returning it to the drain tank since it had been charged.

Before heating the boiler, the loop was exhausted by opening the metal valve connected to the vacuum pumping system, which had been previously evacuated down to 10^{-6} Pa. Thus the loop was rendered almost free of noncondensable gas. By regulating the electric input to the boiler, a desired condition for the potassium vapour was obtained. Before the condensation measurements were begun, all surrounding surfaces, including the condensing test surface, were placed in an almost adiabatic environment. The test surface condition was prepared by placing a small heater on the tail end of the copper block. All other parts were regulated by guard heaters behind insulators. Only the auxiliary condenser was operated. With this method some noncondensable gas which could be mixed with the vapour could be collected in the auxiliary condenser. Then, by removing the electric heater from the tail end of the copper block and adjusting a suitable number of screw in-bolt type fins on that end, one condition for the condensation tests was met.

Almost the same values for σ were obtained for the same conditions of different runs. The situation did not change in other runs without the vacuum pump operation one month later. The effects of changing the vapour velocity by adjusting the condition of the auxiliary condenser could not be measured. Attempts to exhaust the noncondensable gas continuously from the auxiliary condenser have been discussed in previous reports (14, 17). Our loop was also equipped for this function, however, it did not seem necessary to use the function as the loop was sufficiently vacuum tight.

The top of the thermo-well for vapour temperature measurement was positioned close to the condensing test surface. To prevent the radiation of the cold test surface from causing errors in the vapour temperature measurement, we changed the test surface temperature against a constant vapour condition and found the radiation effect to be negligible for the present experiment. It is possible that the heat transfer coefficient by condensation was so large that the minimal heat loss by radiation could be compensated.

RESULTS

Our experimental results are listed in Table 1. The coefficients σ are evaluated by Eq. (1). All measurements were performed in a vapour temperature range of from 580 to 670 K. Temperature differences between the vapour and the condensing test surface were 4 to 15 K, which were larger than those noted in studies on potassium by Wilcox and Rohsenow, and Subbotin et al. Even within this range of temperature differences the σ values were still very close to unity. The σ values are plotted in Fig. 3 with the vapour temperature on the abscissa. With careful observation, the tendency of σ values to

Table 1 Experimental Results

Run	$T_V(k)$	$T_S(k)$	$Q(kW/m^2)$	σ
1-1	608.9	594.5	172.1	0.956
2-1	621.2	609.0	212.3	0.991
3-1	638.4	628.1	238.3	0.938
4-1	622.3	609.1	229.1	0.982
5-1	582.8	570.4	93.0	1.034
5-2	630.2	621.8	175.0	0.958
5-3	654.9	647.1	245.8	0.905
6-1	640.1	628.9	287.9	0.970
6-2	668.2	659.0	391.7	0.920
7-1	635.3	625.8	217.7	0.956
8-1	634.4	627.9	161.2	0.987
8-2	639.6	636.2	97.5	0.991
9-1	669.3	664.9	216.8	0.958
9-2	668.6	663.9	214.2	0.941
10-1	641.6	635.8	163.3	0.968
11-1	637.5	631.8	150.7	0.985
12-1	586.5	574.7	97.7	1.031
13-1	604.4	596.6	95.8	0.986
14-1	617.1	612.0	84.7	0.974
14-2	649.4	644.2	163.2	0.947
15-1	655.7	651.1	163.2	0.929
15-2	656.0	629.0	401.8	0.638*
16-1	654.5	632.0	382.9	0.689*
16-2	658.8	649.9	304.5	0.906

* Large temperature difference

decrease slightly with increasing vapour temperatures can be noted. Because the vapour temperature range of the present measurements was not wide enough, we are not certain whether or not the tendency is the same phenomena for high pressure regions as that reported in previous papers.

In order to examine the recent experimental results by Necmi and Rose, we took measurements under conditions of varying heat flux with constant vapour temperatures. Fig. 4 shows our results compared with those of Necmi and Rose. In this figure, data shown by symbol ◑ were taken with the same status of the apparatus as the data in Fig. 3 were taken. Unfortunately, our number of runs was inadequate, due to some technical difficulties with the apparatus during the course of the measurements. Some other data shown by symbols o and ● were added after the seminar. The latters were taken under situations of exhausting continuously the noncondensable gas. Necmi and Rose have reported increasing tendencies of ξ value with increases of Q. However, our data exhibited an inverse tendency. As mentioned above, the theoretical work on

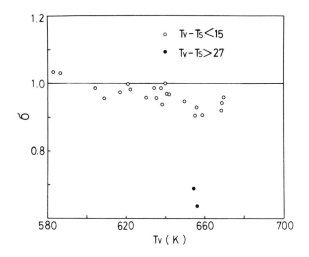

Fig. 3 Condensing Coefficients vs Vapour Temperature

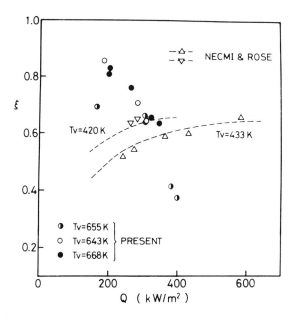

Fig. 4 Dependence of ξ on Condensation Heat Flux

intensive evaporation and condensation by Labuntsov and Kryukov demonstrated the same tendency as that found in the Necmi and Rose data. We have not yet ascertained what caused the inverse tendency.

In most cases, under conditions of intensive condensation and a great temperature difference the effect of noncondensable gas in the vapour, if any, becomes prominent. When this happens, ξ values decrease with the increase of Q. We paid maximum caution to take out the noncondensable gas effect. If our results were affected by noncondensable gas, it appears that the effect was necessary in Necmi and Rose's data in evaluating the ultimate pressure of vacuum of their loop. Another question in conjunction with the correlation of their data to Labuntsov and Krykov's theory is whether both worker's data can be presumed the same under varying conditions.

SUMMARY

Because the condensation coefficient σ must be pursued solely by experimental means for the present, careful measurements of condensation rate under various boundary conditions are of paramount importance. The coefficient cannot be fixed firmly - even when precise experimental data are available - because of the differences in the equations which express the condensation rate between certain theoretical treatments; for example, the $2\sigma/(2 - \sigma)$ factors in Schrage's formula and the $2\sigma/(2 - 0.798\sigma)$ factors in Labuntsov's formula.

In this experiment the condensation coefficient was measured, first of all, under a condition of moderate mass transfer with vapour pressures ranging from 53 to 530 Pa. The σ values fall between 0.9 and 1.0 when calculated by Eq. (1). When the values were evaluated in Eq. (2), they reached a maximum of 1.1, which did not correspond to the recognized values for σ. However our results were almost the same as those obtained by Subbotin et al. in their potassium vapour experiments.

We subsequently conducted experiments to confirm the recent proposals for condensation under intensive conditions and obtained an opposite tendency of ξ values to those reported by Necmi and Rose in their mercury vapour studies. We are not in a position for the time being to conclude whether the cause of the discrepancy is attributed merely to unavoidable errors including the noncondensable gas effect or not.

Examinations of intensive condensations should be continued. We intend to experiment on wider regions of vapour pressure in future studies.

The authors express their sincere appreciations to Dr. T. Kumada who is the associate professor and to Mr. T. Suwa who was a graduate student in the authors laboratory for their helpful assistance in the initial stage of this research.

The authors could have a contact with Professor J. W. Rose, Queen Mary College, University of London, after the seminar. This paper was partly revised with his advice.

REFERENCES

1. Nusselt,W., Z. ver. deut. Ing. 60, 541-569 (1916).
2. Rohsenow,W.M., Trans. ASME, 78, 1645-1648 (1956).
3. Schrage,R.W., A theoretical study of interphase mass transfer, Columbia University Press, New York (1953).
4. Labuntsov,D.A., Teplofiz. Vysok. Temper. 5, 647-654 (1967).
5. Huang,Y.S., Lyman,F.A. and Lick,W.J. Int. J. Heat Mass Transfer, 15, 741-754 (1972)
6. Misra,R. and Bonilla,C.F., Chem. Engng. Prog. Symp. Ser. 52(18), 7-21 (1956).
7. Subbotin,V.I., Ivanovskii,M.N., Sarokin,V.P. and Chulkov,V.A., Teplophiz. Vysok. Temper., 2, 616-622 (1964).
8. Barry,R.E. and Balzhiser,R.E., Proc. 3rd Int. Heat Transfer Conference, 2. 318-328 (1966).
9. Aladyev,I.T., Kondratyev,N.S., Mukhin, V.A., Mukin,M.E., Kipshidze,M.E., Parfentyev,I. and Kisselev,J.V., Proc. 3rd Int. Heat Transfer Conference, 2, 313-317 (1966).
10. Kroger,D.G. and Rohsenow,W.M., Int. J. Heat Mass Transfer, 10, 1891-1894 (1967).
11. Bakulin,N.V., Ivanovskii,M.N., Sorokin,V.P. and Subbotin,V.I., Teplofiz. Vysok. Temper., 5, 930-933 (1967).
12. Ivanovskii,M.N., Milovanov,A.I. and Subbotin,V.I., Atomn. Energ., 24, 146-151 (1968).
13. Sukhatme,S.P. and Rohsenow,W.M., J. Heat Transfer, 88, 19-28 (1966).
14. Wilcox,S.J. and Rohsenow,W.M., J. Heat Transfer, 92, 359-371 (1970).
15. Sakhuja,R.K., Sc. D. Thesis, Mech. Engng., MIT, Sept. (1970).
16. Rohsenow,W.M., Progress in heat and mass transfer - Vol. 7, Pergamon Press. Oxford (1973).
17. Necmi,S. and Rose,J.W., Int. J. Heat Mass Transfer, 19, 1245-1256 (1976).
18. Labuntsov,D.A. and Kryukov,A.P., Int. J. Heat Mass Transfer, 22, 989-1002 (1979).
19. Kobayashi,K. and Kumada,T., Technol. Rep. Tohoku Univ., 33, 169-186 (1968).
20. Touloukian,Y.S.(ed.), Thermophysical properties of high temperature solid materials, Macmillan Co., New York (1967).

HIGH-PERFORMANCE
HEAT TRANSFER: SURFACES

High-Performance Surfaces for Non-Boiling Heat Transfer

ICHIRO TANASAWA

Institute of Industrial Science, University of Tokyo, Japan

ABSTRACT

A state-of-the-art review is presented on the research and development in Japan of the high-performance heat transfer surfaces and the techniques of enhancement of heat transfer. The subject is, however, restricted to convective and condensation heat transfer, since boiling or evaporative heat transfer is mentioned elsewhere.

INTRODUCTION

Improving the performance of heat exchanging devices is the most important problem for the efficient utilization of thermal energies. A solution to this problem is before everything else dependent upon the development of high-performance heat transfer surfaces which are the "hearts" of the heat exchanging devices.

In this brief article the author attempts to give a state-of-the-art review on the research and development in Japan of the high-performance heat transfer surfaces and the techniques of enhancement of heat transfer. However, the subject is restricted to convective and condensation heat transfer since augmentation of boiling or evaporative heat transfer will be mentioned in detail by Professors Nishikawa and Ito. It should also be noted here that because of the limited space only a crude review is presented in the following which may neglect quite a few significant contribution to the latest achievements in Japan.

According to the keynote lecture at the Sixth International Heat Transfer Conference, Toronto, 1978, given by Prof. A.E. Bergles[1] entitled "Enhancement of Heat Transfer," the techniques of heat transfer augmentation can roughly be classified into three categories: (1) Active method, (2) Passive method and (3) Combined method. Active techniques, which require external power, include mechanical aids, surface vibration, fluid vibration, electrostatic fields, injection and suction. Passive techniques, which require no direct application of external power, include treated surfaces, rough surfaces, extended surfaces, displaced enhancement devices, swirl flow devices, surface tension devices, and fluid additives. Two or more of these techniques may be utilized simultaneously to be classified as combined method.

In the chapters which follow a description will be made on the research and development of enhanced heat transfer surfaces in Japan largely following the way of classification proposed by Prof. Bergles.

ENHANCEMENT OF CONVECTIVE HEAT TRANSFER

The coefficient of heat transfer by convection is relatively low. It is especially so when gases are used as the heat transfer media. This necessitates the intensive research and development of high-performance heat transfer surfaces such as extended or finned surfaces.

The recent development seems to show a tendency of utilizing the leading edge effect and boundary layer agitation on top of the effect of surface extension by fins. Also some of the active schemes which had long been neglected have been given attention.

Electrostatic field

The eletrostatic field has been known to increase heat transfer coefficients in free convection. The most typical configuration is a heated wire in a concentric tube maintained at a high voltage relative to the wire, or a fine wire electrode against a horizontal plate.

The research in Japan in this field was started first at the Mechanical Engineering Laboratory, Ministry of International Trade and Industry. Yamaga and the coworkers[2][3][4] applied a high voltage (about 15 kV) between a flat plate (heated surface) and a needle electrode which is maintained vertical to the plate. They found that the flat plate was cooled by the flow of ionized air (corona wind) produced by the corona discharge and that the degree of cooling seemed larger than the one by an impinging air jet having the same approaching velocity with the corona wind. Also shown was a possibility of the application of this phenomenon to the cooling of the tip of bite during the process of cutting a work with a lathe[5][6]. The mechanism of cooling by the corona discharge was not made very clear, however.

Recently Yabe, Mori and Hijikata[7][8] studied theoretically and experimentally the effect of corona wind on heat transfer between a single or plural fine wire electrode(s) and a plate electrode. The outline of the experimental setup is shown in Fig.1. Figure 2 is the result showing the relationship between the heat transfer coefficient and the applied electric tension for the case of a single wire electrode. As the heat transfer coefficient by free convection is shown by the solid circle on the ordinate, augmentation of heat transfer up to a factor of ten is achieved with this configuration. A solid triangle in the figure represents the heat transfer coefficient based on the flow velocity distribution which was

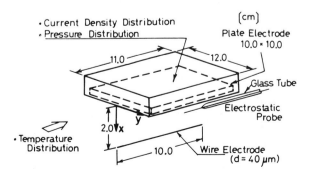

Fig.1 Geometry of wire and plate electrode system

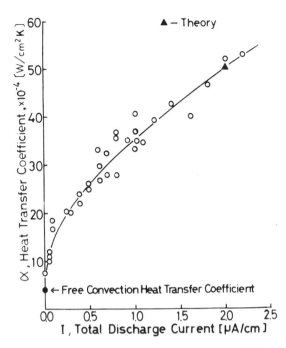

Fig.2 Dependence of heat transfer coefficient
on total discharge current

obtained by solving the electrohydrodynamic (EHD)
equations. This result shows that the mechanism
of heat transfer augmentation by the corona discharge between the wire and plate electrodes can
be fully explained by the velocity distribution
of the corona wind. These authors have also carried
out an   experiment using plural wire electrodes.

Another study has been done by Kikuchi(9)(10)
in which the effect of non-uniform and uniform
electrostatic fields on free convection heat transfer along vertical plate is studied. The details
are not mentioned here, however.

Enhancement of heat transfer by an
electric field seems efficient only in
the case of free convection. Mizushina,
Ueda and Matsumoto(11) state that even
with intense fields, the enhancement disappears as turbulent flow is approached
in a circular tube with a concentric inner electrode. Bergles(1) states that
(even though the heat transfer coefficient could be increased by several hun-
dred percent when sufficient electric
power is supplied,) the equivalent
effect could be produced at lower capital cost and without the hazards of
10,000-100,000 volts by simply providing forced convection with a blower or
fan.

The above statement may be correct
in most of the cases. However, in some
special situations where a blower or fan
cannot be used adequately, the utilization of electric field may become of
great use. Fujikake(12) has proposed an
application of the EHD cooling to automobile equipments, for which the requirement for reduction of noises has
been increasing.

Liquid or solid additives

Aihara and coworkers(13)(14) made

a fundamental experiment on the heat transfer enhancement by addition of water droplets to air
stream. The heat transfer surface was a wedge
placed horizontally in the stream and heated uniformly. A summary of the experimental results is
shown in Fig.3(14). It was found that the addition
of a small amount of water droplets (0.05 to 3% in
weight fraction) enhanced the two-phase heat transfer coefficients from 2 to 14 times the corresponding single-phase convection. It was also found
that the dryout of water film occurred from the upper surface of the downstream part of the wedge and
that at high mass flow ratios the value of the enhancement ratio (the ratio of two-phase heat transfer
coefficient to the single-phase coefficient) increased as the wedge surface temperature increased,
while at low mass flow ratios just the opposite
phenomenon was observed. In the subsequent report
(15) the same authors have analyzed the heat transfer from an isothermal wedge placed vertically upward in the downward air-water mist flow. They
have found that, in the case of the weight fraction
of water mist between 0.1 to 10% and the temperature difference between 10 to 30 °C, the thickness
of water film over the wedge is only 20 to 100 μm,
the dryout occurs from the leading edge, and the
heat transfer enhancement, as large as 4.6 to 57
times the single-phase flow, is achieved.

As for the enhancement of heat transfer by
solid-air two-phase flow, Aihara et al.(16) carried
out an experiment for air flow in a horizontal circular pipe (54.5 mm I.D.). Glass particles with
the mean diameter of 36.6 μm were suspended in the
stream. Spacial distribution of the particles, the
change in the distributions of velocity and temperature, and the local heat transfer coefficient
were measured.

Extended surfaces

Using extended surfaces is one of the traditional ways of heat transfer augmentation. It is
especially effective and indispensable when gases

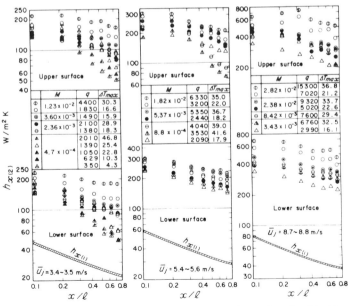

(q in W/m², T_{max} in °C; Symbols ☉ and ● in right figure refer to $\bar{u}_j = 7.1$ m/s).

Fig.3 Local heat transfer coefficient for
air-water mist flow $h_{x(2)}$

are used as the heat transfer media. Figure 4,5 and 6 show some examples of the recently developed finned surfaces. Figure 4* is named the pin-tube element and is used in the heat exchanger for waste heat recovery. To attain compactness the pins are bent so that the envelope of cross section may have a rectangular profile. Figure 5** is called the pine tube as it may remind one of the outer skin of a pineapple. The fins of the pine tube are manufactured by "peeling" the surface of a scored tube. This surface is effective in increasing the air-side thermal conductance of evaporator tube used for the car-cooler. Star insert tubes are shown in Fig.6**. They are a kind of inner fin tubes and

Fig.4   Pin-tube element

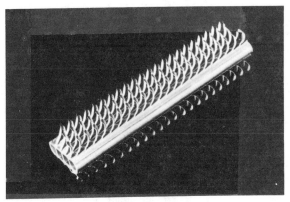

Fig.5   Pine fin tube

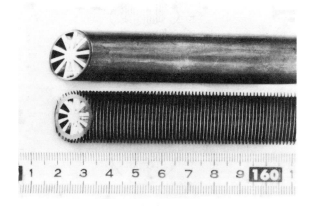

Fig.6   Star insert tubes

are developed for the refrigerator.

Recently the progress in the manufacturing technique has given the fin functions other than that of extending the heat transfer surface area. An example is the strip fin or louver fin which has played an important role in reducing the size of the radiator for automobile use. This type of surface is very effective since a part of the fin is made to protrude into the gas flow, which forces the boundary layer to start anew at the edge of each protrusion. Manufacturers in Japan have spent much effort to develop such kind of surfaces. During these twenty years, the air side heat transfer co-efficient (which accounts for 80 % of radiation capacity) has increased by 80 %, and also the water side coefficient has increased by the same amount. As a result the weight and the volume of the radiator (with the same radiation capacity) are both approaching nearly one-third of those at twenty years ago. This is mainly the result of an improvement of heat transfer surfaces which have changed from plate fins to corrugated fins and then to louver-type fins. Progress in materials and manufacturing techniques have played significant parts at the same time.

Surface roughness or turbulence promoter

Surface roughness or turbulence promoter has long been used extensily to enhance forced convection heat transfer. Various types of surface with roughness or promoter are now available commercially. However, the academic background for these high-performance surfaces seems rather poor.

Mori and Daikoku(17) studied the basic mechanism of heat transfer augmentation expected in the downstream of a circular promoter placed on a flat plate. They measured the distribution of pressure and mean velocity along the wall. Also measured were the turbulence intensity at the wall and the local heat transfer coefficient with the use of electrochemical method. It was found that the heat transfer coefficient behind the roughness took the maximum value at the distance of about ten times the height of roughness element where the separated flow reattached the wall. According to their conclusion the distribution of the local heat transfer coefficient behind the roughness coincided with the distribution of turbulence intensity at the wall and therefore the profile and size of roughness element should be chosen so that the turbulence intensity at the surface might be increased.

Ichimiya et al.(18) studied the effect of a single roughness element placed on the upper wall of a two-dimensional parallel flow channel. Only the lower wall was heated electrically. The heat transfer coefficient, the intensity of turbulence and the velocity distribution were measured along the heated wall. As for the profile of roughness element the authors chose a square and three types of triangle. The results of their measurements show that the position where the maximum heat transfer occurred is dependent on both the height of roughness and the width of flow channel. The maximum heat transfer coefficient obtained for the square roughness was around 210 kcal/m^{2}h°C

* The picture is offered by courtesy from Gadelius Company in Japan.
** The pictures are offered by courtesy from Sumitomo Light Metal Industry Co. Ltd.

when Re=$5.9\times10^4 \sim 1.5\times10^5$ and the width of the channel was 35 mm. This value is about 2.5 times the heat transfer coefficient obtained for the parallel channel without roughness element. A problem pertinent to the roughness type heat transfer promoter is the simultaneous increase of the resistance to flow. The authors have observed a considerable increase in the pressure drop through the channel. The ratio of the increase in the average Nusselt number to that of the friction coefficient lie between 0.45 and 0.75.

The present author has been studying the effectiveness of the repeated rib roughness. Figure 7 shows one of the surfaces tested by the author. The roughness element with rectangular profile is made of bakelite. The elements are placed on a copper plate which is heated electrically. Two sheets of copper plate with rectangular turbulence promoters are faced together to constitute a rectangular channel. The pitch-to-height ratio of the promoter, the channel width, the mass flow rate of the air and the heat input are the variable parameters. It has been found that with this type of turbulence promoter the increase in pressure drop considerably exceeds the gain in heat transfer. The authors are now continuing to experiment with promoter elements having somewhat different geometry.

ENHANCEMENT OF CONDENSATION HEAT TRANSFER

In connection with the utilization of thermal energy source having relatively low temperature, such as solar heat, geothermal and ocean thermal energies and waste heat, the research and development of condensers with excellent performance have become of more importance.

In the past there was a period when little consideration was given to the enhancement of condensation heat transfer because condensation showed a higher heat transfer coefficient than convection. However, as the temperatures of thermal energy sources of interest have decreased, the use of heat transfer media having a lower boiling point than water has been increasing. Since those media (such as Freons) have rather poor thermal conductivity, high heat transfer coefficient cannot be expected. Thus, the demand for enhanced condensation heat transfer surfaces has grown

In the sections which follow, various types of processed surfaces for condensers are mentioned in the first place. Then the review is made on a couple of fields of research activities which are now in progress; dropwise and direct contact condensation.

Application of electrostatic field to condensation process will not be mentioned (though there have been a few reports published hitherto) be-

cause little definite result has been obtained yet.

Finned surfaces

The primary object of finned surfaces is to increase the amount of heat transfer by extending the area of heat transferring surface. The use of extended surface is especially effective in the case of condensation of organic vapors such as Freons because the heat transfer coefficients by condensation of these vapors are considerably lower than that of water vapor.

But today finned surfaces aim at another effect. That is the effect of thinning the condensate film by the action of surface tension force.

When a vapor condenses onto a wavy surface, the pressure in the liquid film where its surface is convex exceeds the pressure inside the concave film, owing to the action of surface tension, thus resulting in the flow of condensate from the convex part toward the concave part. This causes thinning of the liqiud film on convex surface and raises the condensation rate. Such a phenomenon occurring on the condensing fin is sometimes called Gregorig effect after the name of the first researcher(19). Fujii and Honda(20)(21) studied analytically the laminar film condensation on a vertical surface with sinuous fins. But according to Hirasawa et al.(22), they did not take into account the capillary force when calculating the flow of liquid film at the bottom of the wavy fin.

Hirasawa et al.(22)(23) have made both theoretical and experimental investigation on the laminar film condensation along a vertical surface having small leading radius. Their results show that for a plate of a given height the heat transfer coefficient near the leading edge takes a maximum value at a certain optimum leading edge radius and the effect of surface tension is noticeable when the height of the condensing surface is smaller.

Mori et al.(24) have proposed to attach circular discs at a certain spacing to a vertical finned surface (Fig.8) in order to remove accumulated

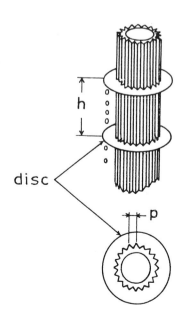

Fig.8 Vertical finned surface having circular discs for condensate removal

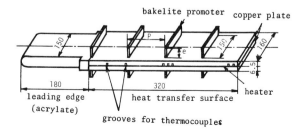

Fig.7 Heat transfer surface with repeated roughness elements

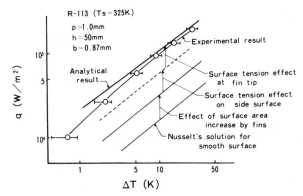

Fig.9 Dependence of heat flux q on surface subcooling ΔT for film condensation of R-113
on vertical finned surface

condensate from the surface before the liquid film
grows too thick. According to the authors, removal
of condensate by the discs also prompts direct
contact heat transfer between the vapor and the
liquid drops which fall from the edges of discs.
They also have shown the optimum design of such
type of finned surface. An example of the results
is shown in Fig.9. The relationship between the
heat flux q and the surface subcooling ΔT is shown
for condensation of R-113. The fin pitch p and the
fin height b are 1.0 mm and 0.87 mm respectively,
and the height of condensing surface h is 50 mm.
When compared with the condensation on a smooth
pipe with the same surface subcooling, which is
calculated by the Nusselt equation, augmentation
of heat flux of a factor of seven is achieved.
Since the increase of surface area by fins is only
by a factor of two, additional enhancement must be
achieved by the capillary effect.

Although it has become evident that the finned
surfaces which make efficient use of the capillary
force are very effective in enhancing filmwise condensation, determination of optimum size and geometry is not always so easy. It is because the
effect of surface tension depends on the geometry of
the fin, the rate of condensation (i.e. heat flux),
and the rate of removal of the condensation heat

Enya et al.(25) measured condensation heat
transfer coefficient of R-113 on various types of
finned tubes to find out the most effective geo-

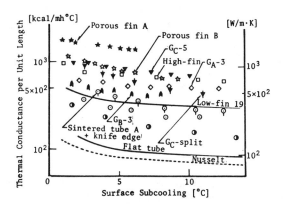

Fig.10 Results of heat transfer coefficient of
R-113 on thirteen different finned tubes

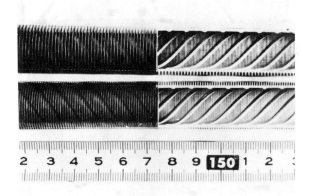

Fig.11 Rifle fin tubes

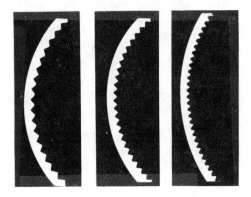

Fig.12 Cross sections of fine-ripple tubes

metry. Figure 10 shows the result. The result reveals that the tube with a sintered metal layer on
the fin surface shows the highest performance. An
economical problem in manufacturing such kind of
tube still remains, however.

Quite a few types of finned tubes for condenser use are commercially available. Figure 11* shows
rifle-fin tubes which have low fins on the outer
surface for enhancement of condensation and spiral
grooves inside for enhancement of forced convection
heat transfer. Fine-ripple tubes, the cross sections of which are shown in Fig.12* enhance the
heat transfer by condensation of R-22 by a factor
of two when compared with the smooth tube.

Shown in Fig.13 is a magnified picture of a
microstructure surface (commercially, Thermoexcel-
C) developed by Hitachi Ltd.) Minute fins which
have saw-shaped tips with pointed ends are machined
on the metal surface. According to the description
in the catalog(26) the thickness of condensate film
is kept thin due to prompt removal of condensate,
resulting in enhanced condensation. Figure 14
shows the performance characteristics of Thermoexcel-C for condensation of R-12(26). Enhancement
of condensation of a factor of about ten is obtained when compared with a smooth tube.

The high-performance condenser surfaces as
described above are all tubular surfaces. A quite
different surface has been developed for the condenser for OTEC use. It is the plate-type condenser surface. For a more detailed description refer
to Uehara(27).

---

* These pictures are offered by courtesy of
Sumitomo Light Metal Industry Co. Ltd.

Cross section of THERMOEXCEL-C

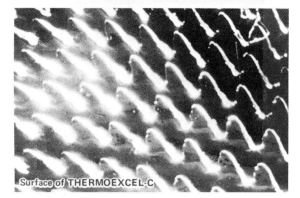

Surface of THERMOEXCEL-C

Fig.13* Microscopic view of the surface of
       Thermoexcel-C

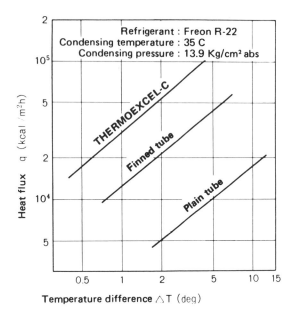

Fig.14* Heat transfer performance of Thermo-
       excel-C (Condensation of R-12)

Dropwise condensation

Dropwise condensation is known as the most
excellent heat transfer process. As for condensation of water vapor at atmospheric pressure, the
heat transfer coefficient obtained by laminar film
condensation remains at best 10 kW/m²K, while that

* These pictures are copied from Hitachi Catalog
  (26) by courtesy of Hitachi Ltd.

by dropwise condensation can easily exceed 200
kW/m²K.

What is the reason why such a high heat transfer coefficient is attained by dropwise condensation? Generally speaking, the heat transfer coefficient by condensation largely depends on the
thermal resistance of liquid film on the surface. In
the case of usual film condensation, the surface is
always covered with a liquid film having a finite
thickness. Because the heat conduction through this
liquid film controls the overall heat transfer, the
heat transfer rate cannot exceed a certain limit (except the top of the condensing surface where the condensate film starts from). The finned surfaces as
mentioned in the previous section are the results
of attempts to reduce the film thickness with the use
of the capillary force effect. The mechanism of heat
transfer by dropwise condensation is quite different.

If we sum up the volume of every individual drop
let on the condensing surface where dropwise condensation of steam is taking place, and then divide
the sum by the surface area, we obtain a value
about 0.1 mm as an averaged thickness of condensate
film(28). This value is equivalent to a thermal
conductance of 6 kW/m²K. The reason why the extremely high heat transfer coefficient by dropwise condensation is obtained in spite of a considerable
amount of condensate on the surface may be as follows: During the process of dropwise condensation
every part of the condensing surface is very
frequently exposed to the vapor due to the incessant
repetition of coalescence, departure and sweeping
of droplets. A high rate of transient condensation
takes places between these exposed parts and the
vapor, thus making the overall heat transfer rate
extremely high. In a sense, dropwise condensation
resembles the microstructure surfaces since it also
makes efficient use of the capillary force. While
the microstructure surface needs some mechanical
treatment of the surface, dropwise condensation
needs surface-chemical treatment.

Transient heat transfer occurring in the process of dropwise condensation can be recognized by
measuring the temperature fluctuation of condenser
surface underneath a drop. The present author and
his coworkers(29) verified it by a different way.
The authors designed a condensing chamber with
a wiper. All the drops on the condensing surface
were swept off by the wiper blade simultaneously
and periodically before they departed the surface
by the action of gravity and steam force. The
changes in surface temperature and heat flux were
measured as shown in Fig.15. A very prompt rise of
the surface temperature occurred immediately after
the surface was cleared off. The heat flux and the
heat transfer coefficient also rose at the same
time and reached values as high as 1.7 MW/m² and 1.1
MW/m²K, respectively.

Because of the high rate of heat transfer by
dropwise condensation, deterioration of heat transfer may occur easily at a lower part of the condensing
surface or at lower stages of a vertically arrayed
tube bundle. However, according to Rose (30) the
effect of height is very small in dropwise condensation. Also the result of experiment by the
present authors(31) conducted at large surface subcooling reveals that the coefficient of heat transfer by dropwise condensation does not decrease much
even when the rate of condensation becomes considerable. Furthermore, Enya(32) carried out an experiment of dropwise condensation for a vertical
row of pipes and found that there was little difference if any in the heat transfer coefficient between the top and bottom(8th) tube. These results

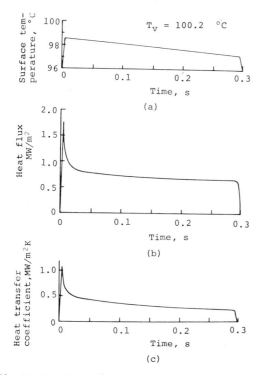

Fig.15 Results of transient dropwise
        condensation experiment

show that the heat transfer deterioration due to
increase in heat flux or condensate is of minor
importance in dropwise condensation.

The more frequent   the movement of drops
on the surface, the higher the heat transfer coefficient of dropwise condensation becomes.There-
fore, reducing the departure size of a drop by some
means or other, such as by raising the vapor velocity, is strongly recommended(33). In addition
to this, high speed vapor flow blows off the non-
condensable gases accumulated on the condensing
surface and hence is doubly effective.

The greatest of the reasons why dropwise condensation has not yet been utilized in commercial
condensers is that the low-energy surface which
can maintain dropwise condensation for a sufficiently long period of time has not yet been obtained. Though a more detailed description is not
presented here, the seemingly most promising way
is coating the condenser surface with a low-energy
polymer such as Teflon(34). In this case the
coating should be as thin as possible because the
polymer has usually a poor thermal conductivity.
Problems arise on   strength and durability at
the same time. Enya(35) calculated the allowable
maximum thickness of Teflon coating and obtained
a value of 10 μm.

Another problem which possibly rises when
the surface is coated with a poor conductive material is a heat transfer deterioration caused by
constriction of heat flow at the condensing surface. The possibility was for the first time
pointed out by Mikic(36). When dropwise condensation occurs on a poor conductive surface, the
heat flow on the surface becomes nonuniform owing
to the random distribution of droplets, which leads
to a decrese in the total heat flow.

A group of researchers(37)(38)(39) support
the opinion of Mikic, while the others(40)(41)(42)

do not. No definite conclusion has been obtained
yet. However, the present author considers that
the effect can be neglected practically when a thin
film is coated on a substrate having high thermal
conductivity.

The way to the practical applications of dropwise condensation seems still distant as mentioned
above, and it is our great regret that its very
high heat transfer ability is left unutilized.
Although the further scope is not clear yet, the
author would like to agree entirely with the opinion
stated by Prof. Westwater(43) at a Round Table
Meeting at the Fifth International Heat Transfer
Conference: "Over the short range, I am pessimistic
and believe that the things we don't know will
require much more effort than we expect. But in
the long range, I am an optimist and expect to see
significant commercial applications of dropwise
condensation in my lifetime."

Direct contact condensation

Conventional heat exchangers are designed such
that two fluids having different temperatures and
separated by a solid wall exchange thermal energy
by the process of overall heat transfer across the
wall. Usually the thermal resistance due to conduction through the wall is negligible when compared with the heat transfer between the wall and
fluid. But the progress in the enhancement technique
has given rise to a demand of reducing the wall
resistance. Also some kind of fluid brings about
the problem of fouling which makes the apparent
wall resistance significant.

Direct contact condensation has become the
object of attention because it is free from the
wall problem. It is considered very efficient for
heat exchangers used in low-temperature thermal
energy systems.

Direct condensation of the vapor onto liquid
film or jet of the same component has already been
studied for other purposes. The one especially
interesting in energy problems is the direct condensation between immiscible components. When one
of the two components is dispersed into the other
as a form of bubbles or droplets, very rapid and
efficient condensation may occur because of direct
contact and large specific surface area. This will
reduce the size and weight of the condenser. Also
the use of immiscible components will make the
separation of two fluids easier.

Mori(44) classified the modes of direct contact condensation of this type into four categories
by comparing interfacial energies. Shown in Fig.16
are the four modes (in which IV cannot exist stably).
Higeta et al.(45)(46) has carried out experiments
for the remaining three modes I, II and III. Vapor
bubbles are injected into a liquid column having a
lower temperature and forced to condense while they

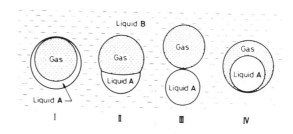

Fig.16 Possible types of configuration of
        two-phase bubble

are moving upward by the buoyant force. A typical result is shown in Fig.17. The combination of two components, i.e. n-pentane vs. glycerin belongs to the mode I. According to the authors the heat transfer performance deteriorates in the order of I, II and III. This means that the heat transfer performance becomes worse as the wettability between two component is reduced. The reason is not clear yet, but it is very interesting when we compare it with the condensation on a solid wall. Dropwise condensation, which takes place when the condensed phase does not wet the wall, predominates the filmwise mode.

The present author would like to add that the study on direct condensation experiment of the reverse combination (that is, the vapor condenses onto falling liquid droplets) is in progress in the author's laboratory.

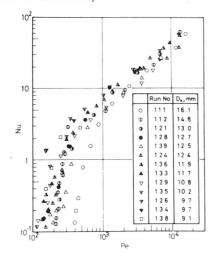

Run No	D_o, mm
111	16.1
112	14.8
121	13.0
128	12.7
139	12.5
124	12.4
136	11.9
133	11.7
129	10.8
135	10.2
126	9.7
134	9.7
138	9.1

Fig. 17 Condensation of n-pentane
         vapor in glycerin

REFERENCES

1. Bergles, A.E., Proc. 6th Int. Heat Transfer Conference, Vol.6 (1978), 89.
2* Yamaga, J., Fukuzawa, K., Jido, M. and Takeya, T., Preprint of JSME, No.720-5 (1972), 1.
3* Yamaga, J. and Jido, M., Proc. 11th National Heat Transfer Symposium of Japan (1974), 177.
4* Jido, M., Fukuzawa, K., J. Mech. Engng. Lab., 31 (1977), 1.
5* Yamaga, J. and Jido, M., Proc. 10th National Heat Transfer Symposium of Japan (1973), 289.
6* Jido, M., J. Mech. Engng. Lab., 31 (1977), 298.
7. Yabe, A., Mori, Y. and Hijikata, K., AIAA J. 16 (1978), 340.
8. Yabe, A., Mori, Y. and Hijikata, K., Proc. 6th Int. Heat Transfer Conference, Vol.3 (1978) 171.
9* Kikuchi, K., Trans. JSME, 43 (1972), 3477.
10. Kikuchi, K., Bull. JSME, 22 (1979).
11. Mizushina, T., Ueda, H. and Matsumoto, T., J. Chem. Engng. Japan, 9 (1976), 97.
12* Fujikake, K., Proc. 17th National Heat Transfer Symposium of Japan (1980), 109.
13* Aihara, T., Taga, M. and Haraguchi, T., Trans. JSME, 44 (1978), 1006.
14. Aihara, T., Taga, M. and Haraguchi, T., Int. J. Heat Mass Transfer, 22 (1979), 51.
15* Aihara, T. and Fu, W-S., Proc. 17th National Heat Transfer Symposium of Japan (1980), 331.

16* Aihara, T., Narusawa, K. and Haraguchi, T., Preprint of JSME, No.790-18 (1979), 22.
17* Mori, Y. and Daikoku, T., Trans. JSME, 38 (1972), 832.
18* Ichimiya, K., Hasegawa, S. and Yamazaki, K., Trans. JSME, 46 (1980), 482.
19. Gregorig, R., Zeitschrift für Angewandte Mathematik und Physik, 5 (1954), 36.
20. Fujii, T. and Honda, H., Proc. 6th Int. Heat Transfer Conference, Vol.2 (1978), 419.
21* Honda, H. and Fujii T., Proc. 14th National Heat Transfer Symposium of Japan (1977), 250.
22* Hirasawa, S., Hijikata, K., Mori, Y. and Nakayama, W., Trans. JSME, 44 (1978), 2041.
23. Hirasawa, S., Hijikata, K., Mori, Y. and Nakayama, W., Proc. 6th Int. Heat Transfer Conference, Vol.2 (1978), 413.
24. Mori, Y., Hijikata, K., Hirasawa, S. and Nakayama, W., Condensation Heat Transfer, Proc. 18th ASME-AIChE Heat Transfer Conference (1979), 55.
25* Kisaragi, T., Enya, S., Ochiai, J., Kuwahara, K. and Tanasawa, I., Preprint of JSME, No.780-1 (1978), 1.
26. Catalogue, Hitachi Cable Ltd. (1978).(see also Arai, N. et al., ASHRAE Trans., 83 (1978), 58.)
27* Uehara, H., Kikai No Kenkyu (Science of Machine), Yokendo, 31 (1979), 93.
28* Tanasawa, I., Progress in Heat Transfer, Vol.4, Yokendo, 300 (1976).
29. Tanasawa, I., Ochiai, J. and Funawatashi, Y., Proc. 6th Int. Heat Transfer Conference, Vol.2 (1978), 477.
30. Rose, J.W., Int. J. Heat Mass Transfer, Int. J. Heat Mass Transfer, 19 (1976), 1363.
31. Tanasawa, I. and Utaka, Y., Condensation Heat Transfer, Proc. 18th ASME-AIChE Heat Transfer Conference (1979), 63.
32* Enya, S., private communication.
33* Tanasawa, I., Ochiai, J., Utaka, Y. and Enya, S., Trans. JSME, 42 (1976), 2846.
34. Tanasawa, I., Proc. 6th Int. Heat Transfer Conference, Vol.6 (1978), 393.
35* Enya, S., JSME RC-SC 47 Report, Vol.1 (1979), 253.
36. Mikic, B.B., Int. J. Heat Mass Transfer, 12 (1969), 1311.
37. Tanner, D.W., Potter, C.J., Pope, D. and West, D., Int. J. Heat Mass Transfer, 8 (1965), 419.
38. Wilkins, D.G. and Bromley, L.A., AIChE J., 19 (1973), 839.
39. Hanneman, R.J. and Mikic, B.B., Int. J. Heat Mass Transfer, 19 (1976), 1309.
40. Aksan, S.N. and Rose, J.W., Int. J. Heat Mass Transfer, 16 (1973), 411.
41. Rose, J.W., Int. J. Heat Mass Transfer, 21 (1978), 835.
42. Stylianou, S.A. and Rose, J.W., Condensation Heat Transfer, Proc. 18th ASME-AIChE Heat Transfer Conference (1979), 69.
43. Tanasawa, I. and Westwater, J.W., Proc. 5th Int. Heat Transfer Conference, Vol.7 (1974), 186.
44. Mori, Y.H., Int. J. Multiphase Flow, 4 (1978), 383.
45* Higeta, K., Mori, Y.H. and Komotori, K., Preprint of JSME, No.780-18 (1978), 181.
46. Higeta, K., Mori, Y.H. and Komotori, K., Chemical Engineering Progress, Symposium Series, No.189 (1979), 256.

[References with asterisks on numbers are written in Japanese.]

High-Performance Heat Transfer Surfaces: Single Phase Flows

WEN-JEI YANG

Department of Mechanical Engineering and Applied Mechanics,
The University of Michigan, Ann Arbor, Michigan 48109, U.S.A.

ABSTRACT

The use of high performance heat transfer surfaces has a reward in terms of economic benefits and energy conservation. In addition, it enables us to accommodate or encourage high heat fluxes. This article presents a brief summary of up-to-date technical information on available high-performance surfaces including performance evaluation criteria and fouling. An attempt is made to outline areas of research that are potentially fruitful scientifically and may contribute to the advancement of industrial technology.

INTRODUCTION

Energy conservation has politically, socially, and to some extent, economically become a catchword in the United States. Both energy and material saving can be achieved by means of more efficient heat transfer surfaces which can also promote utilization of "raw" or "low grade" energy. The goal is to reduce the size of a heat exchange equipment or pumping power required for a specified heat load, to upgrade the capacity of an existing heat transfer unit, or to reduce the mean temperature difference between the working fluids. The use of high performance heat transfer surfaces may also lead to less excess operating temperatures and thus less system destruction under a specified heat transfer duty.

Most of the research effort in heat transfer is devoted to analyzing the standard or normal situation. However, in recent years, a great deal of research effort has been directed toward the development of high-performance thermal systems. In order to understand the manner in which convective heat transfer may be enhanced, it is useful to begin with Newton's law of cooling: The rate of heat transfer between a fluid and a finned surface can be expressed in a general form as

$$q = h(\eta A)\Delta T \qquad (1)$$

Here, h is the heat transfer coefficient; η, overall surface efficiency; A, total heat transfer area; and ΔT, surface-fluid temperature difference. The definition of η is

$$\eta A = (A - A_f) + A_f \eta_f = A - A_f(1 - \eta_f) \qquad (2)$$

in which η_f signifies the fin efficiency and A_f refers to the heat transfer area of the fins. $h\eta A$ represents the surface conductance. Obviously, high performance may be achieved by using a heat transfer surface with high value of ηA, h or both.

The first type of performance enhancement, an increase of ηA, is the function of extended surfaces. Area is paramount, whereas local heat transfer coefficients for finned surfaces are generally lower. Therefore, extended surfaces provide extra heat transfer area, but at an expense of the surface efficiency. This method is justified when h is low, i.e., the surface cannot efficiently transfer heat. The second type of performance enhancement is an increase of the heat transfer coefficient. A surface of high h is known as enhanced-heat-transfer (EHT) surface.

As a reasonable approximation, the heat transfer coefficient is given by

$$h = \frac{k}{\Delta} \qquad (3)$$

where k denotes the thermal conductivity of the fluid and Δ is the thickness of thermal boundary layer. Therefore, h can be increased by alterating the flow pattern near the surface in order to reduce the thermal boundary layer thickness.

A final category of performance enhancement is a simultaneous augmentation of both ηA and h. For example, extended surfaces in compact heat exchangers have increased heat transfer coefficient by means of surface promoters, such as perforations, louvers and corrugations or displaced promotors, such as canted tubes. In addition to the benefit of increased heat transfer area, there are numerous techniques to achieve the goal of enhancing convective heat transfer performance. Various versions of surveys on heat transfer augmentation methods are listed in reference 1 which is a computerized bibliography. Enhancement techniques [1] were classified as passive methods, which need no external power, and active schemes, which require external power. Compound techniques involves combined utilization of the basic techniques. Carnavos [2] classed these techniques into four groups as listed in Table 1. The effectiveness of various techniques is strongly dependent on the mode of heat transfer.

In this presentation, augmentation techniques are broadly classed into two groups. Group I includes external devices or aids that are employed to enhance the performance of heat transfer surfaces such as displaced enhanced devices, swirl flow devices, fluid additives, mechanical aid, surface or fluid vibrations, electric or magnetic fields and injection or suction. They are augmentation techniques in the strictest sense.

On the other hand, Group II consists of heat transfer surfaces which by themselves perform higher heat transfer rates due to surface configuration or characteristics. We may call them "high-performance heat transfer surfaces" or simply "enhanced surfaces" whose performance require no external factors. They are extended surfaces which include textured or formed, treated and rough surfaces. This presentation reports on the current status of "enhanced surfaces" in single-phase flows and prospects for the future in this area. Treated surfaces such as hydrophobic coatings and porous coatings are most effective for phase changes but not applicable to single-phase

Table 1. Classification of Augmentation Techniques

Bergles [1]	Carnavos [2]	Present Work

1. Passive Techniques (No external power needed)
 	
  Treated surfaces
  Rough surfaces
  Extended surfaces
  Displaced enhancement devices
  Swirl flow devices
  Surface tension devices
  Additives for liquids
  Additives for gases
 	
2. Active Techniques (External power required)
 	
  Mechanical aid
  Surface vibration
  Fluid vibration
  Electric or magnetic fields
  Injection or suction
 	
3. Compound Techniques
	1. Augmentative Stimulation is Continuously Supplied
	
	  Fluid additives
	  Electric or magnetic fields
	  Surface or fluid vibrations
	  Injection or suction
	
	2. Turbulence Promotion Swirl flow devices
	
	  Displaced enhancement devices
	  Spirally grooved tubes
	
	3. Extended Heat Transfer Surfaces
	
	  Externally finned surfaces
	  Internally finned surfaces
	
	4. Enhanced Heat Transfer Surfaces
	
	  Hydrophobic or porous coatings
	  Textured or formed surfaces
		1. Group I (requires external devices or aids)
		
		  Displaced enhancement devices
		  Swirl flow devices
		  Fluid additives
		  Mechanical aid
		  Surface or fluid vibrations
		  Electric or magnetic fields
		  Injection or suction
		
		2. Group II (dependant on surface configuration or characteristics)
		
		  Extended surfaces
		  Treated surfaces
		  Rough surfaces

convection. Therefore, effort is directed toward extended and rough surfaces.

One should note that high-performance heat transfer surfaces are the direct result of various heat transfer augmentation techniques and that the use of high-performance surfaces leads to the practicality of high flux heat transfer or high heat flux devices.

CURRENT STATUS

Surveys of the earlier work on heat transfer augmentation techniques are given in numerous articles that are listed in reference 1 [for example, 2-4]. Reference 5 is an additional asset to the knowledge of enhanced heat transfer. Only a brief discussion of some recent technical information is presented here.

A. Extended Surfaces

Finned tube surfaces have been employed in practically all heat transfer processes including laminar and turbulent flows with or without phase change. Various types are commercially available, ranging from longitudinal fins to transverse.
(i) No shroud.
The fin height and pitch that can be used effectively is fundamentally related to the boundary layer thickness. It is desirable that the fin height should be significantly greater than the boundary layer thickness so that the basic objective of increasing the surface area of the boundary layer is achieved. The rule of thumb is that when thermal boundary layer thickness is large, the heat transfer coefficient h is low and thus high fins are employed. Low fins are selected in case of thin boundary layer

where h is high. The fin pitch should be properly spaced depending on the fin height and thermal boundary layer thickness. If the fin pitch is too small, the main stream cannot penetrate the fin spaces, resulting in no increase in thermal boundary area and an excessive pressure drop.

The fins of nominal 1.6 mm height have become standard for the low-fin type of duty. A low-fin tube is ideally suited for shell-and-tube use. The low-fin type of surface is effective with compressed gases, namely in intercoolers and aftercoolers and is also employed in condensing and evaporating duties, particularly with refrigerants and hydrocarbons. The fins of approximately 3.2 mm are referred to as medium fins which are usually employed in shell-and-tube exchangers. Air-cooled exchangers require high fins which have the height of 6.4 mm to 25 mm. Heat recovery systems prefer medium to high fins in tube banks of in-line arrangement.

(ii) Shrouded

When a fin array is shrouded, the flow cross-section is made up of (i) the inter-fin space and (ii) the shroud clearance gap beyond the tips, i.e., the open area which extends outward from the tips of the fins to the shroud. In seeking the path of least resistance, the flowing fluid will favor the open area in reference to the relatively constrained inter-fin spaces. The flow imbalance will grow more severe when the fins are closely packed and when the open area is large. Such an imbalance gives rise to low velocities near the fin base and high velocities near the fin tip. As a consequence, the minimum fin heat loss occurs adjacent to the fin base and the maximum fin heat loss occurs at the

tip [6]. The traditional analyses which assume
a uniform heat transfer coefficient show the base
region to be the most active heat transfer zone
and a minimum heat loss at the tip.

(iii) Thermal Conductivity of Surfaces

Thermal conductivity is usually not crucial in
selecting the material of a primary surface. However, it is quite important in extended surfaces
because of the relatively long and narrow heat-
flow path which affects their heat transfer performance.

A-1. Perforated Surfaces

Flow visualization [7] indicated that at a
low Reynolds number, Re, or a small value of the
interplate spacing-plate length ratio, r, (equivalent to a low-porosity perforate plate), vortices
were seen to shed from the upstream lip. These
vortices exert little effect on the thermal boundary layer on the downstream plate because they are
either disposed into the free stream or weakened
before they reach the succeeding plate. Thus,
perforation exerts little influence on heat transfer performance under this circumstance. In the
transitional flow regime or for an r exceeding a
critical value (equivalent to the perforation di-
ameter-plate thickness exceeding a critical value in
the case of a perforated plate), vortices are formed
behind the upstream lip and weak flow oscillations
are observed in the inter-plate space. These vortices periodically shed from the inter-plate space
and penetrate into the downstream thermal boundary
layer, causing it to either separate from the surface (thermal boundary layer thus becomes shorter)
or become turbulent. Heat transfer performance
is then significantly enhanced. At a high r
(equivalent to a high-porosity perforated plate),
oscillations of free shear layer in the interplate
spacing causes periodic shedding of vortices at
the downstream edge when Re exceeds a critical
value Re_o. These vortices then interrupt the laminar boundary layer that is developed on the downstream plate, resulting in an augmentation in both
heat transfer and friction loss. The critical Re
for the onset of the "second laminar flow" regime
due to perforations was determined through data
curve-fitting procedure with the aid of a digital
computer [8]. The enhancement in the transition
regime is more prominent than in the second laminar flow regime and remains substantial in the
early turbulent flow regime. The penalty of operating in the transition and turbulent flow regimes,
however, is flow-induced noise and vibration.

In addition to the transition-turbulent flow
enhancement and the "second" laminar flow enhancement, the third type called "laminar flow enhancement" was revealed in reference 9. It occurs over
the entire laminar flow region on low-porosity
composite surfaces consisting of a short upstream
section for producing vortices and a main section
for heat transfer. The boundary layer in the main
section (downstream plate) is interrupted by the
vortices that are periodically generated from the
perforations (or interplate spacings) in the upstream inducing section.

Figure 1 compares the heat transfer and friction loss performance of the four different heat
transfer surfaces: non-perforated plate (NP-1
with surface porosity $\sigma_s = 0$), low -porosity
surfaces (S4-31 with $\sigma_s = 0.129$ and S2-1 with

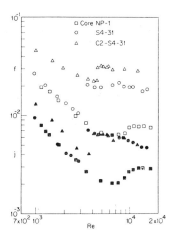

(i) Transition-turbulent flow enhancement and laminar flow enhancement

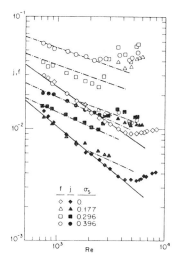

(ii) "Second" laminar flow enhancement

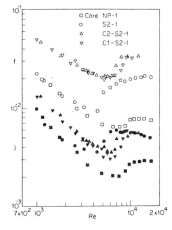

(iii) Effect of inducing section on
laminar flow enhancement

Figure 1. Performance of perforated surfaces

111

σ_s = 0.128), high-porosity plates (σ_s = 0.177, 0.296, 0.396), and composite surfaces (described in Table 2 and Fig. 2). It is interesting to point

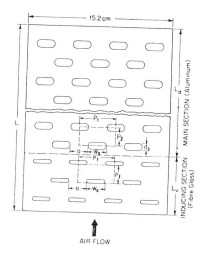

Figure 2. Perforation pattern of composite plate

Table 2. Geometric Parameters of Composite Cores

| | Main Section | | Inducing Section | |
	S2-1	S4-31	C1	C2
L, m	0.61	0.61	0.152	0.152
d_s, cm	0.159	0.318	0.159	0.159
W_s, cm	1.27	1.27	1.27	1.27
a, cm	1.27	1.27	1.27	1.27
P_t, cm	2.54	2.54	2.54	2.54
P_ℓ, cm	0.605	1.17	0.635	0.318
σ_s	0.128	0.129	0.122	0.243

t = 0.0813 cm, W = 15.2 cm, height of test core = 15.2 cm,

b = 0.555 cm, σ_F = 0.872

out that the transition-turbulent enhancement exhibits a very significant increase in both f and j over the non-perforated case only in the transition and turbulent flow regions, Fig. 1 (i). Little change in f or j was detected in the laminar flow regime (up to Re $\cong$ 2800). In the case of high porosity surfaces, there exists a critical Reynolds number Re_o beyond which perforations cause the f and j curves to deviate from those of the non-perforated plate in the laminar flow regime (solid lines) and creates a "second" laminar flow regime (broken lines), Fig. 1 (iii). For Re less than Re_o (corresponding to an intersection of solid and broken lines), perforations exert no effect on the transport phenomena. The main flow is still in a laminar fashion between Re_o and the transition Re (as characterized by a steep increase in both f and j). The presence of an inducing section causes a significant increase in both f and j over the entire laminar flow regime together with a delay in the flow transition, Fig. 1 (ii). This particular feature is very desirable and useful in

application to the plate-fin type heat exchanger which is mostly operated within this Re range. The effect of the inducing section on the transport performance disappears in the turbulent flow regime.

In general, an increase in surface porosity promotes flow transition (to occur at a lower Re) for all three enhancements. It is believed that the manner of vortices interrupting the boundary layers and the mechanisms of enhancement in both f and j that have been disclosed in the perforated plates are applicable to interrupted-wall channels [10]. The interrupted-wall compact surface is referred to as strip fin, serrated fin or offset fin in heat exchanger applications.

When a stack of perforated surfaces are placed normal to the flow, the augmentation in the heat transfer and friction loss performance is caused by the edgetone — self-sustained oscillations due to free shear layer instability [11]. There exists a critical value of the plate spacing-thickness ratio (about 2.5) beyond which the enhancement of j and f begins to level off. Two distinct types of flows are discovered with the surface porosity of approximately 0.28 as the criterion: vortex formation in low porosities and wake instability in high porosities.

Sparrow and his associates studied offset plate transfer surfaces with application to one type of extended surfaces which is referred to as strip fin, serrated fin or offset fin (in A-2).

A-2. Interrupted Surfaces

Sparrow et al [10] obtained solutions for laminar flow and heat transfer in channels whose walls are interrupted periodically along the stream direction. The results demonstrated that such passages can provide superior heat-transfer performance compared to a conventional parallel plate channel. As mentioned earlier, flows in interrupted-wall channels resemble flows through perforated plate stacks. Both flows are strongly influenced by the shedding of vortices. However, the analytical model have neither considered the effect of vortices generated at the leading and trailing edges of the strips nor the effect of unstable laminar wake region downstream of the strips. This defect of their analytical model was also pointed out by Shah [12]. He also suggested one simple way to predict the heat transfer performance of the strip fin using the conventional plain duct thermal entry length solutions. In Fig. 3, the theoretical mean Nusselt

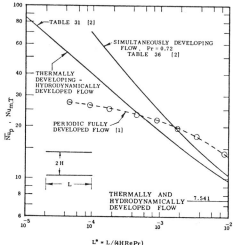

Figure 3. A comparison of laminar mean Nusselt numbers for the plain duct thermal entrance region with those for periodic fully developed flow

numbers $Nu_{m,t}$ for the constant wall temperature boundary condition are plotted against $L^* = L/(4HRePr)$ for parallel plates for two conditions: developing velocity profiles, and fully developed velocity profiles. Superimposed are Nu_p from Table 1 of [10] for the periodic fully developed flow and $Nu_{m,t} = 7.541$ for the fully developed flow through parallel plates. The augmentative characteristics of the interrupted wall is clearly seen in the region of $L^* > 0.0017$. For $L^* < 0.0005$, however, Nu_p is lower than $Nu_{m,T}$ for a thermally developing-hydrodynamically developed flow, since the periodic fully developed region is thermally saturated, while the plain duct thermally developing region is unsaturated. The figure also indicates that Nu_p approaches an asymptotic value for lower values of L^*.

Sparrow and Liu [13] treated a more generalized model of laminar airflow in interrupted-wall channels [10] and obtained solutions for the arrays of in-line or staggered plate segments. The heat transfer for the segmented arrays exceeds that for the parallel-plate channel. Under practical conditions, the staggered array yields better performance than the in-line one.

In order to understand the basic mechanism in interrupted-wall channels, Cur and Sparrow [14] conducted experiments on a two-plate colinear array aligned parallel to air flow. Results showed that under most practical considerations (those commonly encountered in an interrupted-plate heat exchanger), the effect of increasing plate thickness is to increase both the Nusselt number and pressure drop. The same experiments [15] were conducted on a more generalized model, namely an array of numerous, equally spaced colinear plates aligned parallel to the flow in a rectangular duct. The study affirmed the periodically developed regime (at sufficiently downstream distances): identical developing boundary layers on successive plates and identical flow fields in the successive inter-plate spaces [16]. Both the Nusselt number and pressure drop increase with the plate thickness. The presence of the interruptions serves to augment the heat transfer coefficients which are on the order of twice those for a conventional duct flow in the fully turbulent regime. Patankar et al [16] formulated generalized concepts of fully developed flow and heat transfer for ducts of periodically varying cross sections. The concepts were illustrated by applications to laminar flow and heat transfer in a flow field consisting of successive ranks of plate segments placed transverse to the flow direction. Both the heat transfer and pressure drop for the transverse plate array are substantially higher than those of conventional laminar duct flows. The fully developed Nusselt numbers show a marked dependence on the Reynolds number.

Using analysis and numerical techniques, Sparrow and Prakash [17] determined the natural convection enhancement by a staggered array of discrete vertical plates. The maximum enhancement was a factor of two higher than the parallel plate case. The general degree of enhancement compares favorably with that which has been accomplished in forced convection systems.

A-3. Pin Fins

An array of wall-attached cylinders (e.g. rods, wires) mounted perpendicular to the wall is used as pin fins. Sparrow and Ramsey [18] measured heat transfer in a staggered array of circular cylinders situated in a crossflow of air in a flat rectangular

duct. It was disclosed that fully developed conditions prevailed for the fourth and all subsequent rows. The fully developed transfer coefficients are quite insensitive to the cylinder height, increasing only about 20% as the height increases. The array pressure drop, however, increases markedly with increasing cylinder height. In the case of in-line pin fin arrays [19], fully developed conditions also prevailed for the fourth row and beyond. In general, the fully developed heat transfer coefficients for the in-line array are lower than for the staggered array, but the pressure drop is also lower. The deviations between the two arrays increase with increasing fin height. The in-line array transfers more heat than the staggered one under conditions of equal pumping power and equal heat transfer area, while the latter requires less heat transfer surface than the former at a fixed heat load and fixed flow rate. The flow and heat transfer phenomena are too complex to analyze due to the presence of vortices and wake regions behind the cylinders.

A-4. Internally Finned Tubes and Annuli

The gain in heat transfer and the penalty in pressure drop have been measured in airflow for a longitudinal fin with various augmentation treatments (perforated, slitted, screened, laterally wired) which subdivides a rectangular channel into two parts [20]. Sparrow et al [6] theoretically studied the laminar heat transfer characteristics of an array of longitudinal fins with an adiabatic shroud situated adjacent to the fin tips. It was found that (i) the heat loss increases monotonically along the fin from the base to the tip, and (ii) the fin is a more efficient transfer area than the base surface in terms of the overall heat transfer per unit area. Thus, the conventional uniform heat transfer coefficient model is inapplicable to shrouded fin arrays.

Patankar et al [21] analyzed the fully developed turbulent flow and heat transfer performance for tubes and annuli with longitudinal internal fins. The highest heat loss was found to occur in the neighborhood of the fin tip where the velocity was the largest. The heat loss exhibited the smallest value at the fin base. In general, the fins were found to be as effective in heat transfer as the primary surface (wall) on the unit area basis. The analytical predictions agree satisfactorily with the Nusselt number data for air-flow in various internally finned tubes [22]. Carnavos [22] developed an empirical correlation for the turbulent heat transfer and friction characteristics of internally finned tubes in single-phase forced convection. He [23] concluded that internally finned and smooth tubes have the same Prandtl dependence $Nu \propto Pr^{0.4}$ for heating.

Recently, Webb and Scott [24] conducted a parametric analysis of the performance of internally finned tubes in turbulent forced convection. By variation of the fin geometric parameters, the analysis defined the benefits of internally finned tubes for three cases: tube material reduction, UA increase and pumping power reduction, where U is the overall heat transfer coefficient. It also compared the performance of internally finned and internally roughened tubes.

B. ROUGH SURFACES

Kader and Yaglom [25] brought together all the available turbulent heat transfer results for tubes and channels having roughness elements in the form of regularly repeated ridge-like protrusions perpendicular to the stream direction and correlated them satisfactorily. With the gas-cooled reactor as the motivating application, Dalle Donne and Meyer [26] developed a transformation to generalize heat transfer and friction data for a single roughened rod contained in a concentric smooth tube to predict results for a cluster of roughened rods. The rough surface at a salt-brine interface yields a higher natural convection mass transfer rate than a smooth surface does [27]. Han et al [28] conducted turbulent airflow experiments in a parallel plate channel with rib-roughened surfaces. Their results showed that ribs at a 45-degree angle of attack have superior heat transfer performance at a given friction power than the ribs at a 90-degree angle of attack or than sand-grain roughness. A new form of Reynolds analogy for a rough surface was derived by Leslie [29] on the assumption that the eddy Prandtl number is constant. It is purported to supersede existing relationships. Ramakrishna, et al [30] conducted an analysis on turbulent heat transfer from a rough vertical plate. It predicts that the Nusselt number is proportional to the Grashof number raised to the one-half power:

$$Nu_L/f = C_3 \, Gr_L^{0.5}$$

where the constant C_3 is a function of the Prandtl number alone; Gr is the Grashof number; and the friction factor f, is

$$f = 0.044 \, (e/L)^{0.051}$$

where e and L denote the absolute roughness and surface length, respectively. The results agree very well with test data. Moffat et al [31] conducted heat transfer measurements with a turbulent boundary layer on a rough (comprised of small spherical elements arranged in a most-dense array with their crests coplanar), permeable plate with and without blowing. It was disclosed that blowing through the rough surface reduced both the Stanton number and the roughness Reynolds number and also that a correlating equation for smooth walls with blowing was applicable to the rough wall case.

C. HELICAL TUBES AND DUCTS

Masliyah and Nandakumar [32] obtained numerical solutions for laminar flow and heat transfer in internally finned helical coils: fRe is a function of both the Dean number and the radius of curvature, whereas the Nusselt number is dependent upon Dean number alone. Janssen and Hoogendoorn [33] conducted heat transfer experiments in helical coiled tubes. The results indicated that the overall heat transfer coefficients are relatively insensitive to whether the boundary condition is a uniform wall temperature or uniform heat flux.

D. CORRUGATED SURFACES

Flows in corrugated channels are associated with secondary flows, suppression of the secondary flow by counteracting centrifugal forces, and destruction of the secondary flow by the onset of turbulence. Goldstein and Sparrow [34] disclosed that in the laminar range (up to Re of about 1000 to 1200), the average heat transfer coefficient h for the corrugated wall channel were only

moderatley larger than those for a parallel-plate channel. However, in the low turbulent regime (Re of about 6000 to 8000), the wall corrugations caused an increase of nearly a factor of three in h compared with the smooth wall channel. Grosse and Schiestel [35] obtained a finite difference solution for turbulent heat transfer in a wavy-wall tube which shows that positive results of the augmentation can be achieved at moderate and small Pr but not at large Pr. Vajravelu and Sastri [36] used numerical analysis to predict the natural convection across a vertical layer confined between one flat wall and a parallel wavy wall. For an inclined layer between a flat wall and a parallel corrugated wall, ElShirbiny et al [37] obtained a larger heat flux than if both walls are flat.

Al-Arabi and El-Refaee [38] studied natural convection from triangularly corrugated horizontal heated surfaces facing upward to the surrounding air. The heat transfer coefficient decreases with the increase of both corrugation length and angle. In the practical range of corrugation angles, 30 to 60 degrees, Fig. 4 illustrates that the use of

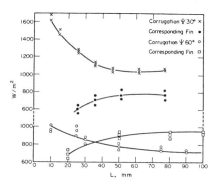

Figure 4. Comparison between corrugated plates and corresponding finned ones.

the corrugated surface as an extended surface is worth considering. Comparison was made under the same pitch, thickness and weight (by selecting a fin height).

E. PERTINENT PROBLEMS

Three problems are intimately related to enhanced surfaces: (i) testing methods, (ii) performance evaluation criteria and (iii) fouling.

E-1. Testing Methods

It is essential to employ a reliable experimental technique to determine the heat transfer and pressure drop performance of the surfaces. Direct heat transfer experiments can be classified into the steady-state and transient (single-blow) techniques [39]. The former technique often requires substantial time in reaching a steady state. Recently, steady state mass transfer experiments were employed rather than steady-state heat transfer experiments and the mass transfer results were converted to heat transfer results via the well-known analogy between the two processes [for example 14]. The naphthalene sublimation technique was often used for the mass transfer experiments.

E-2. Performance Evaluation Criteria

An enhanced surface is more efficient in

transferring heat than what might be called the standard situation. Numerous factors enter into the decision of the merit of surfaces developed to augment heat transfer: heat duty increase, area reduction, initial cost, pumping power or operating cost, maintenance cost, safety and reliability, among others. These factors are to some extent difficult to quantitize. So far, there is no universally accepted criterion for evaluating performance, although numerous criteria have been in practice. Bergles et al [40] reviewed these criteria which were classified into three objectives: increase heat transfer, reduce pumping power and reduce exchanger size.

Kraus et al [41] suggested to evaluate the performance of arrays of extended surfaces by the ratio of total heat dissipated divided by operating temperature excess. Webb [42] recommended the Dipprey-Sabersky correlation for rough surface heat transfer data. A "universally" applicable method has been proposed for the performance evaluation of plate-fin heat exchanger surfaces [43]. Shah [44] summarized the performance comparison methods for compact heat exchanger surfaces. A new approach was due to Bejan [45] who proposed the use of the number of entropy production unit as a means of evaluating the performance of a heat transfer surface directly in terms of the amount of usable energy wasted by the heat exchange device. This criterion is closely related to the current trend towards using the concept of second law efficiency in thermal design.

E-3. Fouling

The stickiest problem in enhanced surfaces is fouling. It is the least understood and has thus been roughly treated. Recently, considerable research effort has been directed toward the mechanism and determination of fouling. There has been no important breakthrough reported and the research status seems to still remain in the state of the art.

PROSPECTS FOR THE FUTURE

The current literature on heat transfer in channel flows reflects a strong interest in flows of complex configurations and in techniques to enhance heat transfer. Efforts to develop new types of extended surfaces are continuing. Rough surfaces and extended surfaces are competing augmentation techniques. As a matter of fact, rough surfaces with very large protuberance may be considered extended surfaces. In order to further enhance heat transfer performance, both the roughened and extended surfaces may be applied with a helix angle — leading to helical surfaces. Corrugated surfaces may be employed as extended surfaces with the advantage of easy manufacturing. The shape of enhanced surfaces that are employed in actual heat exchange devices are quite complex. As a consequence, information about their heat transfer and friction loss characteristics has, in the main, been confined to overall coefficients obtained from tests on either actual or large-scale models of enhanced surfaces. Although this information can be directly applied in the design of heat exchangers with enhanced surfaces, it fails to provide insights into the basic phenomena that occur within the individual flow passages. A better quantitative understanding of the flow phenomena is essential for better data correlations and improved heat exchanger design.

The theoretical studies of complex flows and heat transfer are either non-existing or for overly idealized models at best. Recent efforts by Sparrow and his associates in analyses of several fins are plausible. However, their mathematical models represent oversimplifications, particularly with the neglect of the effects of vortices and unstable wake regions which play an extremely important role in heat transfer augmentation. More reliable information can be obtained through a refinement of the models.

A. EXTENDED SURFACES

A substantial progress has been made in the area of extended surfaces during the past few years. Continuing efforts are currently directed toward two goals: new types of extended surfaces and improvement of heat transfer coefficients. More fundamental research is needed to clarify the performance of finned surfaces. Refined theoretical models can be brought about only through a thorough knowledge on the behavior of vortices and unstable wake regions which can be observed with the use of flow visualization techniques. The "second" laminar flow enhancement has been disclosed in perforated surfaces. This type of augmentation may exist in other kinds of enhanced surfaces. Since the practical flow range in heat exchanger application includes high laminar transition and low turbulent flows, the basic study on the transition or "transition-turbulent" enhancement of extended surfaces is also needed. Only experimental means are feasible in this case, while mathematical treatment will be formidable.

More basic research is needed to clarify the performance of internally finned tubes, externally finned tubes and tube banks.

Further study is needed to clarify the change in performance of extended tubes between single tubes and tube banks.

B. ROUGH SURFACES

Integral roughness may be produced by the process of machining, forming, casting, or welding. Because of infinite numbers of possible geometric variations, there is no unified treatment available. Efforts should be directed to develop more general analogy-based correlations for single-phase flow in order that optimum geometries can be selected without extensive test programs. More information is needed on the Prandtl number effect which varies with the type of roughness (typically, sand grain or repeated rib). The effect of fouling on rough surfaces is virtually unknown. Few studies have investigated the natural convection or laminar heat transfer performance of rough surfaces. Non-Newtonian flows have received practically no attention. When the protuberance is large (but not large enough to be extended surfaces), the selection of the characteristic length and cross-sectional area becomes a problem which is yet to be resolved.

C. METHODS OF INVESTIGATION

More flow visualization studies are encouraged to observe the flow patterns in complex flows. A sophisticated analytical model requires better understanding of the flow phenomena. Flow visualization methods may be tied to the use of image processing techniques since vortex phenomena and unstable wake region are difficult to describe by mathematical expression. Eventually, the information obtained from the image processing of a complex flow may be employed as input data to solve the flow and heat transfer equations for the predictions of the heat transfer and friction loss characteristics. Of course, the task is formidable but not impossible. This innovative approach can also be applied to the flows and heat transfer involving phase change.

D. FOULING

Fouling remains one of the unresolved problems in heat transfer area. Since fouling, even at modest extent, may well wipe out all the benefits of high performance brought about in an enhanced surface, provisions should be made to reduce fouling. It is therefore essential to understand the fouling characteristics of enhanced surfaces. Research in this area is strongly recommended.

REFERENCES

1. Bergles, A.E., Webb, R.L., Junkhan, G.H. and Jensen, M.K., "Bibliography on Augmentation of Convective Heat and Mass Transfer," Engineering Research Institute, Iowa State University, Ames, Iowa (May 1979).

2. Carnavos, T.C., "Some Recent Developments in Augmented Heat Exchange Elements," in Heat Exchangers: Design and Theory Source Book, Scripta, Washington, D.C., 441-489 (1974).

3. Bergles, A.E., "Enhancement of Heat Transfer" in Heat Trasnfer-1978, Vol. VI, Hemisphere, Washington, D.C., 89-108 (1978).

4. Bergles, A.E., Junkhan, G.H. and Webb, R.L., "Energy No. Conservation via Heat Transfer Enhancement", Paper EGY205 presented at 1978 Midwest Energy Conference, Chicago, Ill. (19-21 November, 1978).

5. Chenoweth, J.M., Kaellis, Michel, J. and Shenkman, S.M., (eds.)," Advances in Enhanced Heat Transfer", ASME Symposium, Vol. 100122 (1979).

6. Sparrow, E.M., Balgia, B.R. and Patankar, S.V., "Forced Convection Heat Transfer from a Shrouded Fin Array With and Without Tip Clearnace", J. Heat Transfer $\underline{100}$, 572-579 (1978).

7. Liang, C.Y., "Heat Transfer Flow Friction Noise and Vibration Studies of Perforated Surfaces", Ph.D Thesis, Department of Mechanical Engineering, University of Michigan, Ann Arbor, Michigan (1975).

8. Lee, C.P. and Yang, W.J., "Augmentation of Convective Heat Transfer from High-Porosity Perforated Surfaces", Heat Transfer 1978, Vol. II, Hemisphere, Washington, D.C., 589-594 (1978).

9. Yang, W.J., "Three Kinds of Heat Transfer Augmentation in Perforated Surfaces", Letters J. Heat Mass Transfer, $\underline{5}$, 1-10 (1978).

10. Sparrow, E.M., Baliga, B.R. and Patankar, S.V., "Heat Transfer and Fluid Flow Analysis of Interrupted-Wall Channels, With Application to Heat Exchangers", J. Heat Transfer, $\underline{99}$, 4-11 (1977).

11. Lee, C.P. and Yang, W.J., "Heat Transfer and Friction Loss in Flows Normal to Stacks of Perforated Plates", Heat Transfer 1978, Vol. II, Hemisphere, Washington, D.C., 625-630 (1978).

12. Shah, R.K., "Discussion on Heat Transfer and Fluid Flow Analysis of Interrupted-Wall Channels, with Application to Heat Exchangers by E.M. Sparrow, B.R. Baliga and S.V. Patankar", J. Heat Transfer $\underline{101}$, 188-189 (1979).

13. Sparrow, E.M. and Liu, C.H., "Heat Transfer, Pressure-drop and Performance Relationships for In-Line, Staggered, and Continuous Plate Heat Exchangers", J. Heat Mass Transfer $\underline{22}$, 1613-1625 (1979).

14. Cur, N. and Sparrow, E.M., "Experiments on Heat Transfer and Pressure Drop for a Pair of Colinear , Interrupted Plates Aligned with the Flows", Intern. J. Heat Mass Transfer, $\underline{21}$, 1069-1080 (1978).

15. Cur, N. and Sparrow, E.M., "Measurements of Developing and Fully Developed Heat Transfer Coefficients Along a Periodically Interrupted Surface", J. Heat Transfer, $\underline{101}$, 211-216 (1979).

16. Patankar, S.V., Liu, C.H. and Sparrow, E.M., "Fully Developed Flow and Heat Transfer in Ducts Having Streamwise-Periodic Variations of Cross-Sectional Area", J. Heat Transfer, $\underline{99}$, 180-186 (1977).

17. Sparrow, E.M. and Prakash, C., "Enhancement of Natural Convection Heat Transfer by a Staggered Array of Discrete Vertical Plates", J. Heat Transfer, $\underline{102}$, 215-220 (1980).

18. Sparrow, E.M. and Ramsey, J.W., "Heat Transfer and Pressure Drop for a Staggered Wall-Attached Array of Cylinders with Tip Clearance", Intern. J. Heat Mass Transfer, $\underline{21}$, 1369-1377 (1978).

19. Sparrow, E.M., Ramsey, J.W. and Altemani, C.A.C., "Experiments on In-Line Pin-Fin Arrays and Performance Comparisons with Staggered Arrays", J. Heat Transfer, $\underline{102}$, 44-50 (1980).

20. Tishchenko, Z.V.and Butskii, N.D., "Investigation of Heat Transfer and Drag During Flow of a Gas in a Rectangular Channel with a Single Fin", Intern. Chem. Engg., $\underline{17}$, 498-504 (1977).

21. Patankar, R.V., Ivanovic, M. and Sparrow, E.M., "Analysis of Turbulent Flow and Heat Transfer in Internally Finned Tubes and Annuli", J. Heat Transfer, $\underline{101}$, 29-37 (1978).

22. Carnavos, T.C., "Cooling Air in Turbulent Flow with Internally Finned Tubes", Heat Transfer Engineering, $\underline{1}$, 41-46 (1979).

23. Carnavos, T.C., "Heat Transfer Performance of Internally Finned Tubes in Turbulent Flow", Heat Transfer Engineering, $\underline{1}$, 32-37 (1980).

24. Webb, R.L. and Scott, M.J., "A Parametric Analysis of the Performance of Internally Finned Tubes for Heat Exchanger Appliciations", J. Heat Transfer, $\underline{102}$, 38-43 (1980).

25. Kader, B.A. and Yaglom, A.M., "Turbulent Heat and Mass Transfer from a Wall with Parallel Roughness Ridges", Intern. J. Heat Mass Transfer, $\underline{20}$, 345-358 (1977).

26. Dalle Donne, M. and Meyer, L., "Convective Heat Transfer from Rough Surfaces with Two Dimensional Ribs: Transitional and Laminar Flow", Intern. J. Heat Mass Transfer, $\underline{20}$, 583-620 (1977).

27. Chang, C., Vliet, G.C., and Saberian, A., "Natural Convection Mass Transfer at Salt-Brine Interfaces", J. Heat Transfer, $\underline{17}$, 603-608 (1977).

28. Han, J.C., Glicksman, L.R. and Rohsenow, W.M., "An Investigation of Heat Transfer and Friction for Rib Roughened Surfaces", Intern. J. Heat Mass Transfer, $\underline{21}$, 1143-1156 (1978).

29. Leslie, D.C., "The Form of the Extended Reynolds Analogy for Rough Surfaces", Letters J. Heat Mass Transfer, 5, 99-109 (1978).

30. Ramakrishna, K., Seetharamu, K.N. and Sarma, P.K., "Turbulent Heat Transfer by Free Convection from a Rough Surface", J. Heat Transfer, 100 727-729 (1978).

31. Moffat, R.J., Healzer, J.M. and Kays, W.M., "Experimental Heat Transfer Behavior of a Turbulent Boundary Layer on a Rough Surface with Blowing", J. Heat Transfer, 100, 134-142 (1978).

32. Masliyah, J.H. and NandaKumar, K., "Fluid Flow and Heat Transfer in Internally Finned Helical Coils", Canadian J. Chem. Engg., 55, 27-36 (1977).

33. Janssen, L.A.M. and Hoogendoorn, C.J., "Laminar Convective Heat Transfer in Helical Coiled Tubes", Intern. J. Heat Mass Transfer, 21, 1197-1206 (1978).

34. Goldstein, L., Jr. and Sparrow, E.M., "Heat/Mass Transfer Characteristics for Flow in a Corrugated Channel", J. Heat Transfer, 99, 187-195 (1977).

35. Gosse, J. and Schiestel, R., "Thermal Convection in the Wavy Tubes of a Type of Heat Exchanger", Intern. Chem. Engg., 18, 1-7 (1978).

36. Vajravelu, K. and Sastri, K.S., "Free Convective Heat Transfer in a Viscous Incompressible Fluid Confined Between a Long Vertical Wavy wall and a Parallel Flat Wall", J. Fluid Mech., 86, 365-383 (1978).

37. ElSherbiny, S.M., Hollands, K.G.T. and Raithby, G.D., "Free Convection across Inclined Air Layers with One Surface V-Corrugated", J. Heat Transfer, 100, 410-415 (1978).

38. Al-Arabi, M. and El-Refaee, M.M., "Heat Transfer by Natural Convection from Corrugated Plates to Air", Intern. J. Heat Mass Transfer, 21, 357-359 (1978).

39. Liang, C.Y. and Yang, Wen-Jei, "Modified Single-Blow Technique for Performance Evaluation on Heat Transfer Surfaces", J. Heat Transfer, 97, 16-21 (1975).

40. Bergles, A.E., Blumenkrantz, A.R. and Taborek, J., "Performnce Evaluation Criteria for Enhanced Heat Transfer Surfaces", Heat Transfer 1974, Vol. II, JSME, 239-243 (1974).

41. Kraus, A.D., Snider, A.D. and Doty, L.F., "An Efficient Algorithm for Evaluating Arrays of Extended Surface", J. Heat Transfer, 100, 288-293 (1978).

42. Webb, R.L., "Toward a Common Understanding of the Performance and Selection of Roughness for Forced Convection", ASME Paper No. 78-WA/HT-61 (1978).

43. Soland, J.G., Mack, W.M., Jr. and Rohsenow, W.M., "Performance Ranking of Plate-Fin Heat Exchanger Surfaces", J. Heat Transfer, 100, 514-519 (1978).

44. Shah, R.K., "Compact Heat Exchanger Surface Selection Methods", Heat Transfer 1978, Vol. IV, Hemisphere, Washington, D.C., 193-199 (1978).

45. Bejan, A., "General Criterion for Rating Heat-Exchanger Performance", Intern. J. Heat Mass Transfer, 21, 655-658 (1978).

Augmentation of Nucleate Boiling Heat Transfer by Prepared Surfaces

K. NISHIKAWA and T. ITO
Kyushu University, Fukuoka, Japan

ABSTRACT

The objectives of the research on the augmentation of nucleate boiling heat transfer by prepared surfaces are described first. Next the two basic methods to promote nucleate boiling heat transfer by prepared surfaces are discussed. Then the experimental results from many institutions including the authors' are introduced. They exemplify the above-mentioned basic methods. In citing reports from other sources emphasis is on Japanese research. Fields of investigation open to future studies are enumerated.

NOMENCLATURE

Dimensions are assigned by convention; M = mass, L = length, T = time, Θ = temperature, F = force(= ML/T^2), E = energy (= ML^2/T^2), < = plane angle and - = nondimensional.

d = diameter of sintered particle, L
h = heat transfer coefficient, $E/L^2 T \Theta$
L = latent heat of vaporization, E/M
P = pressure, F/L^2
q = heat flux, $E/M^2 T$
R = radius of curvature, L
ΔT = degree of superheating, Θ
X = length from the vertex along the generatrix of cone, L
δ = thickness of sintered layer, L
ε = porosity, -
θ = contact angle, <
λ = thermal conductivity, $E/LT\Theta$
μ = viscosity, FT/L^2
ρ = density, M/L^3
σ = surface tension, F/L
ϕ = half apex angle of cone, <

SUBSCRIPTS

l = liquid
LV = vapor-liquid interface
m = apparent value for sintered layer saturated with liquid
opt = optimum
SL = solid-liquid interface
SV = solid-vapor interface
v = vapor
∞ = liquid around the cavity

INTRODUCTION

Heat transfer with large temperature differences should be avoided for the efficient use of heat sources, especially when energy of high quality is to be generated most efficiently from two heat sources with a small difference of temperature between them. A large temperature difference will increase the irreversibility by heat transfer, and give rise to the degradation of thermal energy. It also means larger dimensions of the heat exchanger and reduced efficiency of the power plant cycle.

The thermal energy is of lower grade intrinsically. However the effort to make full use of the available temperature difference between heat sources is gaining in importance under the current world situation of the deficiency of an energy supply. Heat transfer by nucleate boiling is one of the most important modes of heat exchange occurring in many constituent devices of a thermal plant. So the augmentation of nucleate boiling heat transfer is sure to go a long way toward the efficient use of thermal energy. The current world-wide research on the augmentation of heat transfer by nucleate boiling suggest many potential measures for it. But only two of them are dealt with in this article. The first is the treatment of the heating surface that reduces the wettability of the surface by the boiling liquid. The second is the fabrication of the heating surface with re-entrant cavities. Namely only the measures applied to the heating surface to promote the heat transfer by nucleate boiling are within the scope of the present article. Though the ideas developed and data referred to might be of use for the flow boiling systems, the following descriptions are concerned exclusively with the nucleate boiling in pool.

The authors have been involved in an investigation to find the general requisites to the high-performance heating surface of nucleate boiling and to examine experimentally some potential high-performance surfaces. A high-performance surface in this context means, of course, a surface that requires a significantly smaller temperature difference for a given heat flux than a commercial smooth surface does. The objective of the authors' work is to develop high-performance surfaces which do not degrade in liquids of commercial purity in long-term operations. The experiment on pool nucleate boiling from horizontal cylinders covered by a sintered layer of metal powder to saturated refrigerant-11 (R-11), refrigerant-113 (R-113) and benzene under the atmospheric pressure is described, followed by a discussion of the results. It also contains the comparison with data from other sources.

THE METHOD TO PROMOTE HEAT TRANSFER BY NUCLEATE BOILING

REDUCTION OF WETTABILITY (METHOD I)

The surface tension between liquid and its vapor exerts the negative effect on heat transfer in most correlations of nucleate boiling heat transfer. That is, the larger the surface tension is, the smaller the heat transfer coefficient for a given heat flux will be. When the surface tension is large,

(i)The diameter of the bubble at the instance of its departure from the heating surface becomes large,
(ii) the surface can hardly get wet, and

(iii) the vapor trapped in the cavity is prone to stay there, thus forming a stabilized site of nucleation.

At the boundary of the three phases formed by a liquid droplet put on a surface of solid (Fig. 1), the balance of forces determines the contact angle θ , the angle between the solid surface and the vapor-liquid interface,

$$\cos \theta = \frac{\sigma_{SV} - \sigma_{SL}}{\sigma_{LV}} \qquad (1)$$

where σ_{SV}, σ_{SL}, and σ_{LV} are the surface tensions at the solid-vapor interface, the solid-liquid interface and vapor-liquid interface respectively. It is apparent from Eq. (1) that the contact angle θ is larger for liquids with larger surface tension σ_{LV}, and the solid surface has a smaller chance to get wet by the liquid. This is the situation that (ii) above-mentioned means. (i) will be easily seen because the Laplace constant include the square root of σ_{LV}. Then let us examine the stability of the small amount of vapor trapped in an inverted conical cavity as shown in Fig. 2. The radius of curvature of the vapor-liquid interface is calculated as follows, the apex angle of the conical cavity being 2ϕ,

$$R = \frac{\sin \phi}{\sin(\theta - \phi \pm \pi/2)} X \qquad (2)$$

$$+ : \theta < \pi/2 + \phi \qquad (a)$$

$$- : \theta > \pi/2 + \phi \qquad (b)$$

where X is the length of the generatrix covered by vapor. The coefficient of X in r.h.s. of Eq. (2) is positive in each case. The difference of pressure between that in the vapor in the cavity P_V and that in the liquid P_∞ is,

$$P_V - P_\infty = \pm \frac{2\sigma_{LV}}{R} \qquad (3)$$

$$+ : \theta < \pi/2 + \phi \qquad (a)$$

$$- : \theta > \pi/2 + \phi \qquad (b)$$

If, whatever the reason, the vapor reduces its volume in case (a) of Fig. 2, that is, if the length X decreases, the radius R decreases also. The pressure of vapor then increases according to Eq. (3) and the saturation temperature becomes higher. The vapor will condense releasing the latent heat, if its temperature is lower than that saturation temperature. The resulting loss of the mass of vapor will cause a reduction of its volume to some extent. Then this process will go on repeatedly unless a new thermodynamic equilibrium is established owing to the heating given rise to by the liberated latent heat. The meniscus will "descend a downward slope" as shown in the lower half of Fig. 3.

In case (b) of Fig. 2, on the other hand, there is little likelihood of the contraction of the volume occurring, because it will lower the pressure of the vapor and the saturation temperature associated with the pressure. In these circumstances the temperature of the vapor will

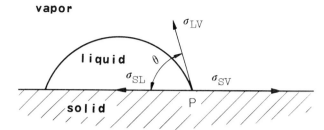

Fig. 1 Force balance at the boudary of three phases

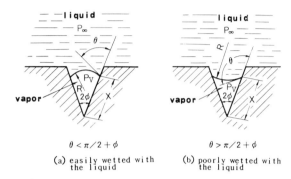

(a) easily wetted with the liquid

(b) poorly wetted with the liquid

Fig. 2 Stability of the vapor trapped in the inverted conical cavity

come up to the saturation temperature sooner or later and the vapor is said to be stable for any imposed disturbances. The meniscus has to "go uphill" to reduce the volume of the trapped vapor as shown in the upper half of Fig. 3.

If the surface tension at the vapor-liquid interface σ_{LV} and that at the solid-liquid interface σ_{SL} are large and/or that at the solid-vapor interface σ_{SV} is small, the contact angle becomes large. Then there is a fair chance that the inequality (b) holds and the vapor trapped in inverted conical cavities is stable. This is what (iii) above-mentioned meant. The poor wettability, namely the large contact angle at the cavity wall is vital to the successful augmentation of the heat transfer by nucleate boiling in the present method. So we should apply such surface treatments that would reduce the wettability by the boiling liquid only at the cavity wall. The poor wettability at the rest of the heating surface is by no means useful for the augmentation, as seen from (i) and (ii) above-mentioned.

SURFACES WITH RE-ENTRANT CAVITIES (METHOD II)

If every conceivable way to reduce wettability fails, the first method discussed above does not apply to the augmentation of the heat transfer by nucleate boiling. This is the case when fluorocarbons or liquefied gases such as liquefied nitrogen are to be boiled. We seem to have no alternative but to rely on the so-called re-entrant cavities for these classes of liquids. This second method applies to every combination of liquid and surface in principle.

Figure 4 depicts the highly idealized re-entrant cavities. The cylindrical mouth is connecting the cavity with the conical "ceiling" to the outer liquid space. The vertex of cone is in the liquid side. By a similar argument the radius of the curvature of the vapor-liquid interface, when the meniscus is at the conical portion of the cavity, is calculated as follows,

$$R = \frac{\sin\phi}{\sin\{\mp(\theta+\phi-\pi/2)\}} X \qquad (4)$$

$$- : \theta < \pi/2 - \phi \qquad (a)$$

$$+ : \theta > \pi/2 - \phi \qquad (b)$$

where X is the distance of the interface from the vertex of the cone along generatrix. The coefficient of X in r.h.s. of Eq. (4) is positive in each case. The difference of pressure between that in the vapor trapped in the re-entrant cavity P_V and that in the liquid P_∞ is,

$$P_V - P_\infty = \pm \frac{2\sigma_{LV}}{R} \qquad (5)$$

$$+ : \theta < \pi/2 - \phi \qquad (a)$$

$$- : \theta > \pi/2 - \phi \qquad (b)$$

The corner connecting the cylindrical portion and the conical surface is a singular point of the pressure variation with the locations of meniscus. The pressure in this context is when the vapor is in equilibrium with the liquid. While the meniscus is at the cylindrical portion, P_V is higher or lower than P_∞ by $|2\sigma_{LV}\cos\theta/X_0\sin\phi|$ depending on whether the contact angle is smaller or larger than $\pi/2$. Where X_0 is the value of X at the corner and $X_0\sin\phi$ is the radius of the cylindrical mouth. While the meniscus is at the conical surface, on the other hand, the pressure is given by Eqs. (4) and (5) above.

Figure 5 shows diagrammatically what is mentioned on the whole. If the contact angle θ is small enough so that the inequality (a) is satisfied, the vapor in the cavity is unstable because of the "well" at the corner. Namely, the vapor depicted in Fig. 4 (a) will shrink spontaneously until a new equilibrium is established. When the contact angle θ is fairly large, on the other hand, the inequality (b) holds. Then there is little likelihood of the vapor turning around the lower end of the cylindrical mouth of the cavity. There is a "barrier" at the corner. The situation shown in Fig. 4 (b) can hardly occur. Both at the inverted conical cavity are discussed in the previous section and at the re-entrant cavity the vapor trapped in the cavity is stable when the meniscus has its center of the radius of curvature in the liquid side. But the most important feature of the re-entrant cavity is the lower threshold of the contact angle for that

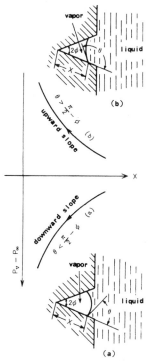

Fig. 3 The variation of vapor pressure when the vapor shown in Fig. 2 shrinks

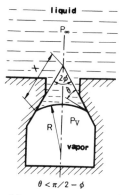 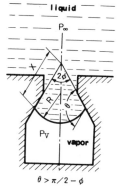

$\theta < \pi/2 - \phi$	$\theta > \pi/2 - \phi$
(a) easily wetted with the liquid	(b) poorly wetted with the liquid

Fig. 4 Stability of the vapor trapped in the re-entrant cavity

geometry. The difference is 2ϕ (the apex angle of the cone) compared with the inverted conical cavity. Furthermore the inequality (b) in Eqs. (4) and (5) is fulfilled always when the apex angle is π, that is, the "ceiling" is horizontal. This means that the vapor trapped in the cavity is stable irrespective of the contact angle.

EXPERIMENTAL RESEARCHES PERTAINING TO METHOD I

Hasegawa et al. /1/ made quite a few interesting observations at the lower heat fluxes when they measured the burn-out heat flux of water at the atmospheric pressure on partly ill-wettable heating surfaces. The experiment was carried out on the electrically heated horizontal strip made of stainless steel, at the central part of which an artificial, ill-wettable portion was installed. The ill-wettable part was obtained by putting a thin layer of Silastic adhesive from DOW CORNING (from 10 to 20 μm), the width being 2a. The contact angle of water on that adhesive is from 90 to 110°/2/ and the adhesive may be said to be hydrophobic. Their boiling curve in the lower heat flux region is reproduced schematically in Fig. 6. Some interesting descriptions which have an immediate connection with the above inference are cited. "(1) Initiation of boiling invariably occurs at the ill-wettable part with Silastic adhesive, and even in the case of the lower heat fluxes where no bubbles are generated on the ordinary surface, there have been found numerous boiling sites at the part with Silastic adhesive. (2) The diameter of bubble on the departure from the surface is unusually large, being about 5 mm. (3) The configuration of the bubbles of such sizes is not constricted at the contact angle of bubbles. (4) There is

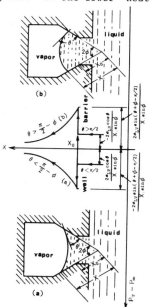

Fig. 5 The variation of vapor pressure when the vapor shown in Fig. 4 shrinks

formed a stable vapor film at the extremely low heat flux (from 5 to 10 W/cm²) on the ill-wettable part, and the heat flux forming the vapor film is inversely proportional to the width of Silastic adhesive ..."

The remarkable point seen from Fig. 6 is that the narrower the width of the ill-wettable part, the more effective the augmentation of nucleate boiling by the reduction of wettability is. It will be inferred that the promotion of boiling by the stabilization mechanism of the cavities is predominant while the width is narrow, but the wider application of the hydrophobic agent might degrade the overall characteristics by the mechanisms (i) and (ii) in the previous section. So the spotted reduction of the wettability, just at the cavity wall if possible, must be the most effective way of the surface treatment to promote nucleate boiling by the method I. Now let's examine two experimental reports that substantiate the idea.

Young and Hummel /3/ tested the effect of pits formed on the heater by pressing emery paper, that of spotted TFE (polytetrafluoroethylene) and the combination of these two processing on the nucleate pool boiling from the upward facing horizontal heater to saturated water at the atmospheric pressure. Figure 7 is a reproduction from their Fig. 3 /3/. The boiling curve 1 was obtained on a heating surface of cleaned stainless steel stock, 3-P on a pitted surface in which the similar surface as in 1 was roughened by pressed emery paper, and 3-PT on a pitted and TFE - processed surface which was prepared by plastic-treating the similar surface as in 3-P. In the plastic treatment, a coat of TFE emulsion was applied first. Then a piece of stiff paper was

used to wipe the TFE, leaving many of the pits partially filled with TFE. Finally the surface was baked to cure the TFE. The boiling curve 7-ST is also for TFE - processed surface. However in that case the surface wasn't pitted. The surface was sprayed with the TFE emulsion and baked.

Though the augmentation by artificial pits is apparent (compare 1 with 3-P and 7-ST with 3-PT), the effect of the TFE - treatment is much more striking, almost altering the order of magnitude of the heat transfer coefficient (compare 1 with 7-ST and 3-P with 3-PT). The effectiveness of the spotted application of layers with low surface energy seems remarkable. The contact angle of water on TFE is 108° at 20°C . The small difference of boiling curves between 3-PT and 7-ST may be inferred from the possibilty that they might have been quite a few pits on the stock surface already or cavities were formed when the coated layer of TFE was cured.

Liquefied nitrogen the small surface tension and spreads on or "wets" TFE. Then no poorly wetted spots would occur when it boils from a TFE coated surface and the promotion of the heat transfer shouldn't be expected. The experiment conducted by Marto et al. /4/ seems to confirm this. Figure 7 of their report is reproduced in Fig. 8. The roughened surface was made by pressing emery paper and the TFE - treatment here is very similar to that in the previous reference /3/. It is quite obvious that there is little point in applying TFE treatment to promote nucleate boiling. Fluorocarbons, important working fluids in the refrigerating cycle also have the small surface tension as liquefied gases do and its order of

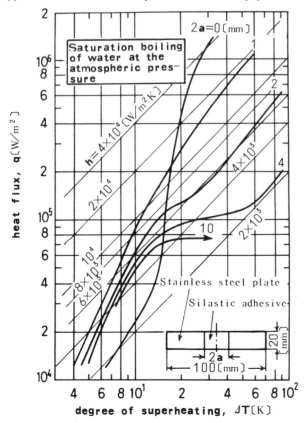

Fig. 6 Promotion of nucleate boiling by the surface with poorly wetted part on it /1/

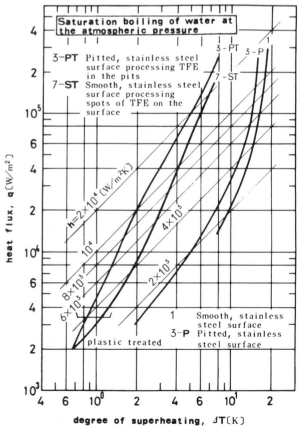

Fig. 7 Promotion of nucleate boiling by spotted application of layer of low surface energy /3/

magnitude is 10^{-2} N/m. No "Freonphobic" substance is known. So METHOD I fails for fluorocarbons. Gaertner /5/ has patented a method of augmenting boiling heat transfer which has a close relation to METHOD I in this article. Before closing this section the requisites to the material applied on the heating surface to enhance the heat transfer by nucleate boiling are cited from the US - Patent gazette. "The material (1) must be insoluble in the boiling liquid, (2) must be chemically and thermally stable, (3) must have a melting point which is sufficiently above the temperature of the heated surface ... , (4) must have good adhesion to the substrate material ... , (5) and must be a low surface energy material highly non-wetted by the liquid, that is, the material yields contact wetting angles of about 80° or greater with the liquid being heated." All the statements seem quite reasonable except the minimum allowable contact angle of 80°. According to Eq. (3) in the previous section, the condition (b) for the stabilized cavity doesn't hold for any contact angles less than 90°.

EXPERIMENTAL RESEARCHES PERTAINING TO METHOD II

The authors' investigation /6,7/ is introduced first in detail. The central part of the apparatus is

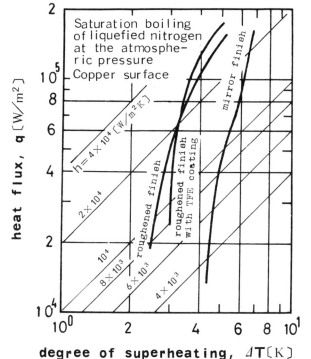

Fig. 8 Experiment verifying that the surface with layer wetted by the liquid does not augment heat transfer /4/

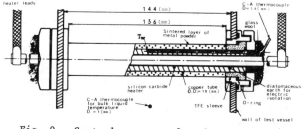

Fig. 9 Central part of the authors' apparatus/6/

shown in Fig. 9. The cylindrical heater is held horizontally in the test vessel. A copper or bronze powder of spherical particles is sintered onto the outer surface of a copper tube of 18 mm in diameter, the nominal diameter of the particle being designated d and the thickness of the sintered layer δ. Diameters ranging from 100 to 1,000 μm and thicknesses from 0(bare tube) to 5 mm were tested. The composition of bronze corresponds to JIS H5111 Class 3 BC3. The measured porosity of the sintered layer ranged from 0.38 to 0.71. By selecting the outer surface of the copper tube (surface of the substrate) as the reference surface, the heat transfer area and the surface temperature are defined.

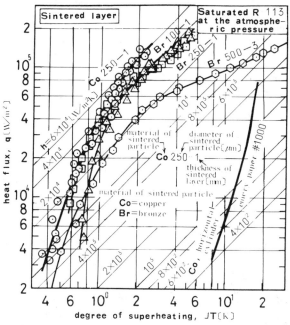

Fig. 10 Some typical examples of the augmentation of nucleate boiling heat transfer by sintered porous layer/7/

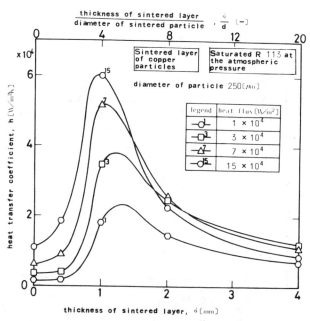

Fig. 11 The effect of the thickness of the sintered layer on heat transfer by nucleate boiling/7/

The experiment was performed on three test fluids of R-11, R-113 and benzene at the saturated temperature under the atmospheric pressure. Figure 10 shows the typical boiling curves of the sintered layer exhibiting a rather favorable performance. The liquid is R-113 saturated at the atmospheric pressure. The curve on the right of the figure was obtained by the authors on a horizontal copper cylinder finished by emery paper #1,000, the diameter of the cylinder being 18 mm. It may be seen that the heat transfer coefficient of these sintered layers for a specified heat flux can be greater than that of the smooth surface by a factor of about 10 at lower heat fluxes, though the performance corrupts at higher heat fluxes. A rough threshold for the heat flux may be around 0.1 MW/m². The optimum geometry of the sintered layer has been searched for experimentally. Figure 11 will afford some ideas. The variation of the heat transfer coefficient with the thickness of the sintered layer is shown at the given heat fluxes. The liquid is again R-113 and the diameter of the powder particle is held constant. The optimum thickness of the sintered layer seems to become thinner as the heat flux increases. The thickness of the sintered layer at the maxima has been

estimated for several combinations of the material of the sintered particle, the diameter of the sintered particle and liquid. They are summarized in Fig. 12. The ordinate is the ratio of the estimated optimum thickness to the diameter of the particle, though the diameter of the particle is held constant on each curve. It will be inferred that the optimum thickness is a slightly decreasing function of heat flux. The layer of bronze layer of 100 μm particle in R-113 (Br100R113) behaves quite differently from others and the optimum ratio is about 10. The physical meaning of the ratio might be open to doubt.

The mechanism of nucleate boiling heat transfer from a porous layer has been only vaguely described in reports from many institutions. It may be too early for constructing a correlating equation. However the authors have tried to propose one based on their simple analysis /6/. In performing the regression analysis the physical quantities included in it have been considered as the relevant physical quantities. The result is as follows,

$$q\delta/\lambda_m \Delta T = 0.639(\sigma^2 L/q^2 \delta^2)^{0.0658}$$
$$X(\delta/d)^{0.626} \quad (qd/\varepsilon L \mu_v)^{0.665}$$
$$X(\lambda_m/\lambda_l)^{-0.692} \quad (\rho_l/\rho_v)^{0.904} \qquad (6)$$

where ΔT = degree of superheating, ε = porosity of the sintered layer, L = latent heat of vaporization, λ = thermal conductivity, μ = viscosity, ρ = density, σ = surface tension at the vapor-liquid interface (σ_{lv}), subscripts l = liquid, m = apparent value for the sintered layer saturated with the boiling liquid, and v = vapor. The l.h.s. represent a Nusselt number, the first non-dimensional quantity in the r.h.s. the effect of the surface tension, and the third the Reynolds number of vapor flow if it fills all of the void.

The proposed equation is compared with the authors' data in Fig. 13, the ordinate being the l.h.s. of Eq. (6) and the abscissa the r.h.s. of the same equation. As entered in the figure, data has been obtained for three liquids, two kinds of material of the sintered layer, the particle diameter ranging from 100 to 1,000 μm and the thickness of the sintered layer from 1 to 5 mm. Most data seem to fall within 30% of the equation.

HITACHI and HITACHI CABLE /8,9,10,11/ have developed a unique tube for nucleate boiling with the product name of THERMOEXCEL. "It has tunnels circumferentially under the outer surface skin, with many openings to the outside. The liquid in the tunnels is heated rapidly and changes to the vapor which leaves through openings as bubbles. A part of the vapor remains always in tunnels, and, therefore, the boiling occurs continuously. The quantity which removes as the vapor phase is compensated with the liquid sucked into tunnels from adjacent openings." The idea of augmentation of boiling is shown in Fig. 14/11,12/, in which li with the same legend except "E" and "S" at the last letter but one should be compared with each other. "E" stands for THERMOEXCEL and "S" for smooth surface. The last letter of the legend represents the geometry of the surface, and the rest of it the liquid; W = water, N_2 = nitrogen,

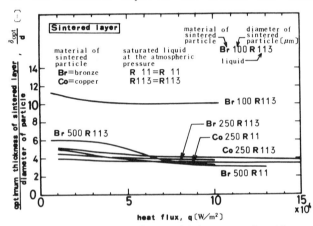

Fig. 12 The optimum thickness of the sintered layer /7/

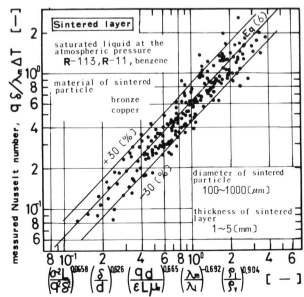

Fig. 13 Authors' correlation of nucleate boiling heat transfer from sintered layers /7/

He = helium 4 and numerals = numbers of fluorocarbons. The reduction of the degree of superheating at the specified heat flux is sometimes in factor of 1/10 and the promotion is more pronounced in the lower heat fluxes as it was the case in the authors' result, Fig. 10. This favorable performance at the lower heat flux operation will also be seen in Figs. 16 and 17. A long term operation as long as 1,000 hours in R-11 and the oil contamination of R-12 up to 3 mass percents have brought no noticable deterioration on the performance of THERMOEXCEL/11/.

Nakayama et al./13,14/ have engaged in the analysis of fluid flow in and around the THERMOEXCEL pores and traced the bubble history to predict the heat transfer characteristics. Their

model is encouraging, as they claim.

UNION CARBIDE, Linde Division/15,16,17,18/ has developed a novel boiling surface consisting of a porous metallic matrix which is bonded to a metallic substrate. The surface layer is about 0.01 to 0.02 in.(0.254 to 0.508 mm) thick with a void fraction of from 50 to 65 % and contains a multiplicity of cavities or pores which function as sites for generation of vapor bubbles. The surfaces are commercially referred to as UC HIGH FLUX. Substantial enhancements of heat transfer have been proved in long-term laboratory as well as field or prototype test. The liquids tested include fluorocarbons, cryogens, ammonia-water solutions, light hydrocarbons, glycol-water solutions, and sea water.

Figure 15 shows comparative performance data taken on a flat heater facing upward for smooth and porous surfaces for propylene, ethanol, R-11, and water. The data clearly illustrate the marked improvements of heat transfer over smooth surface. Futhermore the deterioration of the performance in the higher heat flux region is not recognized in

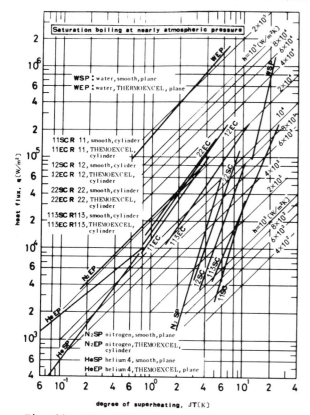

Fig. 14 The idea of the promotion of nucleate boiling heat transfer by THERMOEXCEL/11,12/

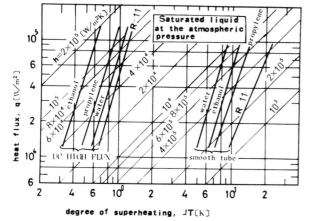

Fig. 15 The idea of the promotion of nucleate boiling heat transfer by UC HIGH FLUX/17,18/

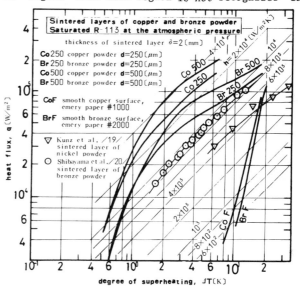

Fig. 16 Comparison of the authors' result of R-113 with other publications

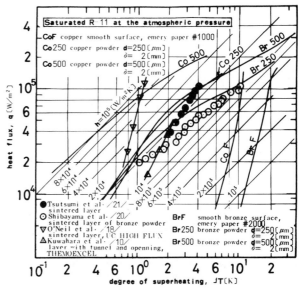

Fig. 17 Comparison of the authors' result of R-11 with other publications

their data. This was not the case in Figs. 10 and 14.

A comparison of the authors' studies with others is made below. Figure 16 includes experiments for R-113. Data marked by inverted triangles and circles are cited from the publications by Kunz et al./19/ and by Shibayama et al./20/ respectively. They are for horizontal plane surface facing upward, covered with a sintered layer of spherical powder particles. In the former the diameter of the nickel powder particle ranges from 150 to 297 μm, the thickness of the sintered layer is about 2.2 mm(estimated), and the porosity is 0.66. The free surface of the test liquid was flush with the upper surface of the sintered layer. This type of nucleate boiling may be named as the heat-pipe mode or wicking mode. In the experiment by Shibayama et al., bronze powder particles of 318 μm in average diameter were sintered onto the heating surface to form a porous layer with a porosity of 0.45. The thickness of the sintered layer is not reported in their paper, but they insist the thickness has no effect upon heat transfer. However, this may not be the case as discussed in Figs. 11 and 12.

In Fig. 17, data from several authors for R-11 are collected and compared with the present authors'. Tsutsumi et al./21/ reported an investigation on a horizontal cylinder covered with a sintered layer of powdered metal particles, but the particle material was not stated. The points marked by filled circles show their results. The average diameter of the powder particle is 108 μm, and the thickness of the sintered layer is 0.5 mm. They say that this geometry is optimum for R-11 under atmospheric pressure. Data from Shibayama et al./20/ are plotted by open circles, the conditions being the same as those cited in Fig. 16 except for the test fluid. O'Neil et al./18/ reported an experiment on a horizontal cylinder covered by a sintered layer of powdered metal particles of irregular shape, UC HIGH FLUX mentioned above. Their data is shown by inverted triangles in the figure. The material of the powder particles and the geometry of the sintered porous layer are not described. Kuwahara et al./10/ published data on the THERMOEXCEL tube and they are also included in Fig. 17, with data points marked by triangles. The authors emphasize the importance of the nominal dimension of the opening, which is the diameter of the inscribed circle of the triangular opening. The most favorable value of it for R-11 under the atmospheric pressure seems to be in the range between 0.06 to 0.08 mm.

CONCLUDING REMARKS

The two basic methods to promote nucleate boiling by prepared surfaces have been discussed, and publications illustrating and proving them have been cited. The scope of the present article has been limited to the methods applied to the surface side of the system of nucleate boiling.

Important fields of future study would include (1) the proposition of flow model in porous structures and its verification by experiments, (2) the preparation of heat transfer correlation from the prepared surfaces, (3) the confirmation of resisitivity of specific surface for fouling by long-term operation and for contamination of liquid, by impurities such as oils, and (4) the assessment of the prepared surface when applied in specific thermal device, which the overall performance of the surface, the substrate and the other side of the surface, the total efficiency of the tube bundle, and the final merit brought by the device with the prepared surface might belong to.

Thanks are due to HITACHI and HITACHI CABLE who supplied the authors with their technical publications and some unpublished data, and Messrs. T. Tanaka and T. Kuroki of the authors' department who helped them prepare the manuscript.

REFERENCES

1. Hasegawa, S., Echigo, R. and Koga, K., Bulletin of JSME, 20, 873-882(1969).
2. Takegawa, T., Hasegawa, S. and Echigo, R., Heat Transfer-Japanese Research, 2, 1-17(1973).
3. Young, R. K. and Hummel, R. L., Chem. Eng. Prog., 60, 53-58 (1964).
4. Marto, P. J., Moulson, J. A and Maynard, M. D., Trans. ASME, Ser. C, 90, 437-444 (1968).
5. Gaertner, R. F., US Patent 3,301,314 (Jan. 31, 1967).
6. Nishikawa, K., Ito, T. and Tanaka, K., Heat Transfer-Japanese Research, 8, 65-81(1979).
7. Ito, T., Nishikawa, K., Tanaka, K. and Yasumoto, K., Proc. of 17th National Heat Transfer Symposium of Japan, Kanazawa, 241-243(1980).
8. Nakayama, W., Daikoku, T., Kuwahara, H. and Kakizaki, K., Hitachi Review, 24, 329-334(1975).
9. Arai, N., Fukushima, T., Arai, A., Nakajima, T., Fujie, K. and Nakayama, Y., ASHRAE Trans., 83, 58-70(1977).
10. Kuwahara, H., Nakayama, W. and Daikoku, T., Proc. of 14th National Heat Transfer Symposium of Japan, Tokyo, 121-123(1977).
11. A Sale Catalogue of HITACHI CABLE, CAT. No. EA 501, (1978).
12. private communication with HITACHI CABLE.
13. Nakayama, W., Daikoku, T., Kuwahara, H. and Nakajima, T., to be published in May issue of Trans. ASME, Ser. C, 102, in two parts (1980).
14. Nakayama, W., Daikoku, T. and Nakajima, T., presented at the 19th National Heat Transfer Conference, Orlando, Florida, July(1980).
15. Czikk, A. M., Gottzmann, C. F., Ragi, E. G., Withers, J. G. and Habdas, E. P., ASHRAE Trans., 76, 96-109(1970).
16. O'Neil, P. S., Gottzmann, C. F. and Terbot, J. W., Chem. Eng. Prog., 67, 80-82(1971).
17. Gottzmann, C. F., O'Neil, P. S. and Minton, P. E., Chem. Eng. Prog., 69, 69-75(1973).
18. O'Neil, P. S., Gottzmann, C. F. and Terbot, J. W., Advances in cryogenic engineering, 17, 420-437(1976).
19. Kunz, H. R., Langston, L. S., Hilton, B. H., Wyde, S. S. and Nasbick, G. H., NASA CR-812, 1-182(1967).
20. Shibayama, S., Hajioka, S., Kitagawa, R. and Ishikawa, K., Preprint of JSME, No. 750-20, 65-68(1975).
21. Tsutsumi, M., Kawai, M., Fujikake, J. and Kumagai, G., Preprint of JSME, No. 710-17, 121-124(1971).

High Performance Heat Transfer Surfaces for Boiling and Condensation

RALPH L. WEBB

The Pennsylvania State University,
University Park, Pennsylvania 16802, U.S.A.

ABSTRACT

This paper presents a survey of special surface
geometries for boiling and condensing on the inner
and outer surfaces of tubes. Future research
needs are also addressed.

INTRODUCTION

Bergles and Webb (1) have defined 13 enhancement
techniques which are catagorized into two group-
ings: passive techniques (special surface geometries or fluid additives) and active techniques(e.g.
electric fields or vibration). Reference 2 provides 1,967 bibliographic journal citations for
all heat transfer modes, 731 of which are for two-
phase processes. A second report which surveys
the U.S. Patent literature will be issued soon,
(3). This survey will be limited to patent and
journal literature on special surface geometries
for enhancement of boiling and condensing processes.

NUCLEATE BOILING OUTSIDE TUBES

Low, integral-fin tubing has been used for many
years. Beginning in the early 1960's, rapid advances were made in the development of special
nucleate boiling geometries. Today, six nucleate
boiling geometries are commercially available, or
reported in the U.S. Patent literature. Four of
these have been reported after 1973. Two basic
approaches are employed to form a high area density
of nucleation sites:

1. Porous Boiling Surfaces (PBS): This consists
of a sintered, porous, metallic matrix bonded to a
base surface. Figure 1 shows the performance improvement provided for several fluids (4). The
sintered porous coating is approximately 0.25 mm
thick, has a 50 to 65% void fraction, and may be
made in several materials. The most important
dimension is the average pore radius, which ranges
from 0.01 mm to 0.1 mm, measured by a capillary
rise test. Smaller pore sizes are preferred for
low surface tension fluids. O'Neill et al. (4,5)
have developed a theory to predict the performance
of the porous surface, and to establish the preferred pore size. An alternative method for
making a porous boiling surface was proposed by
Dahl and Erb (7) in a U.S. Patent. The porous
coating is formed by flame spraying aluminum particles in an oxygen-rich atmosphere. The oxygen-
rich flame produces an oxide film on the particle
and alleviates particle flattening upon impact.
Fuji et al. (6) describe a similar PBS in which
the copper particles are fixed to the base surface by electroplating.

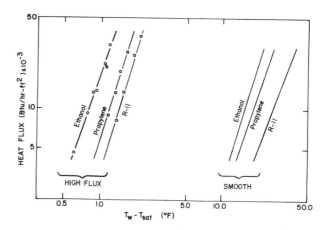

Figure 1   Performance of smooth and
porous boiling surface (4).

Janowski et al. (8) have developed an innovative
process for making a porous surface (Figure 2).
A tube is wrapped with an open cell polyurethane
foam, which is then copper plated. The copper
plating provides structural integrity and good
thermal conduction. The polyurethane is removed
by pyrolysis, which forms additional very small
pores within the skeletal structure. The structure
has 1.5 mm coating thickness, 97% void volume, 4
pores/mm and 0.12 mm pore size, augmented by 0.02
mm pores within the skeletal structure. Because
the flame spray and polyurethane foam coating processes are performed at relatively low temperature,
compared to 960 C for the sintering process, the
mechanical properties of the base tube are not al-
tered.

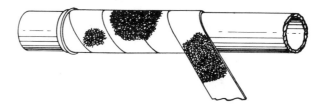

Figure 2   Porous boiling surface (8) formed
by wrapping 1.5 mm thick polyurethane foam, followed by electroplating.

2. Integral "Roughness": The metal is cold worked
to form re-entrant nucleation sites which are interconnected below the surface. Figure 3 shows the
cross section of three commercially used boiling
surfaces of this type (9,10,11). Figure 3a
from Webb's U.S. Patent (9) consists of integral
fin tubing (13 fins/cm, 0.8 mm fin height) in which
the fins are bent to form a re-entrant cavities.
The groove opening at the surface is a critical
dimension, 0.0038-to-0.0089 mm for Refrigerant 11.

The performance is sharply reduced for gap widths outside of this range. The preferred gap spacing is expected to be higher for high surface tension fluids.

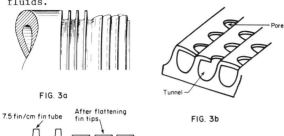

FIG. 3a

7.5 fin/cm fin tube   After flattening fin tips

FIG. 3b

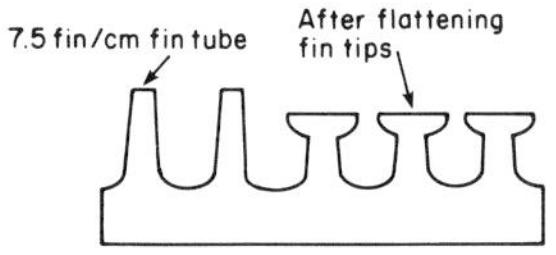

FIG. 3c

Figure 3   Three commercial rough boiling surfaces

A variation of the Webb patent developed by Nakayama et al. (10) is shown in Figure 3b. It is formed from an integral-fin tube, which has small spaced cutouts at the fin tips. These "saw tooth" fins are bent to a horizontal position to form tunnels having spaced pores at their top. Nakayama et al. (12) propose a model for boiling in the enclosed tunnels. In 1979, the Weiland Werke A.G. (11) announced a third nucleate boiling roughness geometry, illustrated by Figure 3c. It is basically a variation of Figure 3a and may be formed from standard 7.5 fin/cm integral-fin tubing.

Kun and Czikk (13) describe a re-entrant grooved surface consisting of a series of closely spaced grooves (11 to 90 grooves/cm) followed by a second set of cross grooves, which are not as deep as the first set.

The performance exhibited by the several types of coated and rough surfaces is nominally comparable. The key to the high performance of the different structures is attributed to three factors: 1) a pore or re-entrant cavity within a critical size range, 2) interconnected cavities, and 3) nucleation sites of a re-entrant shape. When the cavities are interconnected, one active cavity may activate adjacent cavities. The dominant fraction of the vaporization occurs at very thin liquid films within the subsurface structure. The re-entrant cavity shape provides a very stable vapor trap, which will remain active at low liquid superheat values. Coated surfaces offer the opportunity for a duplex tube material construction. The boiling surface may be made of less expensive material than required for the tube-side fluid.

EVAPORATION OF THIN FILMS

As an alternate to pool boiling, a liquid may be vaporized by distributing it as a thin film on the heat transfer surface. This method may be preferred if the static head in a flooded evaporator, causes increased saturation temperature. Because heated films are susceptible to rupture, the liquid rate must be sufficient to assure surface wetting. Heat exchanger geometries for thin film evaporation include sprayed horizontal bundles and vertical tube evaporators (VTE).

The doubly fluted tube VTE was developed for sea-water distillation with condensation on the outer surface (14). The augmentation mechanisms on fluted surfaces are different for condensation and evaporation. In evaporation, the liquid is pulled into the valleys and dewetting may occur on the convex surface, unless it is wetted by waves or other film instabilities. Nucleation within the film will aid surface wetting. Johnson et al. (15) propose possible enhancement mechanisms. Carnavos (16) reports a three-fold increased evaporating coefficient (total area basis) and Alexander et al. (17) report overall coefficient increases of 2-to-3 evaporating fresh water. Johnson et al. (15) found 100% higher coefficients are obtained with seawater, due to interfacial motion caused by surface tension gradients. The addition of a surfactant to water provided a 100% higher overall coefficient in an upflow VTE, due to the formation of thin, foamy liquid films which wet the entire fluted surface (17).

Conti (18) tested low-finned tubing (7.9 and 13.4 fin/cm) with ammonia in a horizontal sprayed film evaporator. The best performance was given by the 7.9 fin/cm tubing with 0.50 mm high fins. This tube has an 80% area increase, and provides a 200% evaporating side enhancement based on plain tube area. Edwards et al. (19) present an analytical model and experimental results for thin film evaporation on triangular threaded horizontal tubes. Their enhancement levels for 3.9 and 6.3 grooves/cm were lower than measured by Conti with 7.9 fin/cm integral fins. Fricke and Czikk (20) report test results on a sprayed horizontal tube evaporator having the Linde porous coating on the outer tube surface. The ammonia heat transfer coefficients were approximately equal to those obtained with the same surface geometry operated as a flooded evaporator (21).

FLOW BOILING INSIDE TUBES

The special surface geometries used for enhancement of nucleate pool boiling generally cannot be applied to forced convection vaporization inside tubes. A notable exception is the Linde PBS (4). Commercial applications are described (22) in which the porous boiling surface is on the inner tube surface with condensation occurring on a fluted condensing surface. An oxygen reboiler-condenser exhibited "a tenfold performance advantage", and a "fivefold improvement over a smooth tube" was achieved in a thermosyphon reboiler for a C_2 splitter with ethane boiling on the tube side. Murphy and Bergles (23) found that the PBS did not yield improvement in high flux saturated flow boiling of a refrigerant.

Special internal roughness geometries have been developed for commercial use. Withers and Hadbas (24) tested tubes having "integral internal helical ridging" (Figure 6a) using R-12 in a direct expansion water chiller. Configurations were tested which gave a 200% increased boiling coefficient. Although the pressure drop was substantially increased, which reduced the LMTD, the best geometry provided a 100% increase in evaporator capacity. Ito et al. (25) and Lord et al. (26) describe other roughened tubes for direct expansion refrigerant evaporators. The tube of Ito et al. has spiraled triangular thread-type grooves approximately 0.2 mm deep. This tube provides boiling heat transfer coefficients 1.5-to-2 times higher than a smooth tube, with very small increased pressure drop.

Kubanek and Miletti (27) and Lavin and Young (28) tested several commercially available internally finned tubes using R-22. The tubes were nominally 15 mm inside diameter with 16 fins (27) and 32 fins (28), and included axial and spiraled fins. Kubanek and Miletti found that fin spiral yielded increased heat transfer coefficients, and the enhancement was equal to or greater than the area increase. The heat transfer coefficient based on total surface area were 80 to 300% greater than the plain tube values. Lavin and Young found that the greatest enhancement (100%) occurred at low vapor qualities, and the axial-fin tube yielded heat transfer coefficients 25% smaller than the smooth tube in the mist flow regime.

Bryan and Seigel (29) tested an R-11 evaporator using wire coil inserts in a smooth tube. Their tests of five different wire coil geometries provided enhancement up to 200%.

Heat transfer coefficients in two-phase flow vary with vapor quality, due to the different flow regimes. Therefore, a given augmentation technique may not be equally effective in all flow regimes. One example is the use of twisted tapes or swirl devices to delay the critical heat flux in sub-cooled boiling (30) or to delay dryout in the mist flow region (31).

CONDENSATION ON HORIZONTAL TUBES

Low, integral-fin tubing has found wide commercial acceptance for condensation on horizontal tubes. Because thin condensate films are formed on the short fins, the condensation coefficient is higher on the finned surface than on a plain tube of equal diameter. Beatty and Katz presents a theoretical model for this geometry (32). The permissable fin spacing is limited by the condensate surface tension and vapor shear (33) as shown in Figure 4. If the condensate bridges the spacing between the fins, the extended surface will not be effective. Fin densities of 7-to-14 fins/cm are used for refrigerants or low surface tension fluids. Externally finned tubes are not normally used for steam condensation due to the probability of condensate bridging, and because the limiting thermal resistance is usually on the tube side.

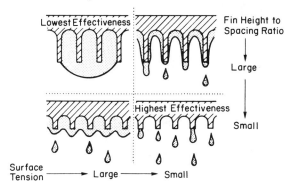

Figure 4 Condensate retention on horizontal low finned tubing.

Figure 5a shows a recently developed modification of the integral-fin tube (10). This commercially available surface, known as "THERMOEXCEL-C", provides a substantially higher coefficient (total area basis) than the 7.5 fin/cm finned tube for Refrigerants 11, 12 and 22 (10). This tube has

14 fin/cm and 1.2 mm fin height. Figure 5b shows a spine-fin array proposed by Webb and Gee (34). Analytical predictions for condensing R-11 and R-12 show that the spine-fin geometry will provide the same condensing side performance as integral-fin tubing and permit a 60% reduction of fin material. Carnavos (35) presents comparative test data on five basic types of enhanced tubes with R-11.

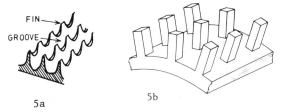

5a   5b

Figure 5 Recent developments in extended surfaces for condensation on horizontal tubes.

6a   6b   6c

Figure 6 Corrugated Tube Geometries

Figure 6 shows three corrugated tube geometries which provide coolant and condensing side enhancement. Figure 7 shows the test results of Mehta and Rao (36) for a corrugated tube which has the Fig. 7a geometry. The maximum enhancement is modest, relative to that of the extended surface geometries. Marto et al. (37) performed comparative tests of the Figure 6 geometries for steam condensation. Their horizontal tube tests showed that the dominant enhancement occurred on the water-side, and that the condensing coefficients on the tubes of Figure 6a and 6b are very close to the smooth-tube values. These results imply that such tubing would be of value only if the water-side is the controlling thermal resistance.

Thomas et al. (21) have tested a smooth tube helically wrapped with wire. Surface tension pulls the condensate to the base of the wires, which act as condensate run-off channels. Tests with condensing ammonia on a 38 mm O.D. tube indicated a condensing coefficient about three times the value predicted by the Nusselt equation for condensation on a smooth tube.

CONDENSATION ON VERTICAL TUBES

In 1954, Gregorig (38) proposed a method of using surface tension forces to enhance laminar film condensation on a vertical surface. Figure 8 illustrates the tube cross section. Surface tension draws the condensate from the convex surface into the concave region. A high condensation rate occurs on the convex portions of the fluted surface and the concave portions serve as condensate drainage channels. Webb (39), Panchal and Bell (40) and Mori et al. (41) present a general summary

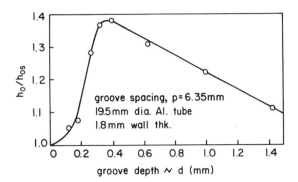

Figure 7 Steam condensation on single helix
corrugated tube, as a function of
groove depth (36).

of the theory of the fluted condensing surface,
defined optimum flute geometries and present equations for construction of the surface profile.
The size of the flutes is selected such that the
drainage channel will be filled to capacity at the
bottom of the vertical surface. Therefore, as the
tube length is varied, the flute size should be
changed accordingly. Mori et al. (41) show that
vertically spaced condensate runoff disks are
effective for condensate removal.

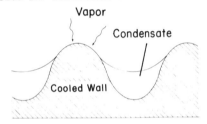

Figure 8 Horizontal cross section through
the wall of a vertical fluted tube.

The same effect produced by fluted surfaces can
be obtained by loosely attached, spaced vertical
wires on a vertical surface. The enhancement
occurs due to the thinned film between the wires.
Thomas (42) gives an analytical model and experimental data for this augmentation technique. In a
later publication (43), he shows that square wires
have a greater condensate carrying capacity than
circular wires of the same diameter. Soviet researchers (44) have also worked with loosely
attached vertical wires for steam condensation.

A 1979 U.S. Patent by Notaro (45) describes a new
concept rough surface geometry for enhanced film
condensation. It consists of an array of small
metal particles bonded to the tube surface. The
particles are .25-to-1.0 mm high covering 20-to-
60% of the surface. Condensation occurs on the
particle array and drains along the smooth base
surface. High condensation rates occur on the
convex surfaces of the particles, due to surface
tension forces. Figure 9 shows a photograph of
the surface, and illustrates the thinned condensate films on the particles.

FORCED CONDENSATION INSIDE TUBES

The augmentation techniques used on the outside
of vertical tubes are applicable to condensation
inside vertical tubes. Augmentation requirements
in horizontal tubes are different because gravity

force drains the film transverse to the flow
direction.

Internal fins yield high augmentation levels.
Vrable et al. (46) and Reisbig (47) used R-12 in
internally finned tubes. In terms of the total
surface area, they measured condensation coefficients 20-to-40% greater than the smooth-tube value.
Accounting for surface area increases of 1.5-to-2.0,
the heat transfer coefficient, based on the nominal
tube inside diameter, is 2-to-3 times the smooth
tube value. Bergles and co-workers condensed steam
(48) and R-113 (49) in four internally finned tubes,
three of which had spiraled fins. The best internal
fin geometry provided 20-to-40% higher heat transfer
coefficients (total area basis) than the smooth
tubes. Bergles et al. (48,49) also tested twisted
tapes with steam and R-113. The twisted tape performed distinctly poorer than the internal fins.
They gave only 30% higher coefficients and exhibited
pressure drops equal or higher than the internally
finned tubes.

Luu (50) found that a repeated-rib roughness increased the R-113 condensing coefficient more than
100%. For the same augmentation level, roughness
may be of greater interest than internal fins, since
a roughness requires less tube material.

Fenner and Ragi (51) tested an attached particle
roughness with condensing R-12. The spaced metal
particles provide extended surface at high vapor
qualities, and turbulence of the condensate film
at lower vapor qualities. The tests were performed
in a 14.5 mm diameter tube having a 50% area density of e/D = 0.031 particles. This tube provided
enhancement levels (smooth tube area basis) of 2.4
for low exit qualities (25-to-60%) and 4.0 for high
exit qualities (60-to-90%). The accompanying
pressure drops were only 6% and 105% larger than
the smooth-tube values for the two exit qualities.

PHOTO OF PARTICLE
COATED SURFACE

ILLUSTRATION OF
SURFACE PARTICLES

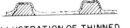

ILLUSTRATION OF THINNED
CONDENSATE FILM ON PARTICLES

Figure 9 Condensing surface formed by
attached metal particles (45).

PROSPECTS FOR THE FUTURE

Several key areas are suggested for future R&D
activity:

1. Enhancement on the Outer Surface of Tubes:
The central need is for lower cost surface geometries, and availability in a wider range of materials. Future technology advances are needed for
condensation of high surface tension fluids, particularly on horizontal tubes.

2. Enhancement on the Inner Surface of Tubes: Advances in tube-side enhancement lag those for shell-side enhancement. Additional R&D activity is needed, as well as the development of manufacturing methods, which allow processing the inner surface of a tube.

3. Doubly Enhanced Tubes: Future activity should consciously consider techniques applicable to doubly enhanced tubes. Preferably, the concepts should allow independent selection of inside and outside surface enhancements. The use of surface coatings (PBS and attached particles) offer interesting possibilities for tubes having internal fins or roughness. Virtually no progress has been made on doubly enhanced vertical condensing tubes having internal fins or roughness.

4. Cost effective manufacturing technology is probably the most significant barrier yet to be overcome. This is particularly true for doubly enhanced tubes. Greater communication is needed between heat transfer researchers and manufacturing technologists. A future cooperative effort is recommended.

5. Analytical design information, applicable to a wide variety of fluids, must be made available to the designer.

REFERENCES

1. Bergles, A. E., and Webb, R. L., "Energy Conservation via Heat Transfer Enhancement," Energy, 4:193-200, 1979.

2. Bergles, A. E., and Webb, R. L., "Bibliography on Augmentation of Convection Heat and Mass Transfer," Report No. HTL-19, Eng. Inst., Iowa State Univ., Ames, IA, May 1979.

3. Webb, R. L., G. A. Junkhan and A. E. Bergles, "Bibliography of U.S. Patents on Augmentation of Convective Heat and Mass Transfer," To be published by Eng. Res. Inst., Iowa State Univ., Ames, IA, 1980.

4. Gottzman, C. F., J. B. Wulf and P. S. O'Neill, "Theory and Application of High Performance Boiling Surfaces to Components of Absorption Cycle Air Conditioners," Proc. Conf. on Natural Gas Res. and Tech., Sess. I, Pap. 3, 1971.

5. Czikk, A. M. and P. S. O'Neill, "Correlation of Nucleate Boiling from Porous Metal Films," Advances in Enhanced Heat Transfer, Ed., J. M. Chenoweth, et al., ASME, pp. 103-113, 1979.

6. Fujii, M., E. Nishiyama and G. Yamanaka, "Nucleate Pool Boiling Heat Transfer from a Microporous Heating Surface," Advances in Enhanced Heat Transfer, Ed., J. M. Chenoweth, et al., ASME, NY, pp. 45-52, 1979.

7. Dahl, M. M. and L. D. Erb, "Liquid Heat Exchanger Interface Method," U.S. Patent 3,990,862, Nov. 9, 1976.

8. Janowski, K. R. and M. S. Shum, "Heat Transfer Surface," U.S. Patent 4,129,181, Dec. 12, 1978.

9. Webb, R. L., "Heat Transfer Surface Which Promotes Nucleate Ebullition," U.S. Patent 3,521,708, Oct. 10, 1972.

10. Arai, N., T. Fukushima, A. Arai, T. Nakajima, K. Fujii and Y. Nakayama, "Heat Transfer Tubes Enhancing Boiling and Condensation in Heat Exchangers of a Refrigerating Machine," Trans. ASHRAE, 83(2):58-70, 1977.

11. Anon., "GEWA-T-Tubes: High Performance Tubes for Flooded Evaporators," Brochure SAE-15e-06.78, Wieland-Werke AG, Metabuerke, Ulm, West Germany, 1978.

12. Nakayama, W., T. Daikoku, H. Kuwahara and T. Nakajima, "Dynamic Model of Enhanced Boiling Heat Transfer and Porous Surfaces," Advances in Enhanced Heat Transfer, Ed., J. M. Chenoweth, et al., ASME, NY, pp. 31-44, 1979.

13. Kun, L. C. and A. M. Czikk, "Surface for Boiling Liquids," U.S. Patent 3,454,081, July 8, 1969.

14. Alexander, L. G. and H. W. Hoffman, "Performance Characteristics of Corrugated Tubes for Vertical Tube Evaporators," ASME Paper 71-HT-30, 1971.

15. Johnson, B. M., G. Jansen and P. C. Owzarski, "Enhanced Evaporating Film Heat Transfer from Corrugated Surfaces," ASME Paper 71-HT-33, 1971.

16. Carnavos, T. C., Chapt. 17 in "Heat Exchangers" Design and Theory Sourcebook, N. Afgan and E. U. Schlunder (Ed.) McGraw-Hill, 1974.

17. Sephton, H. H., "Vertical Tube Evaporation with Double Fluted Tube and Interface Enhancement," ASME Paper 75-HT-43, 1975.

18. Conti, R. J., "Heat Transfer Enhancement in Horizontal Ammonia Film Evaporators," Proc. 6th OTEC Conference, Washington, D.C., V-2, 11.8, 1979.

19. Edwards, D. K., K. D. Gier, P. S. Ayyaswamy and I. Catton, "Evaporation and Condensation in Circumferential Grooves on Horizontal Tubes," ASME Paper 73-HT-25, 1973.

20. Fricke, H. D. and A. M. Czikk, "Enhanced Sprayed Bundle Evaporator Performance Studies," Advances in Enhanced Heat Transfer, Ed., J. M. Chenoweth, et al., ASME, NY, pp. 23-30, 1979.

21. Thomas, A., J. J. Lorenz, D. A. Hillis, D. T. Young and N. F. Sather, "Performance Tests of 1 Mwt Shell and Tube Exchangers for OTEC," Proc. 6th OTEC Conf., Washington, D.C., V-2, 11.1, 1979.

22. O'Neill, P. S., E. G. Ragi and M. L. Jacobs, "Effective Use of High Flux Tubing in Two-Phase Heat Transfer," AIChE Paper presented at Session 83, 86th National Meeting, Houston, April 1979.

23. Murphy, R. W. and A. E. Bergles, "Subcooled Flow Boiling of Flourocarbons: Hysterisis and Dissolved Gas Effects on Heat Transfer," Proc. 1972 Heat Trans. and Fluid Mech. Inst., Stanford Univ. Press, Stanford, CA, pp. 400-416, 1972.

24. Withers, J. G. and E. P. Habdas, "Heat Transfer Characteristics of Helical Corrugated Tubes

for In-tube Boiling of Refrigerant-12," *AIChE Symp. Ser.*, 70(138): 98-106, 1974.

25. Ito, M., H. Kimura and T. Senshu, "Development of High Efficiency Air-Cooled Heat Exchangers," *Hitachi Review*, Vol. 26, No. 10, pp. 323-326, 1977.

26. Lord, R. G., R. C. Bussjager and D. F. Geary, "High Performance Heat Exchanger," U.S. Patent 4,118,944, Oct. 10, 1978.

27. Kubanek, G. R. and D. C. Milletti, "Evaporative Heat Transfer and Pressure Drop Performance of Internally-Finned Tubes with Refrigerant 22," *Jour. Heat Trans.*, 101:447-452, 1979.

28. Lavin, J. G. and E. H. Young, "Heat Transfer to Evaporating Refrigerants in Two-Phase Flow," *AIChE Jour.*, 11:1124-1132, 1965.

29. Bryan, W. L. and L. G. Seigel, "Heat Transfer Coefficients in Horizontal Tube Evaporators," *Refrigeration Eng.* pp. 36-45 and 126, May 1955.

30. Gambill, W. R. and N. D. Green, "Boiling Burnout with Water in Vortex Flow," *Chem. Eng. Prog.* 54(10): 58-76, 1958.

31. Bergles, A. E., W. O. Fuller and S. W. Hynek, "Dispersed Flow Film Boiling of Nitrogen with Swirl Flow," *Int. Jour. Heat-Mass Trans.*, 14: 1345-1354, 1971.

32. Beatty, K. O. and D. L. Katz, "Condensation of Vapors on Outside of Finned Tubes," *Chem. Eng. Prog.* 44(1): 55-70, 1948.

33. Taborek, J., "Design Methods for Heat Transfer Equipment," Chapter 3 in Heat Exchangers: *Design Theory and Sourcebook*, N. Afgan and E. U. Schlunder (ed.), p. 69, McGraw-Hill, 1974.

34. Webb, R. L. and D. L. Gee, "Analytical Predictions for a New Concept Spine-Fin Surface Geometry," *ASHRAE Trans.*, 85(2), 274-283, 1980.

35. Carnavos, T. C., "An Experimental Study: Condensing R-11 on Augmented Tubes," ASME Paper 80-HT-54, 19th Natl. Heat Trans. Conf., Orlando, FL, 1980.

36. Mehta, M. H. and M. R. Rao, "Heat Transfer and Frictional Characteristics of Spirally Enhanced Tubes for Horizontal Condensers," *Advances in Enhanced Heat Transfer*, J. M. Chenoweth, et al., Ed., ASME, New York, pp. 11-22, 1979.

37. Marto, P. J., D. J. Reilly and J. H. Fenner, "An Experimental Comparison of Enhanced Heat Transfer Condenser Tubing," *Advances in Enhanced Heat Transfer*, J. M. Chenoweth, et al., Ed., ASME, pp. 1-10, 1979.

38. Gregorig, R., "Hautkondensation on feingwellten oberflachen bei Beruksichtigung der Oberflachenspannungen," *Zeitschrift fur Angewandte Mathematik und Physik*, 5:36-49, 1954.

39. Webb, R. L., "A Generalized Procedure for the Design and Optimization of Fluted Gregorig Condensing Surfaces," *Jour. Heat Transfer*, 101: 335-339, 1979.

40. Panchal, C. B. and K. J. Bell, "Analysis of Nusselt-Type Condensation on a Vertical Fluted Surface," *Condensation Heat Transfer*, P. J. Marto and P. G. Kroeger, Ed., ASME, New York, pp. 45-54, 1979.

41. Mori, Y., K. Hijikata, S. Kirisawa, and W. Nakayama, "Optimized Performance of Condensers with Outside Condensing Surface," *Condensation Heat Transfer*, P. J. Marto and P. G. Kroeger, Ed., ASME, New York, pp. 55-62, 1979.

42. Thomas, D. G., "Enhancement of Film Condensation Rate on Vertical Tubes by Longitudinal Fins," *AIChE Jour.*, 14: 644-649, 1968.

43. Thomas, D. G., "Enhancement of Film Condensation Rates on Vertical Tubes by Vertical Wires," *Ind. and Eng. Chemistry-Fundamentals*, 6(1): 97-103, 1967.

44. Butizov, A. I., V. G. Rifert and G. G. Leont'yev "Heat Transfer in Steam Condensation on Wire-Finned Vertical Surfaces," *Heat Transfer-Soviet Research*, 7(5): 116-120, 1975.

45. Notaro, F., "Enhanced Condensation Heat Transfer Device and Method," U.S. Patent 4,154,294, May 15, 1979.

46. Vrable, D. A., W. J. Yang and J. A. Clark, "Condensation of Refrigerant-12 Inside Horizontal Tubes with Internal Axial Fins," *Heat Transfer 1974, Fifth Int'l Heat Trans. Conf.*, Vol. III, pp. 250-254 (Tokyo, Japan Soc. Mech. Engrs., 1974).

47. Reisbig, R. L., "Condensing Heat Transfer Augmentation Inside Splined Tubes," *Paper 74-HT-7, AIAA/ASME Thermophysics Conf.*, Boston, July, 1974.

48. Royal, J. H. and A. E. Bergles, "Augmentation of Horizontal In-Tube Condensation by Means of Twisted Tape Inserts and Internally Finned Tubes," *Jour. Heat Transfer*, 100:17-24, 1978.

49. Luu, M. and A. E. Bergles, "Experimental Study of the Augmentation of the In-Tube Condensation of R-113," *ASHRAE Trans.*, 85(2), 1979 (in press).

50. Luu, M. "Augmentation of In-Tube Condensation of R-113," Ph.D. Thesis, Dept. of Mech. Engr., Iowa State University, 1979.

51. Fenner, G. W. and E. Ragi, "Enhanced Tube Inner Surface Heat Transfer Device at Method," U.S. Patent 4,154,293, May 15, 1979.

HIGH-FLUX HEAT TRANSFER:
APPLICATIONS

High Flux Heat Transfer in Gaseous Solid Suspension Flow

S. HASEGAWA and R. ECHIGO
Kyushu University, Fukuoka, Japan

ABSTRACT

This paper discusses a few prominent features embodied in the gaseous solid suspension medium which is affiliated with the heat transfer system at high temperature and high thermal load. Firstly, an analytical approach developed by the authors is introduced concerning the combined radiative and convective heat transfer in the circular tube at a high temperature environment. Examinations on the heat transfer mechanism show that a remarkable increase of the over-all heat transfer coefficient is achieved on account of absorbing and reemitting behaviors of the particulate medium. Some specifications of the experimental facility on the "Heat Transfer Loop-Gaseous Solid Suspension at High Temperature" installed in Kyushu University are depicted together with a few typical experimental results. Alternatively a consecutive variation of the (main) flow direction of suspension medium yields a high heat transfer enhancement due to the drastic interaction between the particulate and fluid media. In this respect the analytical and experimental studies have been performed and discussed in some detail for the heat transfer of gaseous solid suspension flow in a curved tube and also jet impingement.

NOMENCLATURE

C^* = specific heat ratio ($= c_p / c_f$)
D = tube or nozzle diameter
Da = Darcy number
d_p = particle diameter
Fr = Froude number = u_o^2 / gD
$F(\tau_0), H(\tau_0 \eta), K(\tau_0 \eta, \tau_0 \eta_1)$ = radiation extinction functions
f_w = coefficient of friction
g_f = total diffusivity function for gas
h_p = heat transfer coefficient from particle to gas
k = thermal conductivity
M = dimensionless heat transfer parameter between gas and particle
N_R = conduction-radiation parameter
$Nu_{\xi,c}$ = local Nusselt number by convection
$Nu_{\xi,R}$ = local Nusselt number by radiation
n_p = dispersed number density of particle
p = pressure
q_r^R, q_x^R = radial and axial components of radiative heat flux vector
r, x = coordinates in radial and axial directions
T = temperature
U = dimensionless axial velocity
u, v = velocities in axial and radial directions
Γ = solid loading ratio
$\bar{\Gamma}$ = thermal loading ratio
μ = viscosity
ρ = density
ρ_{dp} = dispersed density of particles
θ = dimensionless temperature
τ_0 = optical thickness
ξ = dimensionless axial coordinate
η = dimensionless radial coordinate

Subscripts

f	= fluid	p	= particle
m	= mean	w	= wall
o	= reference or nozzle exit		

INTRODUCTION

The flowing gaseous solid suspension media have been applied to many industrial purposes and discussed extensively to further new applications such as nuclear reactor cooling[1], solid-propellant rocket[2], MHD generator[3], high performance heat exchanger[4] and so forth. These studies have been summarized and discussed in the monographs[5][6] in perspective of fluid dynamics and heat transfer and are not repeated here. These studies have all been more or less concerned with the limited problems of moderate temperature range, wherein the effects of thermal radiation are negligible. From the viewpoint of heat transfer coolant, however, the basic features of solid suspensions are large heat capacities without pressurization and also relevant to the radiative heat transfer. Further, it has to be emphasized that the flowing solid suspensions may sometimes lead to the unexpected effects on heat transfer due to the drastic change of the flow structure except for the tractable flows in straight channels with constant cross section which have been most extensively studied but not often encountered in practical problems. Noticing that the inertia force of the particulate medium is two to four orders of magnitude larger than that of gas, it is easily understood that the flow field, particularly, turbulent structure is changed to a great extent, if the main flow direction of gas-solid mixture is changed successively as being found in a curved channel and jet impingement.

In this paper analytical[7~11] and experimental[12] studies on the combined radiative and convective heat transfer of multiphase flow are proposed in order to evaluate the heat transfer characteristics at high temperature.

Thereafter the heat transfer enhancement of gaseous solid suspension medium is discussed in the fully developed laminar flow in a curved circular tube[13] and also in the jet impingement[14] in connection with the high flux heat transfer problem of first wall and blanket cooling of the controlled thermonuclear reactor.

ANALYSES ON HEAT TRANSFER AT HIGH TEMPERATURE SYSTEM

Simultaneous Radiative and Convective Heat Transfer by Flowing Multiphase Media in a Circular Tube[7~10]

The analyses are concerned with the heat transfer

of fully developed laminar or turbulent flow of the multiphase media in a circular tube and based on the following assumptions:

(1) The surface of the tube is isothermal and black for thermal radiation.

(2) The flow field of the two phases is hydro-dynamically fully developed and has a laminar or turbulent velocity profile.

(3) The physical properties are constant.

(4) The spherical particles that are the same size are distributed uniformly throughout the tube cross section,and emit and absorb thermal radiation in local thermodynamic equilibrium, but do not scatter it.* The gas is transparent for radiation.

(5) The time-mean velocities of the two phases are equal and the presence of the particles has no effect on the velocity profiles and the friction factor of the gas.

(6) The agglomeration, chemical reaction of the particles and viscous dissipation are not considered.

(7) The one-dimensional propagation of radiation is to be valid. ($\frac{\partial q_x^R}{\partial x} \ll \frac{1}{r}\frac{\partial}{\partial r}(rq_r^R)$)

On the foregoing postulations, a heat balance in tube flow yields the following basic equations governing the temperature fields in the dimensionless form.

$$U\frac{\partial \theta_f}{\partial \xi} + M(\theta_f-\theta_p) = \frac{2}{\eta}\frac{\partial}{\partial \eta}(g_f\eta\frac{\partial \theta_f}{\partial \eta}) \qquad (1)$$

$$\left.\begin{array}{l} U\bar{\Gamma}\frac{\partial \theta_p}{\partial \xi} + M(\theta_p-\theta_f) = \frac{2\tau_0}{N_R}H(\tau_0\eta) \\[2mm] + \frac{2\tau_0^4}{N_R}\int_0^1 K(\tau_0\eta,\tau_0\eta_1)\eta_1\theta_p^4 d\eta_1 - \frac{2\tau_0^2}{N_R}\theta_p^4 \end{array}\right\} \quad (2)$$

where
$$\left.\begin{array}{l} H(\tau_0\eta) = \int_1^\infty K_1(\tau_0 y)I_0(\tau_0\eta y)\frac{dy}{y} \\[4mm] K(\tau_0\eta,\tau_0\eta_1) = \begin{cases} \int_1^\infty K_0(\tau_0\eta y)I_0(\tau_0\eta_1 y)dy & \eta_1<\eta \\[2mm] \int_1^\infty K_0(\tau_0\eta_1 y)I_0(\tau_0\eta y)dy & \eta<\eta_1 \end{cases} \end{array}\right\} \quad (3)$$

For the turbulent velocity distribution of two phases, the expressions developed by Reichardt are utilized here. The boundary conditions for eqs. (1) and (2) are taken as follows.

$$\eta=1;\ \theta_f=\theta_p=1, \quad \eta=0;\ \partial\theta_f/\partial\eta=\partial\theta_p/\partial\eta=0, \quad \xi=0;\ \theta_f=\theta_p=\theta_0 \quad (4)$$

Eqs. (1) and (2) are solved numerically by an implicit finite difference method as regarding the radiation term as an iterative one. Once the temperature fields for both phases are prescribed, heat transfer characteristics are readily evaluated as follows.

$$Nu_{\xi,c}=2\frac{\partial \theta_f}{\partial \eta}\Big|_{\eta=1}\Big/(1-\theta_m) \qquad (5)$$

$$Nu_{\xi,R}=2\{\frac{\tau_0^2}{N_R}F(\tau_0)-\frac{\tau_0^3}{N_R}\int_0^1\theta_p^4\eta_1 H(\tau_0\eta_1)d\eta_1\}\Big/(1-\theta_m) \qquad (6)$$

$$Nu_\xi=Nu_{\xi,c} + Nu_{\xi,R} \qquad (7)$$

where
$$\theta_m=(\int_0^1\{\theta_f+\bar{\Gamma}\theta_p\}Ud\eta)\Big/(1+\bar{\Gamma})\int_0^1 Ud\eta \qquad (8)$$

and
$$F(\tau_0)=\int_1^\infty K_1(\tau_0 y)I_1(\tau_0 y)\frac{dy}{y^2} \qquad (9)$$

Fig.1 shows the effect of varying parameter ξ on the temperature profiles. The prominent feature embodied in these figures is the fact that in the vicinity of the heating wall the temperature gradient becomes steeper due to the presence of fine particles and as a result the convective heat

---

*) Recently, Modest et al.[15] have made an analytical study on the same subject by taking account of scattering due to the particulate phase.

---

transfer is promoted, while in the central core of a tube the gas temperature is increased by the heat transfer from fine particles, which absorb thermal radiation. Fig.2 indicates the relations between Nu_ξ and ξ for the typical values of parameters τ_0, N_R and Γ. Nu_ξ decreases to a certain minimum and turns to increase beyond this point. This behavior implies that there exists no asymptotic Nusselt number. In addition, if the comparison is made with the single phase heat transfer, it is noted that the heat transfer characteristic is enhanced prominently by the flowing multiphase media, especially, at high temperatures (at small N_R).

Simultaneous Radiative and Convective Heat Transfer in a Packed Bed with high Porosity[11]

The present model (Fig.3) is considered to be a limiting case of the multiphase flow system; that is, the particulate phase is quiescent in a duct. The assumptions to be introduced here are almost the same as those in the previous section except for the one-dimensional propagation of radiation, which is only admissible in the case of slowly varying temperature field along the flow direction and is not valid in the case when the radiation becomes predominant over other modes of energy transfer in the finite heating section or the boundary conditions at the heating surface yield the varying surface temperature along the flow direction. Instead of the one-dimensional approximation to radiative transfer, the propaga-

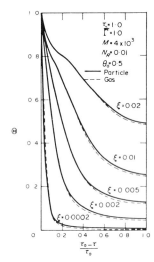

Fig.1 Temperature profiles (laminar case)

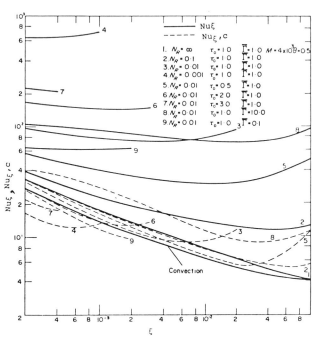

Fig.2 Nu_ξ, $Nu_{\xi,c}$ vs ξ (laminar case)

tion of radiation is exactly treated as being two-dimensional in the current study.

The basic equations pertinent to the present system are written as follows in the dimensionless form.

$$\frac{1}{2} U \frac{\partial \theta_f}{\partial \xi} = \frac{1}{\eta} \frac{\partial}{\partial \eta} (\eta \frac{\partial \theta_f}{\partial \eta}) + M(\theta_p - \theta_f) \qquad (10)$$

$$M(\theta_p - \theta_f) = \frac{\tau_0^3}{2N_R} \int_{-\infty}^{\infty} \theta_w^4 (\tau_{\chi_1}) \frac{1}{\pi} \int_0^{\pi} \frac{\exp(-\tau_{sw})}{\tau_{sw}^3} (\tau_0 - \tau \cos\phi) d\phi d\tau_{\chi_1}$$
$$+ \frac{\tau_0^4}{2N_R} \int_0^1 \eta_1 \int_{-\infty}^{\infty} \theta_p^4 (\tau_{\chi_1}, \eta_1) \frac{1}{\pi} \int_0^{\pi} \frac{\exp(-\tau_s)}{\tau_s^2} d\phi d\tau_{\chi_1} d\eta_1 - \frac{\tau_0^2}{N_R} \theta_p^4 (\tau_\chi, \eta) \qquad (11)$$

where
$$U(\eta) = \{I_0(1/\sqrt{Da}) - I_0(\eta/\sqrt{Da})\}/I_1(1/\sqrt{Da}) \qquad (12)$$

The boundary conditions are taken as follows.

$$\eta = 0; \quad \partial\theta_f/\partial\eta = 0, \quad \partial\theta_p/\partial\eta = 0$$

$$\eta = 1; \quad \theta_p = \theta_f = \theta_w(\xi) \quad \text{(constant)} \quad \text{(heating section)}$$

$$\eta = 1; \quad \frac{\partial\theta_f}{\partial\eta}\Big|_{\eta=1} + \frac{\tau_0}{4N_R} \theta_w^4(\tau_\chi)$$
$$- \frac{\tau_0^4}{2N_R} \int_{-\infty}^{\infty} \theta_w^4(\tau_{\chi_1}) \frac{1}{\pi} \int_0^{\pi} \frac{\exp(-\tau_{sw})}{\tau_{sw}^4} (1-\cos\phi)^2 d\phi d\tau_{\chi_1}$$
$$- \frac{\tau_0^4}{2N_R} \int_0^1 \eta_1 \int_{-\infty}^{\infty} \theta_p^4(\tau_{\chi_1}, \eta_1) \frac{1}{\pi} \int_0^{\pi} \frac{\exp(-\tau_s)}{\tau_s^3} (1-\eta_1\cos\phi) d\phi d\tau_{\chi_1} d\eta_1 \qquad (13)$$
$$= 0 \quad \text{(adiabatic section)}$$

The controlled volume subject to numerical analysis must include regions both upstream and downstream of the heating section because thermal radiation propagates into those regions from the heating section.

The governing equations (10) and (11) are solved at once in this controlled volume by an implicit finite difference method as correcting the radiation term iteratively.

Heat transfer characteristics are evaluated by eqs.(5), (14) and (7).

$$Nu_{\xi,R} = \frac{2}{\theta_w - \theta_{mf}} \{ \frac{\tau_0}{4N_R} \theta_w^4 - \frac{\tau_0^4}{2N_R} \int_{-\infty}^{\infty} \theta_w^4 \times \frac{1}{\pi} \int_0^{\pi} \frac{\exp(-\tau_{sw})}{\tau_{sw}^4} (1-\cos\phi)^2 d\phi d\tau_{\chi_1}$$
$$- \frac{\tau_0^4}{2N_R} \int_0^1 \eta_1 \int_{-\infty}^{\infty} \theta_p^4 \frac{1}{\pi} \int_0^{\pi} \frac{\exp(-\tau_s)}{\tau_s^3} \times (1-\eta_1\cos\phi) d\phi d\tau_{\chi_1} d\eta_1 \} \qquad (14)$$

Fig.4 shows the variation of temperature profiles with parameter ξ. The profiles at $\xi=0.0$ and 0.01 correspond to those at the inlet and outlet of the heating section, respectively. As for the case of two-dimensional radiative heat transfer, the temperature profile of porous bodies at $\xi=0.0$ is characterized by the peak in the vicinity of the surface.

The results of heat transfer are shown in Fig.5. It is recognized that the remarkable heat transfer enhancement is attained by packing the porous materials in a duct at high temperatures and also that the one-dimensional analysis of radiation predicts lower values than those obtained by the two-dimensional one near the inlet and outlet of the heating sec-

tion and this trend becomes more prominent with decreasing N_R, particularly, at the outlet of the heating section.

EXPERIMENTS BY HTL-GSS(HEAT TRANSFER LOOP-GASEOUS SOLID SUSPENSION)[12]

Experimental Facility

A schematic diagram of HTL-GSS(Heat Transfer Loop-Gaseous Solid Suspension) is shown in Fig.6. This experimental closed loop mainly consists of three flow lines, i.e. gas stream (along the arrow ⟶), particle stream (--➤) and gaseous solid suspension stream (=➤). Gas is circulated by a diaphragm-type compressor ①. The gas flow rate is adjusted by by-pass circuit valve and is measured by Venturi tube ⑥. Solid particle feed system consists of ribbon blender ⑯, two ball valves ⑱, feed tank ⑳ and screw feeder ⑧. Leaving a heat transfer section ⑩, the suspension flow enters a gas-solid separator system (cyclone ⑫ and bag filter ⑬). Particles are captured there and sent to the particle weight measuring equipment ⑮ through two rotary valves ⑭ and accumulate in a conical hopper within it. The time variation of the weight of the hopper is recorded automatically and the slope of the recorded curve gives the solid flow rate.

Various measures are taken in the heat transfer section to cope with the situation at high temperature and high thermal load. The main heater is made of an Inconel pipe of 18 mm inner diameter (designated by D), 0.75 mm thickness and 1000 mm length, which is equally divided into five segments (200 mm length regional pipe sections) by water-cooled electrodes in order that regional variations of heating can be controlled. The pipe is wound by a fibrous insulator of outer diameter about 90 mm and is covered with asbestos. This heater assembly is contained in a vessel with a water jacket. A turbulence promoter (twisted tape) is inserted into the heating pipe in some cases. This tape is made of stainless steel plate of 2300 mm total length (2000 mm twisted part), 1 mm thickness, 10 mm width and 400 mm pitch. The tape end extends upstream by 500 mm from the starting point of the heating pipe. Asbestos blocks (3 × 10 × 15 mm³) are attached to the twisted part with separation distance 100 mm for electrical insulation.

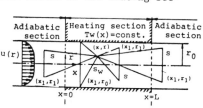

Fig.3 Physical model and coordinate system

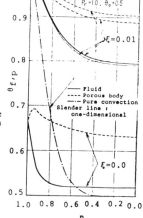

Fig.4 Temperature profiles

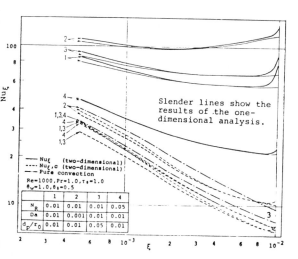

Fig.5 Nuξ, Nuξ,c vs ξ

Experimental Results and Discussions

The results of forced convection heat transfer of helium-graphite suspension (average particle diameter d_p=18 μm) in a straight tube without a turbulence promoter are shown in Fig.7 and 8 in case of constant wall temperature condition up to 1173 K. The Reynolds number based on gaseous phase at inlet Refi ranges from 1.0×10^4 to 2.0×10^4, particle loading ratio Γ reaches about 4. Prior to the experiments the heat losses are estimated through the preliminary measurements by realizing a similar thermal condition in the insulator vessel. The heat transfer coefficient and Nusselt number are locally evaluated as follows,

$$h_x = q_x/(T_w - T_b) \qquad (15)$$

$$Nus_x = h_x \cdot D/k_{fbx} \qquad (16)$$

where q is the net heat flux, k_{fb} is the thermal conductivity of fluid phase at the suspension bulk temperature T_b and the suffix x means a length from the starting point of the heating pipe. The variation of Nus at the location x/D = 50 with Γ for Refi = 1.0×10^4 is shown in Fig.7(a). It turns out that Nus increases only slightly with increase in wall temperature T_w. However, it should be noted that high heat fluxes reduce forced convective Nusselt number for a single phase flow as indicated by the correlation, for example, $Nub = 0.021Re_b^{0.8}Pr_b^{0.4}(T_w/T_b)^{-0.5}$. Therefore, it is considered that both radiative heat transfer mechanism of particulate phase owing to the rise in wall temperature and fluid turbulence enhanced by fine particles might be compensated by the foregoing reduction effect. As shown in Fig.7(b), Nus for Refi=2.0×10^4 decreases to a certain minimum with increasing Γ and beyond this point Nus becomes greater. It is also clear that its rising gradient becomes steeper with increase in T_w.

In Fig.8, the heat transfer results are reproduced in the form of conventional Nusselt ratio Nus/Nug, where Nug is the Nusselt number for a single gas flow obtained experimentally for the same x, T_w and Refi. The effects of the wall temperature T_w on Nus/Nug is emphasized, because Nug is considerably reduced at high thermal load.

The results obtained in a straight tube might be explained by the suppressive effect on fluid turbulence by fine particles. To recover this defect and make use of larger inertia forces of particles, the above mentioned turbulence promoter (twisted tape) is used in flowing gaseous solid suspension systems and its augmentation effect is experimentally

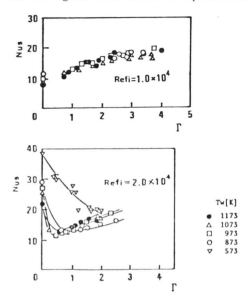

Fig.7 Variation of Nus with Γ (effect of T_w)

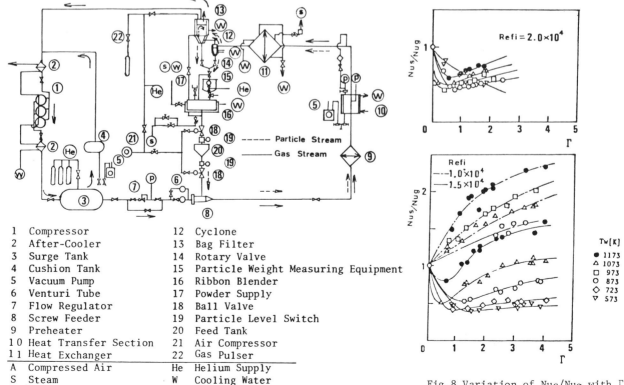

1	Compressor	12	Cyclone
2	After-Cooler	13	Bag Filter
3	Surge Tank	14	Rotary Valve
4	Cushion Tank	15	Particle Weight Measuring Equipment
5	Vacuum Pump	16	Ribbon Blender
6	Venturi Tube	17	Powder Supply
7	Flow Regulator	18	Ball Valve
8	Screw Feeder	19	Particle Level Switch
9	Preheater	20	Feed Tank
10	Heat Transfer Section	21	Air Compressor
11	Heat Exchanger	22	Gas Pulser
A	Compressed Air	He	Helium Supply
S	Steam	W	Cooling Water

Fig.6 Diagram of experimental facility

Fig.8 Variation of Nus/Nug with Γ (effect of T_w)

examined as shown in Fig.9. It becomes evident that the local heat fluxes with the twisted tape increase remarkably with increase in T_W.

ENHANCED HEAT TRANSFER OF GASEOUS SOLID SUSPENSION FLOW IN A CURVED TUBE [13]

Flow Characteristics

An analysis has been performed for a fully developed laminar flow of gaseous solid suspension medium in a curved circular tube with constant curvature. In contrast. to the single phase flow[16] an inherent difference of inertia forces between solid and gas phase will affect the flow structure to a great extent so that a single phase approach (taking account of an increment of apparent density) is not permissible. Fig.10 shows the coordinate system and the prevalent postulations and assumptions are introduced and then the coupled governing equations for respective phases are formulated in the following,

$$V_f \frac{\partial \tilde{\omega}}{\partial R} + \frac{W_f}{R} \frac{\partial \tilde{\omega}}{\partial \phi} = \frac{2}{De} \nabla^2 \tilde{\omega} - \frac{2}{Re} U_f \{ \frac{1}{R} \frac{\partial U_f}{\partial \phi} \cos\phi + \frac{\partial U_f}{\partial R} \sin\phi \}$$
$$- \frac{2\Gamma}{\Lambda De} \frac{1}{R} \{ G_p (\frac{\partial V_p}{\partial \phi} - W_p - R \frac{\partial W_p}{\partial R} + R\tilde{\omega}) + (V_p - V_f) \frac{\partial G_p}{\partial \phi} - R (W_p - W_f) \frac{\partial G_p}{\partial R} \} \quad (17)$$

$$\tilde{\omega} = -\nabla^2 \psi \quad (\text{where } \nabla^2 = \frac{\partial^2}{\partial R^2} + \frac{1}{R} \frac{\partial}{\partial R} + \frac{1}{R^2} \frac{\partial^2}{\partial \phi^2} \text{, } V_f = \frac{1}{R} \frac{\partial \psi}{\partial \phi} \text{, } W_f = -\frac{\partial \psi}{\partial R}) \quad (18)$$

$$V_f \frac{\partial U_f}{\partial R} + \frac{W_f}{R} \frac{\partial U_f}{\partial \phi} = \frac{16}{De} DP + \frac{2}{De} \nabla^2 U_f + \frac{2\Gamma}{\Lambda De} G_p (U_p - U_f) \quad (19)$$

$$\frac{\partial}{\partial R}(G_p R V_p) + \frac{\partial}{\partial \phi}(G_p W_p) = 0 \quad (20)$$

$$V_p \frac{\partial V_p}{\partial R} + \frac{W_p}{R} \frac{\partial V_p}{\partial \phi} - \frac{W_p^2}{R} - \frac{\cos\phi}{Rc} U_p^2 = \frac{2}{\Lambda De}(V_f - V_p) \quad (21)$$

$$V_p \frac{\partial W_p}{\partial R} + \frac{W_p}{R} \frac{\partial W_p}{\partial \phi} + \frac{V_p W_p}{R} + \frac{\sin\phi}{Rc} U_p^2 = \frac{2}{\Lambda De}(W_f - W_p) \quad (22)$$

$$V_p \frac{\partial U_p}{\partial R} + \frac{W_p}{R} \frac{\partial U_p}{\partial \phi} = \frac{2}{\Lambda De}(U_f - U_p) \quad (23)$$

where the dimensionless variables and parameters are defined as

$$U = u/u^*, \quad V = v/u^*, \quad W = w/u^*, \quad Rc = r_0/r_c, \quad R = r/r_0, \quad R_p = d_p/2r_0$$
$$DP = (\partial p/\partial x)/(8\mu u^*/r_0^2), \quad Re = 2\rho_f r_0 u_{fm}/\mu, \quad De = Re\sqrt{r_0/r_c} \quad (24)$$
$$\Gamma = (\rho_p U_p)/(\rho_f U_f), \quad G_p = (\rho_p/\rho_f)/\Gamma, \quad \Lambda = (\rho_p/\rho_f)(d_p/r_0)^2/18$$
$$\xi = (x/r_0)/DePr, \quad u^* = u_{fm}\sqrt{r_0/r_c}$$

While the boundary conditions of fluid are similar to the former study[16], special attention is paid to the behavior of solid particles in close vicinity of the wall. If the radial force exerted on a particle contacting to the wall is positive, the friction drag is taken into account for the equation of motion which constitutes one of the boundary conditions for particulate phase.

$$\phi = 0, \pi : \quad \partial U_f/\partial \phi = \partial V_f/\partial \phi = W_f = 0, \quad \partial U_p/\partial \phi = \partial V_p/\partial \phi = W_p = \frac{\partial G_p}{\partial \phi} = 0 \quad (25)$$

$$R = 1 : \quad U_f = V_f = W_f = 0 \quad (26)$$

$$\frac{W_p}{R} \frac{\partial W_p}{\partial \phi} + \frac{\sin\phi}{Rc} U_p^2$$
$$= \frac{2}{\Lambda De}(W_f - W_p) - f_W \frac{W_p}{\sqrt{U_p^2 + W_p^2}} \{ \frac{W_p^2}{R} + \cos\phi \frac{U_p^2}{Rc} + \frac{2}{\Lambda De} V_f \} \quad (27)$$
$$\frac{W_p}{R} \frac{\partial U_p}{\partial \phi} = \frac{2}{\Lambda De}(U_f - U_p) - f_W \frac{U_p}{\sqrt{U_p^2 + W_p^2}} \{ \frac{W_p^2}{R} + \cos\phi \frac{U_p^2}{Rc} + \frac{2}{\Lambda De} V_f \}$$

Numerical solutions have been obtained by S.O.R. method based on the finite difference scheme, in which the grid spacing is nonuniform with a higher density of nodal points near the wall. In Fig.11, typical results on the flow field are illustrated in which the parameters are set as $R_c = 100$, $R_p = 10^{-3}$, $\Lambda = 10^{-3}$, $f_W = 0.1$. Examination on the figures shows that the particle loading deforms the velocity profile appreciably and

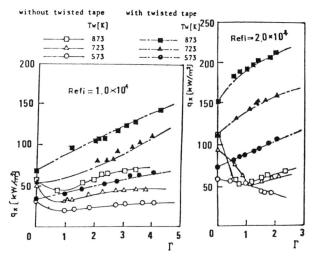

Fig.9 Variation of q_x at $x/D = 50$ with Γ

enhances the secondary flow and it is worthy to note that higher particle density is observed in the vicinity of the wall due to the centrifugal and Coriolis forces. The higher population of particles near the wall will play an important role in the heat transfer mechanism which will be discussed briefly in the next section. Detailed behaviors of particulate phase in a close vicinity of the wall are depicted in Fig. 12 for Dean number De = 30 and particle loading ratio $\Gamma = 1$, respectively. For a set of parameters shown in the figure, the particles contacting the wall ($R = 1 - R_p$) detach at 20 deg. due to the Stokes' drag of the secondary flow and yield a steep increase of Gp at

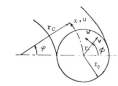

Fig.10 Coordinate system

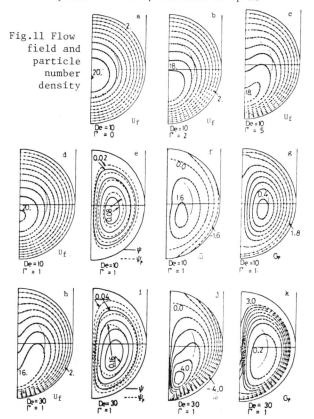

Fig.11 Flow field and particle number density

$R = 1 - 2R_p$. The number density of particles at $R = 1 - 2R_p$ is gradually decreasing (becomes zero at 170 deg.), while it comes to increase at $R = 1 - 4R_p$ ($\Delta R = 0.025$, $\Delta R_p = 0.001$). The limit of the most populated condition of attached particles to the wall in a single layer is expressed as

$$Gp \leq (\rho_p/\rho_f)\pi/6\Gamma \qquad (28)$$

The calculation results executed in this study clear the foregoing conditions. These behaviors are intuitive for understanding the flow and heat transfer mechanism in this system.

Heat Transfer Analysis and Discussions

Once a flow field is established it is comparatively easy to get a solution of energy equations for fluid and particle medium. The energy equations in the dimensionless forms state

$$PrDe\{V_f\frac{\partial\theta_f}{\partial R} + \frac{W_f}{R}\frac{\partial\theta_f}{\partial\phi}\} + U_f\frac{\partial\theta_f}{\partial\xi} = 2\nabla^2\theta_f + \frac{2}{3}\frac{Nu_p}{\Lambda}\Gamma Gp(\theta_p-\theta_f) \qquad (29)$$

$$PrDe\{V_p\frac{\partial\theta_p}{\partial R} + \frac{W_p}{R}\frac{\partial\theta_p}{\partial\phi}\} + U_p\frac{\partial\theta_p}{\partial\xi} = \frac{2}{3}\frac{Nu_p}{\Lambda C^*}(\theta_f-\theta_p) \qquad (30)$$

with the boundary conditions of

$$\phi=0,\pi : \partial\theta_f/\partial\phi=\partial\theta_p/\partial\phi=0, \quad R=1 : \theta_f=1 \qquad (31)$$

where $\theta=(T-T_0)/(T_w-T_0)$, $Pr=\mu c_f/k_f$, $Nu_p=h_p d_p/k_f$ (32)

For the numerical calculations some of the dimensionless parameters related to the energy equations are fixed as $Pr = 0.7$, $C^* = 1$ and $Nu_p = 2$. Fig.13 shows the typical temperature profiles along the $\phi = 0-\pi$ axis, where the temperature of the particle is not drawn in the figures because the differences from the fluid are minor. However, a detailed comparison of the particle temperature with the fluid indicates that the fluid temperature exceeds the particle temperature to a certain extent over an entire cross section and this fact is important from a viewpoint of heat transfer mechanism. In order to discuss the heat transfer characteristics, the mixing mean temperatures and Nusselt numbers are defined in the following,

$$\bar{\theta}_m = \frac{\int_A \theta_f W_f dA + \int_A C^* Gp\theta_p W_p dA}{\int_A W_f dA + \int_A C^* Gp W_p dA} \qquad (33)$$

$$\bar{\theta}_f = \frac{\int_A \theta_f W_f dA}{\int_A W_f dA} \qquad \bar{\theta}_p = \frac{\int_A \theta_p W_p dA}{\int_A W_p dA} \qquad (34)(35)$$

$$Nu(\xi,\phi)=2(\frac{\partial\theta_f}{\partial R})_{R=1}/\{1-\bar{\theta}_m\}, \quad Nu(\xi)=\frac{1}{\pi}\int_0^\pi Nu(\xi,\phi)d\phi \qquad (36)(37)$$

$$\overline{Nu}(\xi) = -(1+C^*\Gamma)\sqrt{R_c}\ \ln(1-\bar{\theta}_m)/(2\xi) \qquad (38)$$

The mixing mean temperatures ($\bar{\theta}_m$, $\bar{\theta}_f$ and $\bar{\theta}_p$) and mean Nusselt number $Nu(\xi)$ against ξ are shown in Fig.14 for $De = 1$, 10 and 30 with loading ratio $\Gamma = 1$. Though the local temperature of fluid is always exceeding the particle temperature, the mixing mean temperature of fluid $\bar{\theta}_f$ is rather lower than that of particles $\bar{\theta}_p$. Noticing the local excursion of distribution of Gp, the physical meaning is self-explanatory.
It is important to gain insight into the circumferential distributions of the local Nusselt number $Nu(\xi,\phi)$. The difference between the outside ($\phi = 0$) and inside ($\phi = \pi$) of the curvature is prominent.

ENHANCED HEAT TRANSFER OF GASEOUS SOLID SUSPENSION JET IMPINGEMENT

Basic Characteristics of Free Jet

In order to grasp the outline of the impinging jet heat transfer, the characteristic features of the free jet should be clarified in the first place. In the analytical studies of gaseous suspension jet, it is a commonly accepted procedure to regard the suspended particulate phase only as a passive contaminant. It makes a whole equation system quite simple and economizes the necessary computational labor. However, momentum equations for the particulate phase are retained in our free jet analysis, because our final goal is to clarify the heat transfer of the impinging region where inertia difference of both phases is expected to play a decisive role in heat transfer. Modelling of the suspended phase is the same one as in the former section. Moreover, the suspended particles are assumed to be coarse enough so that they do not follow the turbulent fluctuation of the fluid. Accordingly, the turbulence terms are neglected in the equations for particulate phase. In describing the turbulent motion of the fluid phase, Prandtl's turbulent energy model is applied. Based on these assumptions, conservation equations for continuity, momentum and turbulence energy are written as follows.

$$\frac{\partial u_f}{\partial x} + \frac{1}{r}\frac{\partial}{\partial r}(rv_f)=0 \quad (39) \qquad \frac{\partial}{\partial x}(\rho_{dp}u_p)+\frac{1}{r}\frac{\partial}{\partial r}(r\rho_{dp}v_p)=0 \qquad (40)$$

$$\rho_f(u_f\frac{\partial u_f}{\partial x}+v_f\frac{\partial u_f}{\partial r})$$
$$=\frac{1}{r}\frac{\partial}{\partial r}\{r(\mu+\mu_t)\frac{\partial u_f}{\partial r}\}-\frac{1}{8}C_D(Re_{p\mu})\pi d_p^2\rho_f n_p(u_f-u_p)\cdot|u_f-u_p| \qquad (41)$$

$$\rho_{dp}(u_p\frac{\partial u_p}{\partial x}+v_p\frac{\partial u_p}{\partial r})=\frac{1}{8}C_D(Re_{p\mu})\pi d_p^2\rho_f n_p(u_f-u_p)\cdot|u_f-u_p|\pm\rho_{dp}g$$

(+ ; downward, - ; upward) (42)

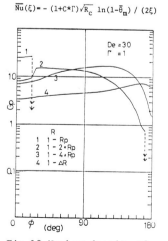

Fig.12 Number density of particle near the wall

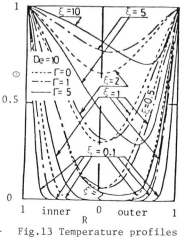

Fig.13 Temperature profiles along $\phi = 0-\pi$ axis

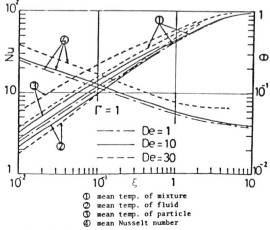

① mean temp. of mixture
② mean temp. of fluid
③ mean temp. of particle
④ mean Nusselt number

Fig.14 Mixing mean temperature and Nu

$$\rho_{dp}(u_p\frac{\partial v_p}{\partial x}+v_p\frac{\partial v_p}{\partial r})=\frac{1}{8}C_D(Re_{ep,v})\pi d_p^2\rho_f\eta_p(v_f-v_p)\cdot|v_f-v_p| \quad (43)$$

$$\rho_f(u_f\frac{\partial k}{\partial x}+v_f\frac{\partial k}{\partial r})=\frac{1}{r}\frac{\partial}{\partial r}[r(\mu+\frac{\mu_t}{\sigma_k})\frac{\partial k}{\partial r}]+\mu_t(\frac{\partial u_f}{\partial r})^2-C_{dis}\frac{k^{3/2}}{\ell}\rho_f$$
$$-\eta_p K_{dis}+\eta_p K_{prod} \quad (44)$$

$$k=\frac{1}{2}(\overline{u_f'^2}+\overline{v_f'^2}+\overline{w_f'^2}), \quad \mu_t=C_\mu\rho_f\ell\sqrt{k}, \quad \ell=C_\ell(r_{0\cdot 1}-r_{0\cdot 9})$$

$$Re_{ep,u}=\frac{\rho_f d_p|u_f-u_p|}{\mu}, \quad Re_{ep,v}=\frac{\rho_f d_p|v_f-v_p|}{\mu}$$

ℓ is the length scale and $\gamma_{0\cdot 1}$ and $\gamma_{0\cdot 9}$ indicate the radial positions the velocities of which are 10% and 90% of the center line velocity. C_D in the momentum equations indicates the drag coefficient of a single sphere and is written as,

$$C_D(Re)=\frac{24}{Re}(1+0.158Re^{\frac{2}{3}}) \quad (45)$$

in order to cover the coarser particle case. K_{dis}

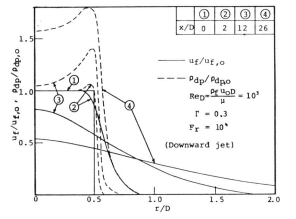

Fig.15 Fluid velocity and dispersed particle density

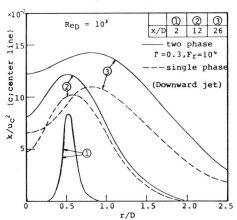

Fig.16 Turbulent energy distribution

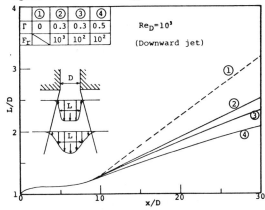

Fig.17 Jet half width

and K_{prod} are added as dissipation and production terms of turbulent energy due to the presence of a particle. They are formulated as follows. K_{dis} is originated from the interaction between a particle and fluctuating component of the fluid motion. Similar to Danon et al.[17] it is written as

$$K_{dis}=\frac{\sqrt{2}}{4}C_D(Re_{ep,k})\pi d_p^2\rho_f k^{\frac{3}{2}}[1-\exp\{-C_1 Re_{ep,k}^{\frac{3}{2}}(\frac{\rho_p}{\rho_f}\frac{d_p}{\ell})^{\frac{1}{2}}\}] \quad (46)$$

$$Re_{ep,k}=\frac{\rho_f d_p\sqrt{k}}{\mu}$$

Meanwhile, K_{prod} is assumed to be originated from the difference in mean velocities of both phases.

$$K_{prod}=\frac{1}{8}C_D(Re_{ep,u})\pi d_p^2\rho_f|u_f-u_p|^3\times C_2\{1-\exp(-C_3 Re_{ep,u})\} \quad (47)$$

Eqs.(39)~(47) are solved simultaneously by finite difference procedure with the necessary boundary conditions at the round nozzle of finite diameter and at the radial infinity. The following empirical constants are tentatively used.

σ_k	C_{dis}	C_μ	C_ℓ	C_1	C_2	C_3
1.0	1.0	0.078	0.625	1.0	1.0	1.0

Fig. 15~17 reproduce the numerical results. Small peaks of ρ_{dp} in Fig.15 are due to the deceleration of the particles in the shear layer. In Fig.16, the turbulent energy for suspension jet is more than that for single phase jet. Moreover, Fig.17 indicates that the half width of the suspension jet is narrower than that of the single phase jet. This tendency coincides with the experiment by Laats[18].

Experiment of Impinging Jet Heat Transfer[14]

Presented here is an experimental research of impinging jet heat transfer of gaseous suspension. Fig.18 shows a brief sketch of the experimental set-up which is set within a two phase line of HTL-GSS. Details of this experiment are described in (14). Fig.19 shows a typical result of suspension jet wherein the heat transfer around the stagnation point is seen to be remarkably enhanced through the addition of suspended phase. Fig.20 summarizes the stagnation point heat transfer coefficient h_s normalized by those of the single phase jet of the same Reynolds number. It illustrates that the suspension impinging jet is a powerful means of heat transfer enhancement even for a small loading ratio. A direct incentive that urged

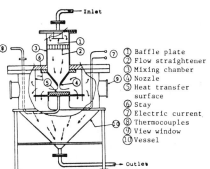

① Baffle plate
② Flow straightener
③ Mixing chamber
④ Nozzle
⑤ Heat transfer surface
⑥ Stay
⑦ Electric current
⑧ Thermocouples
⑨ View window
⑩ Vessel

Fig.18 Test section

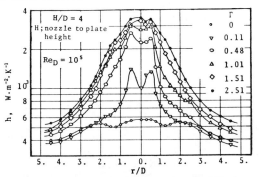

Fig.19 Impinging jet heat transfer coefficient

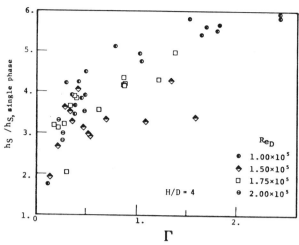

Fig.20 Heat transfer at the stagnation point

the authors into the investigation of this novel heat transfer problem is explained in Fig.21, where a new concept of suspension-cooled fusion reactor blanket system is illustrated. The selection of coolant material of the fusion reactor blanket is still left to further synthetical examinations. Two main coolant materials that have been proposed so far have their inherent shortcomings. Namely, liquid metal such as Li is inevitably influenced by the MHD effect of plasma confining magnetic field, although it has preferable heat transfer properties. Conversely, single phase helium gas has inferior heat transfer properties, although it is liberated from the MHD effect. If we compare these features with those of the gaseous suspension, following remarks can be pointed out. Firstly, the MHD effect of the suspension is of little significance if the particle number density is not so high that the collision frequency of the particles is negligible. Secondly, its heat transfer characteristics are desirable particularly for high thermal load condition of the blanket. Furthermore, as is pointed out by Sze[19], if a certain chemical compound of Li is selected as a suspended material, tritium breeding and heat removal may possibly be achieved simultaneously. In our concept, the first wall of the blanket, the thermal condition of which is severest among others, is cooled by the impinging jet of gaseous suspension. For the cooling of the tritium breeding region, suitably constructed helical tubes of the suspension flow may be possibly applied. The heat transfer characteristics of it have been already discussed in the former section.

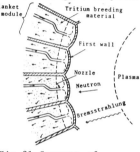

Fig.21 Concept of suspension-cooled fusion blanket

SUMMARY

In this paper we have discussed a few particular problems of gaseous solid suspension flow associated with the "High-Flux Heat Transfer" and shown that the remarkable increases of heat transfer coefficient have been realized on account of the radiative behaviors of particulate media and, in some cases, may be attributed to the inherent difference of inertia forces between gas and particles in flowing multiphase media. However, a relevant exposition on this subject has not been found in literature and in view of the lack of experimental evidence and theoretical prediction regarding the effect of solid addition on the turbulent structure, one has to develop further elaborate studies in order to stimulate and facilitate this technique to the extensive applications of high flux and high temperature heat transfer system.

ACKNOWLEDGEMENT

Finally the major part of the paper has been performed by the present authors along with the coworkers who are found in the literatures cited and the authors wish to express their gratitude to Dr. K. Kamiuto, Dr. K. Kanemaru and Mr. A. Shimizu who helped to prepare the current manuscript.

REFERENCES

1. Schulderberg, D.C. et al., Nucleonics, 19, 67-76 (1961).
2. Rudinger, G., AIAA J., 8, 1288-1294 (1970).
3. Honma, T., Res. Elect. Lab., No.666 (1966) (in Japanese).
4. Babcock & Wilcox Co., "Final Report on the Gas-Suspension Coolant Project" AECBAW-1159 (1959).
5. Soo, S.L., "Fluid Dynamics of Multiphase Systems" Blaisdell Pub. Co. (1967).
6. Boothroyd, R.G., "Flowing Gas-Solids Suspensions" Chapman Hall (1971).
7. Hasegawa, S. and Echigo, R., J. JSME, 75, 679-688 (1972) (in Japanese).
8. Echigo, R. and Hasegawa, S., Int. J. Heat Mass Transfer, 15, 2515-2534 (1972).
9. Echigo, R., Hasegawa, S. and Tamehiro, H., Int. J. Heat Mass Transfer, 15, 2595-2610 (1972).
10. Tamehiro, H., Echigo, R. and Hasegawa, S., Int. J. Heat Mass Transfer, 16, 1199-1213 (1973).
11. Kamiuto, K., Kanemaru, K., Ando, T., Echigo, R. and Hasegawa, S., Trans. JSME, 44, 4234-4242 (1978) (in Japanese).
12. Kanemaru, K., Hasegawa, S., Echigo, R., and Ichimiya, K., J. High Temperature Society, 6, 25-33 (1980) (in Japanese).
13. Echigo, R. et al., 17th National Heat Transfer Symp. Japan, 307-309 (1980) (in Japanese).
14. Shimizu, A., Echigo, R. and Hasegawa, S., Advances in Enhanced Heat Transfer, 18th Nat. Heat Transfer Conf., San Diego, 155-160 (1979).
15. Modest, M.F., Meyer, B.R. and Azad, F.H., Joint ASME/AIChE National Heat Transfer Conf., Orlando, Paper No.80-HT-27 (1980).
16. Mori, Y. and Nakayama, Trans. JSME, 30, 977-988 (1964) (in Japanese).
17. Danon, H., Wolfshtein, M. and Hetsroni, G., Int. J. Multiphase Flow, 1, 715-726 (1974).
18. Laats, M.K., Inzh. fiz. Zh., 10, 11-15 (1966).
19. Sze, D.K., Larsen, E.T. and Clemmer, R.G., ANS Trans., 22, 21-22 (1975).

High Flux Heat Transfer: Applications to Power and Propulsion Systems

F. W. STAUB

Corporate Research and Development, General Electric Company,
Schenectady, New York, U.S.A.

ABSTRACT

The current status of and future perspectives in heat transfer research in the subject area is summarized to provide a first foundation for further discussion and input by other workers in the field. Emphasis is placed on two-phase diabatic processes with brief reference to high velocity single phase systems and the need for the coupling of technologies. Attention is focused on areas where much work has been reported and on those that require further effort.

INTRODUCTION

A brief comment on the areas considered and those not considered here is in order. The obvious high flux systems considered involve phase change or heat transfer enhancement due to high single phase fluid velocities or two-component flow behavior. It is recognized that a high two-phase flux with one fluid, such as a fluorocarbon refrigerant, is a small two-phase flux with water or liquid metals. The maximum heat flux capability of the fluid of interest is thus implied here. The above high flux systems involve high coefficients at generally moderate temperature differences. High flux systems also involve those with lower coefficients at high temperature difference. Since the latter is the subject of a separate summary, high temperature systems are not discussed here except in one area that is becoming increasingly important.

Highly specialized areas are not considered in this brief summary. For example, heat transfer in cryogenic systems, liquid metal systems and light water nuclear reactor operation are not discussed. Such areas require in-depth understanding related to specific geometries, specific operating or accidental transients or data on fluids with unique properties that have less broad applications at this time. The status and recommended high priority further work in these areas should be proper subjects for specialized symposia dedicated to these applications.

The purpose of this summary is to highlight the status of, and perspectives on, several high flux areas of general application to power and propulsion systems as a first foundation for further discussion and additions by other workers in this field. This summary is neither meant to be all inclusive nor is it meant to be an in-depth review. Where references are given they are only meant as examples of those commonly used in areas where work has been published.

STATUS AND FURTHER WORK NEEDED

Vaporization

Liquid vaporization can take place via nucleate boiling in pools or under forced flow conditions, or via evaporation from liquid films or droplets. Maximum heat flux conditions are obtained on transition from nucleate to film boiling and on breakup or depletion of liquid films and droplets. While steam generators, refrigerant evaporators and waste heat recovery boilers generally employ forced convection vaporization inside tubes, shell-side boilers require pool boiling correlations and two-phase cooling in rotating fields requires flow regime and thin film evaporation correlations. In electrical equipment cooling, such as vaporization cooled transformers and solid state power conversion systems, all of the above vaporization modes are encountered depending on the equipment power rating and the particular application. These systems usually require a dielectric coolant or highly demineralized water. A knowledge of pressure drop and heat transfer coefficients in the above applications must be supplemented with understanding of the material behavior and the fouling rates where applicable. Before outlining the need for further work in this area, as it relates to power and propulsion systems, some comments are first given below on vaporization configurations on which considerable information has been published.

A large amount of data on pool boiling heat transfer exists with generalized correlations, such as those of (1) and (2), which can be employed for reasonable first estimates. The performance of new fluid-surface combinations must generally be checked experimentally since the nucleation characteristics of surfaces effect the heat transfer coefficients. The maximum pool boiling heat flux can be reasonably well predicted for standard geometries using generalized correlations such as those of (3) and (4) including the effect of high gravity forces such as those encountered in rotating equipment.

For forced flow vaporization inside tubes with wetted walls several useful correlations exist, such as that of (5), which are often satisfactory especially when the overall heat transfer coefficient is not controlled by the vaporization side. The maximum heat flux, or dry-out, condition inside tubes is less predictable although much data exist, as summarized by (6) for example, that allow first estimates.

Evaporation across liquid films is reasonably well understood, up to film break-up, and can be predicted from the film behavior work discussed below under condensation. Evaporation from droplets, as these exist in dry-wall evaporators and "fog-cooled" ducts, is much less predictable because one usually does not know the droplet size and number density. Fortunately, in most evaporators and cooling systems one needs to avoid this dry wall condition. Nevertheless in steam reheaters, wet steam turbines and once through boilers with superheat, the dry wall vaporization process becomes important.

The need for further application information in pool boiling is primarily related to the maximum, or critical, heat flux and to the minimum heat flux required to establish pool boiling. The effect of non-standard geometries, such as closely spaced equipment in immersed power conversion systems, on the critical heat flux is required especially at low reduced pressures. Under part load conditions it is also necessary to more accurately predict the onset of pool boiling especially if large amounts of non-condensable gases are dissolved in the coolant. While the latter is not of concern in most water cooled systems, the solubility of many gases in typical dielectric liquids is considerable. In addition, while of more restricted interest to the cooling of electrical equipment in cold climates, there is almost no data on pool boiling very close to the freezing (or slush point) of dielectric coolants. Of increasing interest in shell-side boilers, the question of local dry-out requires more attention since both process steam evaporators and some waste heat recovery units are performance limited by this condition.

For both shell-side boilers and forced flow in-tube vapor generators the minimum temperature difference required for established nucleate boiling is of increasing concern as efficient heat recovery limits the available temperature difference to small values. Cyclic vapor generation due to excessively low temperature differences is highly undesirable in such applications. The critical heat flux, or dry-wall, condition in forced flow evaporators, as well as improved prediction of the post-dryout heat transfer, is also not well predictable from the literature especially in view of the very real effects of bends in ducts and tubes that redistribute the flow. Even in long evaporator ducts without bends, more information on transition from stratified to slug flow is needed. Without slug flow the heat transfer is reduced and with very limited slugging wall corrosion can result especially in water systems. The latter will be of concern in combined cycle power boilers, fluidized bed combusters and coal gasifiers. In these applications the heat transfer in direct contact evaporating flows will also require greater predictive reliability although it is understood that this is primarily a droplet fluid mechanics problem. The same will hold true in some heat pump or heat recovery systems that can employ direct contact evaporators.

Single phase cooling of ducts in rotating equipment, such as generators, motors and gas turbine blades has been in use for some time. Two-phase cooling of such components has received increasing consideration as the heat removal density in such equipment has increased. The two-phase fluid mechanics of radial or axial cooling ducts in rotating equipment has received very little attention. Emphasis on wetted wall vs. dry wall conditions is needed as a function of the volumetric vapor-liquid flow rates for both rotating

radial and axial ducts. Such work should include the determination of minimum liquid film wetting rates which has received some attention in non-rotating systems.

Condensation

While condensation at a surface can take place under droplet or film conditions, droplet or drop-wise condensation on a surface is not discussed here in view of its fairly rare occurrence in power or propulsion applications. Free convection film condensation takes place on the shell side of power condensers while forced convection in-tube condensation takes place in tube-side power condensers, in nuclear steam reheaters and in refrigerant condensers employed in electric power cooling systems. In air cooled condensers, without the presence of non-condensable gases, the air side is usually the controlling side while in water cooled condensers the accurate prediction of the condensing side performance is more important. The presence of enough non-condensable gases to effect condenser performance is possible in power generation or large propulsion systems and it is quite likely in some two-phase cooling systems for electrical equipment since dielectric coolants can dissolve a large amount of any cover gas. While direct contact spray or film condensers have so far seen very limited use in power and propulsion systems, the need to decrease the condenser-coolant temperature difference in bottoming cycles, waste heat recovery systems or other "low temperature engines" has increased the application interest in direct contact spray condensers.

The direct contact condensation of flowing steam in droplet form during expansion is of particular interest to 'wet' power or propulsion turbines due to its effect on turbine performance. The direct condensation of steam bubbles in liquid takes place in submerged condensers and in the suppression pool of some light water nuclear reactors. Neither the expanding steam condensation nor the direct contact bubble condensation applications are further considered here in view of their limited application in power and propulsion systems. Comments on condensation configurations on which considerable information has been published are followed below by some identified areas for further work.

The correlations for liquid film thickness and condensation heat transfer are given for flat plates and outside tubes, by (7) for example, and for the inside of tubes by (8), (9), (10) and (11). A reasonably satisfactory method for predicting the effect of non-condensable gases on the condensing heat transfer is given by several workers such as (12). In general the available information for the prediction of the condensing heat transfer coefficients with low non-condensables present and in the high velocity region of power and propulsion condensers is satisfactory for application purposes. In view of the need to limit the pressure drop in such condensers, however, the accuracy of the condenser pressure drop correlations, especially at low reduced pressures, needs improvement. The latter is especially true for tube bends and manifold losses for in-tube condensers and for the steam inlet region loss for shell-side condensers.

The primary need for improved predictive methods for condensation in power and propulsion systems are in the area of non-condensable effects in power condensers, in electrical equipment cooling systems employing dielectric coolants and in direct contact spray condensers. Both the volumetric heat transfer rates and the non-condensable gas effects in direct contact condensers are primarily given by inadequate ad hoc data in the literature.

As a general guide, the areas in condensation heat transfer that need the most attention are those that can sharply reduce heat transfer, such as non-condensable gases or the flooding of bare or enhanced tubes in multi-row shell-side condensers, and those that can achieve a large increase in the overall heat transfer rate, such as direct contact condensation where the 'condensation side' performance is governing.

Thermohydraulics

The undesirable effect of flow transients in two-phase flow caused by the interaction of the fluid dynamics and the heat addition or removal process deserves a brief comment. The multipassage and multicomponent two-phase systems that are typical in power handling and propulsion equipment almost always involve local flow transients that, if severe enough, can damage equipment via mechanical excitation or via premature dry-out in evaporators. It is thus important to further study such realistic configurations from the standpoint of thermohydraulic stability. While this point does not directly fall under the subject of this discussion, its indirect effect on the real performance of two-phase systems may be as great or greater than the effect of improved information on the several heat transfer mechanisms above.

Two-Component Flow and Heat Transfer

Gas-solid particulate flow and heat transfer has more recently become of increasing interest in power generation. Fluidized bed boilers and coal gasification systems, now under development, employ solids handling systems and various forms of fluidized beds in which heat is generated and removed. While much information has been generated by the chemical process industry in this two-component area, the increasing scale, greater reliability and plant availability required by power generation systems create a demand for improved understanding and generalized design correlations involving gas-solid particulate flow and heat transfer. These systems result in an order of magnitude increase in the heat flux compared to that in either single phase gas flow or moving bed (packed) solids flow. While most of the related applications would operate at high source temperatures, their consideration here is warranted in view of the recent power application possibilities in this area.

Most of the co-current gas-solids flow and heat transfer data and correlations, such as those of

(13) and (14), have been generated using particles smaller than 200μm flowing in pipes. Heat transfer correlations and mechanisms in low velocity, bubbling, fluidized beds have been published, see (15) for particle sizes less than 500μm and for heat transfer to tubes and tube banks immersed in the fluidized beds. The power applications of fluidized beds have generally evolved into higher velocity systems employing particle sizes larger than 500μm.

In view of the above, there is a need for flow and heat transfer data in large particle (500μm - 3000μm) systems operating at high temperatures (to 950°C) at pressures up to about 10 atmospheres. Some limited work on large particle, high velocity, turbulent fluidized beds and on the effect of particle size and thermal radiation on heat transfer in fluidized beds, (16) and (17) for example, has been published, but, in general, there is a continuing need for more flow and heat transfer research in high velocity fluidized bed and circulating or fast beds. The latter should be extended to the consideration of the radiation component of heat transfer from high temperature gas-solid particle systems over a large particle number density range and gas velocity range. Finally, material erosion information in these two-component systems is an important parallel consideration that requires equal attention.

High Velocity Single Phase Flow and Heat Transfer

The occasional need for high velocity liquid or gas cooling systems in large power generation equipment, large magnets or gas turbine components requires a brief comment here since high heat flux values can be sustained, without vapor generation with liquids, at high fluid velocities. While the heat transfer and pressure drop is reasonably well understood in most of these applications, except perhaps for complex cooling duct geometrics in some turbines, the primary need for improved information is that relating to material erosion or cavitation damage. As the local wall temperature approaches the saturation temperature, the interaction between the evolution of dissolved gases and the onset of local cavitation damage can use more attention than that available in the literature.

SUMMARY OF RECOMMENDATIONS

The following summarizes the priority items which require increased attention, as outlined in the discussion, for application to power handling and propulsion equipment.

Vaporization

Pool Boiling - The effect of non-standard geometries, such as close vertical and horizontal clearances, and of operation near the freezing point on the critical heat flux needs effort. The local critical heat flux in shell-side boiling of shell and tube exchangers and the effect of non-condensable gases on the initiation of nucleate boiling need attention. Emphasis should be on demineralized water, dielectric coolants and refrigerants over a broad reduced pressure range.

Forced Flow Vaporization – The minimum temperature difference required to initiate wall nucleation with dissolved non-condensable gases present needs attention especially for dielectric coolants that can dissolve a large quantity of dielectric gases. The effect of stratified flow on the onset of wall dryout in near horizontal tubes and near return bends as well as the stratified-to-slug flow regime transition needs to be better determined with water to define both heat transfer and corrosion limits. The effect of entrained liquid on dry wall heat transfer including the effect of return bends on this flow regime needs effort with emphasis on water and a lower surface tension coolant. The effect of cooling duct rotation on the two phase flow regime and heat transfer needs attention with reference to cooling ducts in rotating power and propulsion machinery. The minimum liquid film wetting rates under the effect of large centrifugal acceleration are also needed for these applications.

Direct Contact Vaporization – The rate processes involve in spray evaporators over a broad range of spray generation devices must be better correlated together with increased effort on mist removal techniques usually necessary to minimize the droplet content of the generated vapor.

Condensation

Surface Condensation – The effect of increasing non-condensable gas content on shellside and on inside-tube forced flow condensation needs better definition as a function of fluid velocity. Emphasis should be on water and on dielectric coolants that contain dissolved non-condensable dielectric gases. The pressure drop for inside tube condensation at very low reduced pressures and with realistic tube bends and manifolds requires better understanding and correlation with water as the working fluid. The effect of multiple bare and finned horizontal tube rows on local shell side condensation performance needs to be better established.

Direct Contact Condensation – The available correlations and models of direct contact spray condensers are inadequate especially as these are effected by non-condensable gases and the various devices available for spray generation. In view of their increased consideration in waste heat recovery or bottoming power cycles the present design base needs to be improved.

Thermohydraulics

As a more general and very necessary area for further work, the effect of multiple tube evaporators and condensers, coupled to transfer lines and throttling valves, on the dynamic system flow oscillations and local (tube-to-tube) flow oscillations requires selected in-depth study to isolate the oscillation modes and their cause. More unanticipated two-phase flow equipment malfunction and failures can be traced to inadequate understanding of transient thermohydraulics than to the inability to predict steady state heat transfer or pressure drop. It is thus recommended that selected work in this area be carried forward in parallel with the other activities outlined here.

Two Component Flow and Heat Transfer

Greater concentration on the gas-solid particulate flow behavior and heat transfer to immersed and containing surfaces is required especially for particles in the 500μm – 3000μm size range at temperatures above about 300°C.

The thermal radiation component requires further work above about 500°C in these systems. Concurrent and countercurrent flowing systems as well as fluidized beds operating at gas velocities larger than those in bubbling beds need attention. Circulating or fast fluidized beds should be included as these are also under study for power generation systems involving coal combustion or gasification.

As the cost of some of the experiments involved in the above research areas becomes large, increasing effort is required to couple other high priority technologies into the experimental plans. Where erosion, corrosion or other material incompatibilities become possible limits to successful application, for example, measurements relating to such limiting phenomena may have to take place at the same time to obtain meaningful application data.

Periodic updates of new research needs should take place in organized form, either at special symposia or as part of the International Heat Transfer Conference. As new technologies or discoveries alter our priorities the benefit of their rapid communication becomes obvious.

REFERENCES

1. Rohsenow,W.M., Trans. ASME, 74, 969 (1952).

2. Forster,K. and Zuber N., AICHE Journal, 1, 531 (1955).

3. Kutateladze,S.S., "Fundamentals of Heat Transfer," 362, Academic Press Inc., New York, (1963).

4. Zuber,N., Tribus,M. and Westwater, "International Developments in Heat Transfer," pt. II, 230-236, ASME, (1961).

5. Chen,J., ASME Paper 63-HT-34, (1963).

6. Tong,L.S., "Boiling and Two-Phase Flow," 163, John Wiley & Sons Inc., New York, (1965).

7. McAdams, "Heat Transmission," 3rd ed. McGraw Hill Co., New York, (1954).

8. Chato,J.C., J. Am. Soc. Heating Refrig. Air Cond. Engrs., 52, Feb. 1962.

9. Rohsenow,W.M., Webber,J.H. and Ling, A.T., Trans. ASME, 78, 1637 (1956).

10. Altman M., Staub, F. W., Norris,R.H., 3d Nat. Heat Transfer Conf., ASME/AICHE, Storrs, Conn., Preprint 115, Aug. 1959.

11. Akers,W.W., Crosser,O.K. and Deans,H.A., Proc.
     2nd Nat. Heat Transfer Conf. ASME/AICHE,
     Aug. 1958.

12. Colburn,A.P., Hougen,O.A., Ind. Eng. Chem.,
     26, 1178 (1934).

13. Hawes,R.J., United Kingdom Atomic Energy
     Authority Report AEEW-R244, (1964).

14. Jepson G., Transactions of the Inst. of Chem.
     Eng., 41, 207-211, (1963).

15. Botterill,J.S.M., "Fluid Bed Heat Transfer,"
     Academic Press, New York, (1975).

16. Staub,F.W., ASME Jour. of Heat Transfer, 101,
     No. 3, 391-396, (1979).

17. Baskakov,A.P., Powder Technology, 8, (1973).

HIGH–TEMPERATURE HEAT TRANSFER: APPLICATIONS

Design, Construction, and Operation Experience of He-He Intermediate Heat Exchanger

H. UKIKUSA, T. NAKANISHI, T. NAKADA, M. ITOH, and K. WATANABE

Ishikawajima-Harima Heavy Industries Co., Ltd., Tokyo, Japan

ABSTRACT

In Japan, the research and development of direct steelmaking technology with the use of high temperature reducing gas, that is, nuclear steelmaking, has been taken up as one of the large-scale national projects by the Agency of Industrial Science and Technology of the Ministry of International Trade and Industry (MITI). In this project, Ishikawajima-Harima Heavy Industries Co., Ltd. (IHI), as a member of the Engineering Research Association of Nuclear Steelmaking (ERANS), is taking charge of the research and development of the high temperature heat exchanger, which is selected as one of the six major items. In this research and development, the 1.5 MWt He-He intermediate heat exchanger and the high temperature helium test loop were designed, constructed and operated. The intermediate heat exchanger has been operated satisfactorily without any structural trouble and has shown the estimated performance as expected. The experimental results on the heat transfer characteristics of the intermediate heat exchanger were confirmed to be in good agreement with the theoretical results based on the fundamental experiments.

NOMENCLATURE

A = heat transferring area, m^2/m $\bar{A}_i = A_i/A_1$
C_p = specific heat at constant pressure, J/kgK
d = diameter, m
Fi-j = shape factor from surface i to surface j
G = flow rate, kg/s
H = distance between radiation plates, m
h = heat transfer coefficient, $W/m^2 K$ $\bar{H}_i = h_i A_i/h_1 A_1$
M = radiation convection parameter = $\sigma T_{m2in}^3/h_1$
Nu = Nusselt number, $N_{u1} = h_1 d_1/\lambda$, $N_{u2} = h_2 d_2/\lambda$, $N_{uR} = h_R d_1/\lambda$
qi = heat flux, W/m^2 $Q_i = q_i/h_1 T_{m2in}$
Re = Reynolds number, $Re_1 = U_{m1} d_1/\nu$, $Re_2 = U_{m2} d_2/\nu$
Smi = free flow area, m^2
S = longitudinal pitch of tube bundle, m
T = temperature, K $\bar{T}_i = T_i/T_{m2in}$
U = overall heat transfer coefficient, $W/m^2 K$
x = distance along the axis of helical coil tube bundle from the inlet of sec. helium, m $X = x/x_0$
x_0 = height of helical coil tube bundle, m
ε = emissivity of surface
λ = thermal conductivity, W/mK
ρ = density, kg/m^3
σ = Stefan-Boltzmann constant, $W/m^2 K^4$

Suffix

c = for convection
i, j = surface wall, i = 1: outer surface of tube, i = 2: inner surface of tube, i = R: radiation plate
m_1 (m_1 in) = primary fluid (inlet)
m_2 (m_2 in) = secondary fluid (inlet)
Rin = incoming radiation
$Rout$ = outgoing radiation
tot = for convection and radiation

INTRODUCTION

The multipurpose utilization of nuclear heat is now attracting public attention from the view point of improving and strengthening the energy supply structure. By the multipurpose utilization of nuclear heat it means utilization of nuclear energy not for generating electric power but as a source of process heat for steelmaking, chemical industry, sea water desalination, district heating and so forth. Now that high temperature heat energy is required for industries such as steelmaking, chemicals etc., a high temperature helium gas-cooled reactor has been brought into focus of attention as one type of nuclear reactor for its application. In Japan, the research and development of direct steelmaking technology with the use of high temperature reducing gas, that is, nuclear steelmaking, has been taken up as one of the large-scale national project by the Agency of Industrial Science and Technology of the Ministry of International Trade and Industry. The six major items of the project are shown in Figure 1 and the research and development has been kept in steady progress. In May 1973, the Engineering Research Association of Nuclear Steelmaking was established as an organization, whose role was to implement and promote this project in the most efficient way. At present, fifteen companies and one association are the members affiliated with ERANS. Also at the Japan Atomic Energy Research Institute, the 50 MWt experimental multipurpose high temperature gas-cooled reactor project is under research and development aiming for 1000°C helium temperature and pressure of 3.92 MPa (40 kgf/cm²) at the reactor outlet. In the nuclear steelmaking system, the reducing gas is produced with the use of nuclear thermal energy obtained from a nuclear reactor, and iron ore pellets are reduced through direct reduction by means of the high temperature reducing gas.

Ishikawajima-Harima Heavy Industries Co., Ltd., as one member of ERANS, has been responsible for one of the research themes adopted for this project, that is, "Research and Development of High Temperature Heat Exchanger" and has been involved and been proceeding with studies and experiments from the very beginning [1,2,3]. And also in this project, we designed, constructed and operated the 1.5 MWt He-He intermediate heat exchanger [3,4]. The inlet primary helium temperature and the outlet secondary helium temperature of the intermediate heat exchanger are designed to be 1000°C and 925°C respectively. This paper describes the outline and progress of the project, the

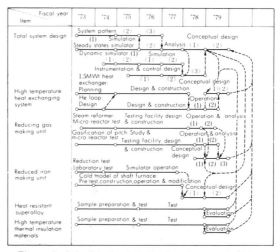

Figure 1. Schedule for R & D of nuclear steelmaking

results of the intermediate heat exchanger.

RESEARCH AND DEVELOPMENT OF HIGH TEMPERATURE HEAT EXCHANGER

Purpose of research and development

One of the important factors to be considered for the utilization of nuclear thermal energy as process heat is to prevent the fission product from mixing into the processed products. For this reason, in Japan, safety considerations have led to the adoption of the system generally known as the indirect heat exchanging system, which is characterized by an intermediate heat exchanging loop connecting the nuclear reactor with the utilization systems. Research and development work on the high temperature heat exchanger is directed to the subject matters of the intermediate heat exchanging system and the intermediate heat exchanger, which play important roles in the whole process, as mentioned previously. The conditions required for the primary helium, which is the cooling medium for the high temperature gas-cooled reactor, are described as 1000°C in temperature and 3.92 MPa (40 kgf/cm²) in pressure at the outlet of the reactor, while the conditions required for the secondary helium at the outlet of the intermediate heat exchanger are 925°C in temperature and 4.41 MPa (45 kgf/cm²) in pressure. This means that relatively high temperature and high pressure are required for the system.

The purpose of the present research and development project is to establish design method, safety evaluation technique, and safety-securing measures to be applied to an intermediate heat exchanger and to the overall heat exchanging system, which will satisfy those strict operating conditions as mentioned above. Research and development works on the high temperature heat exchanger have been carried out on each and every item listed in Figure 2. The objectives of the research and development are divided into two categories: the first concerning the development of the high temperature heat exchanger itself, and the other, the heat exchanging system including thermal insulation piping,

helium circulator, etc. In order to develop an ideal type of heat exchanger, we have established the design criteria and performed a number of experimental tests on the characteristics of thermal insulation, augmentation of heat transfer, degradation of heat transfer by high heat flux, vibration of heat transfer tubes and so on. In addition, safety problems including hydrogen permeation and research on wear and self-welding have been studied to give a sound basis to the desing and construction of a high temperature heat exchanging system. On the basis of the results obtained from these basic researches, the design and construction of the 1.5 MWt He-He intermediate heat exchanger had been brought to completion by the end of 1977. Further, the design, fabrication and installation of the high temperature helium test loop, which is required for the operation tests of the intermediate heat exchanger and the other high temperature heat exchanging equipments, had satisfactorily been completed and the operation of the test loop had been conducted in 1978 and 1979.

High temperature helium test loop

The high temperature helium test loop was designed and constructed for the purpose of carrying out examinations of overall functions and performance of the intermediate heat exchanger and the high temperature heat exchange system, and also for the purpose of ascertaining the functions of the individual equipments, which were the main objectives of development under the "Research and Development of High Temperature Heat Exchanger" project. To be more specific, the individual equipments were designed and constructed by incorporating the results of various kinds of basic researches, and thereafter the operating tests have been conducted on those equipments integrated into the test loop. The flow diagram and the major plant parameters of the high temperature test loop are shown in Figure 3 and Table 1 respectively. The arrangement of the test loop is also shown in Figure 4. The circuit system of the high temperature helium test loop consists of a primary helium circuit, secondary helium circuit and heat utilization system in its main circuit and the auxiliary system. This test loop has been put into operation since the latter half of 1978, and in December 1979 all of the operation tests,

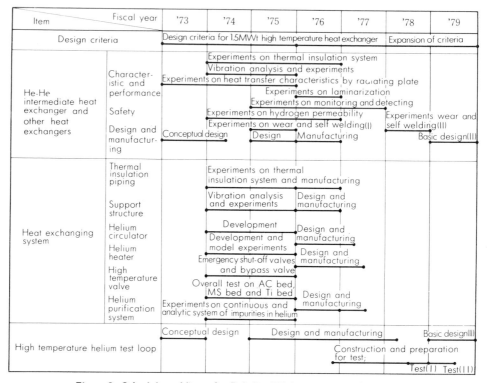

Figure 2. Schedule and items for R & D of high temperature heat exchanger

which were initially scheduled, were completely finished. Furthermore, the operation of the test loop is continued this year for the purpose of the endurance test of the intermediate heat exchanger.

DESIGN AND CONSTRUCTION OF He-He INTERMEDIATE HEAT EXCHANGER

The high temperature heat exchanger which has been taken up under the present project is placed under research and development for the purpose of obtaining basic data required for the design and construction of a high temperature heat exchanger to be used in an experimental plant for direct steelmaking. This experimental plant will be connected into a 50 MWt multipurpose high temperature helium gas-cooled reactor. As was explained previously, the He-He intermediate heat exchanger is to be used under conditions of extremely high temperature, specifically 1000°C in terms of primary helium temperature and 925°C in the secondary helium. In order to design a high temperature heat exchanger which will meet these strict requirements, as many results obtained from basic researches as possible have been incorporated into the design studies carried out from a technical standpoint. As regards the type of intermediate heat exchanger, a helical coil counterflow type has been adapted upon comparison and evaluation made on several different types such as straight tube type, U-tube type and others. A Helical coil type is considered to have advantages from the viewpoint of its size, heating surface, axial symmetry of temperature and flow distribution, feasibility of scale-up and adequacy of height and diameter. The general arrangement and the principal specifications of the He-He intermediate heat exchanger are shown in Figure 5

and Table 2 respectively. The high temperature primary helium is led at 1000°C into the heat exchanger from the inlet nozzle located in the lowest part of the center of the heat exchanger, and then it heats up the secondary helium as it rises in the helical coil tube bundle. The primary helium, after transferring heat to the secondary helium, reverses its flow-direction at the upper part of tube bundle, flows down the outer annular flow path provided between the inner duct and the inner surface of the shell, and is led to the outside of the intermediate heat exchanger through the outlet nozzle located at the bottom of the heat exchanger. The secondary helium is conducted to the helical coil tube bundle through the upper connecting tube from the secondary helium inlet nozzle located at the uppermost part of the shell of the heat exchanger. The helical coil tube bundle is composed of a total of 30 tubes with three layers of helical coils. The secondary helium is heated up to 925°C flow-

Table 1. Plant parameters of high temperature helium test loop

System	Parameter	Value
Primary helium system	Primary helium flow rate	0.483 kg/s
	Temp. entering IHX	1,000°C
	Press. entering IHX	3.92 MPa
	Temp. leaving He circ.	400°C
	Heat transferred, IHX	1558 kW
	Pressure drop	0.20 MPa
Secondary helium system	Secondry helium flow rate	0.494 kg/s
	Temp. leaving IHX	925°C
	Press. leaving IHX	4.41 MPa
	Temp. entering SH	910°C
	Temp. leaving SH	750°C
	Temp. entering RGH	910°C
	Temp. leaving RGH	650°C
	Heat transferred, steam superheater	400 kW
	Heat transferred, reducing gas heater	250 kW
	Heat transferred, steam generator	1132 kW
	Pressure drop	0.24 MPa
Water steam system	Steam flow, steam superheater	0.429 kg/s
	Steam generator	0.446 kg/s
	Steam pressure, superheater outlet	3.92 MPa
	Steam generator outlet	4.12 MPa
	Steam temp., superheater outlet	850°C
	Steam generator outlet	450°C
Reducing gas system	Reducing gas composition	H₂ : 90% CO : 10%
	Flow	0.047 kg/s (0.010 kmol/s)
	Temp. leaving RGH	850°C

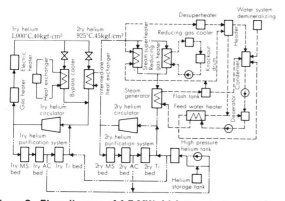

Figure 3. Flow diagram of 1.5 MWt high temperature test loop

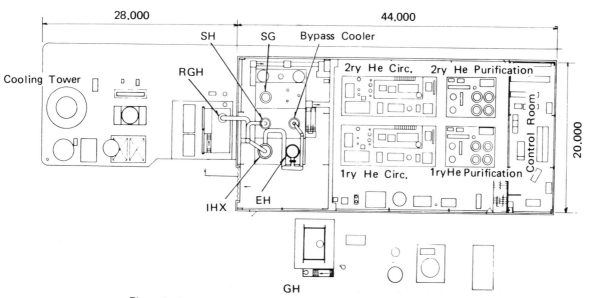

Figure 4. Arrangement of 1.5 MWt high temperature helium test loop

ing down inside of the helically coiled tube bundle by the primary helium, collected into the manifold located at the lower part of the centerpipe and flows upward through the centerpipe to go out from the outlet nozzle provided at the top of the shell.

As shown in Figure 5, the component parts to be exposed to extremely high temperature are arranged along the axis of the heat exchanger and the overall structure is designed so as to form the axial symmetry for the purpose of decreasing the thermal stress and thermal expansion uniformly. As for the selection of the material to be used for the heat transfer tube in the high temperature section, one kind of heat-resistant alloy was decided to be selected from a list of seven such kinds of alloys which were designated as "candidates". Inconel 617 was the one chosen as a result of detailed examination and deliberations. Incoloy 800 and 2 1/4Cr-1 Mo steel are also used for the medium and low temperature zones depending on the required temperature of each tube.

In the intermediate heat exchanger, the heat transfer augmentation by means of radiation is applied. As the radiation plates, the thin cylindrical plates are installed between the three helical coil tube layers. Also these radiation plates inserted between the layers of tube bundles has the effect of restraining the flow induced vibrations [5]. The external appearance of radiation plates and helical coil tube bundle are shown in Photo 1.

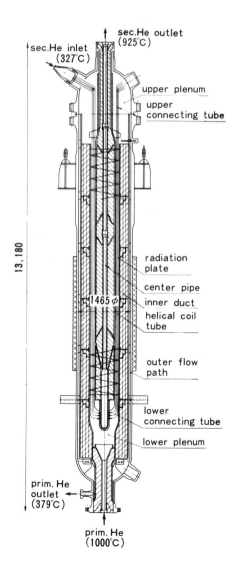

General arrangement of He-He intermediate heat exchanger

For the construction of the He-He intermediate heat exchanger, technical investigations and preparations were conducted with particular attention on the fabrication of the heat transfer tube bundle. The major investigations are:
  - examination of the welding method to be applied to the welding joints of the same material as well as of different materials,
  - trial construction of a mock-up in the scale of an actual helical coil tube bundle, and
  - trial fabrication of a supporting structure for the heating tube bundle.

Moreover for the operation test of this intermediate heat exchanger, a total of 239 kinds of temperature, pressure and flow rate were measured, and these data were acquired and processed by the mini computer system.

Table 2. Specification of He-He intermediate heat exchanger

Type		Helical coil, one-through type
No. of unit		1
Tube side	Fluid	Helium
	Flow rate	0.494 kg/s
	Temp. ent./leav.	327/925°C
	Press.	4.41 MPa
Shell side	Fluid	Helium
	Flow rate	0.483 kg/s
	Temp. ent./leav.	1,000/379°C
	Press.	3.92 MPa
Heat transferred		1558 kW
Tube	Material	Inconel 617, Incoloy 800, STBA24
	Size	25.4 mm O.D. x 4.0 mm t
	Number	30
	Pitch	45 mm (trans.) x 40 mm (long.)
Helical coil	No. of layer	3
	Dia. max./min.	780/600 mm
Shell	Material	2-1/4 Cr. - 1 Mo
	Outside dia.	1,555 mm
	Height	13.180 m

Photo 1. Helically coiled tubes and radiation plate

EXPERIMENTAL RESULTS

Outline

The operation tests of 1.5 MWt He-He intermediate heat exchanger were conducted to confirm the performance of IHX at the rated condition with the inlet primary helium temperature of 1000°C and the performance at the partial load conditions. The tests also confirmed the efficiencies of thermal insulation of the internal insulation structure and the function and stability of the components by extracting and inspecting the helical coil tube bundle. The operation tests were conducted from 1978 to 1979 for two years, and all of the scheduled experiments were completed as expected in December 1979. Moreover the operation of the test loop is continued this year for the endurance test of the intermediate heat exchanger. The total operation time consumed and start-up/shut-down frequencies as of the end of 1979 are as follows:

Helium circulation operation : 5059 hours
Heat up by gas heater : 4430 hours
High temperature test
operation : 2370 hours
(primary helium temperature
at IHX inlet above 900°C)

Full load test operation : 1172 hours
(primary helium temperature
at IHX inlet 1000°C)
Frequencies of start-up and shut-down : 31

Heat exchanger performance

When the primary helium temperature entering the intermediate heat exchanger was 1000°C, the desired secondary helium temperature of 925°C was attained at the outlet of the heat exchanger with a small adjustment in the amount of primary helium flow rate. According to the full load operation condition, shown in Table 2, the secondary helium temperature leaving the intermediate heat exchanger was slightly lower than the designed value of 925°C. So the flow rate of the primary helium was increased by 5% of the estimated value in order to gain the desired secondary helium temperature. The difference between the experimental result and the design value of the primary helium flow rate is caused by the decrease of the heat transfer coefficient outside of the tube and the increase of the heat loss from the heat exchanger shell surface. The decrease of the heat transfer coefficient is mainly due to the eccentricity of the heat transfer tubes from the center between the radiation plates, and the other manufacturing tolerances of the tube bundle. The heat loss from the heat exchanger shell surface is less than 5% of the heat transferred. The heat received by the secondary helium coincides within the range of 1.2% error with the heat transferred from the primary helium from which the heat loss is subtracted. At the above mentioned helium flow condition, the heat exchanger performance was examined varying the inlet primary helium temperature from 700°C to 1000°C. Also at 75% helium flow rate, the experiments were conducted. The performance test results in the case of 100% flow rate are shown in Figure 6. The abscissa shows the primary helium temperature entering the intermediate heat exchanger. The experimental results of the temperature and pressure are in good coincidence with the estimation. However the primary helium flow rate is about 5% higher than the predicted value as mentioned before.

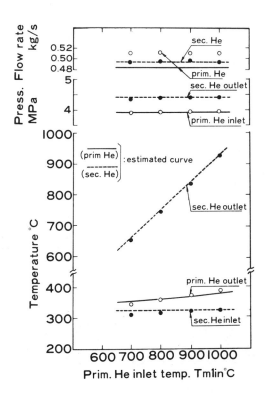

Figure 6. Experimental results of He-He intermediate heat exchanger performance (100% flow)

Axial temperature distribution in helical coil tube bundle

Experimental result of axial temperature distribution of primary helium, radiation plate, tube wall and secondary helium at the helical coil tube bundle is shown in Figure 7, when the primary and secondary helium temperature entering the heat exchanger is 999°C and 325°C respectively and the primary and secondary helium flow rate is 0.510 kg/s and 0.494 kg/s. The measured primary helium temperatures are compensated for the radiation from the surrounding structures [6] and the tube wall temperature is measured on the outer surface.

Secondary helium temperatures of the helical coil tube bundle are calculated from the measured temperatures of primary helium and the secondary helium inside of the centerpipe and the heat loss from the heat exchanger surface. The solid lines are the analytical results which are explained afterward. The temperature distribution outside of tubes becomes lower by the following order: primary helium, radiation plate and heat transfer tube. From this fact, it is known that the heat of the primary helium is transferred directly to the heat transfer tube by convection, and also to the radiation plate. The heat transfer from the primary helium to the heat transfer tube is affirmed to be augmented by the radiation from the radiation plate to the heat transfer tube wall.

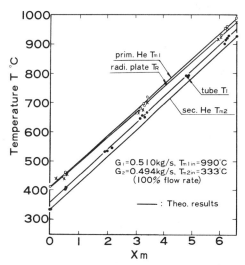

Figure 7. Distribution of temperature in helically coiled tube bundle of He-He intermediate heat exchanger

Heat transfer characteristics in helical coil tube bundle

Overall heat transfer coefficient Utot in the helically coiled tube bundle is calculated by Eq. (1) using the logarithmic mean temperature difference. Here, q is the average heat flux for the heat received by the secondary helium based on the outer surface area of the tubes.

$$U_{tot} = q/\Delta T_m \qquad (1)$$

The experimental result of Utot increases from 262 W/m² K to 280 W/m² K, when the inlet primary helium T_{m1} in is varied from 700°C to 900°C at 100% flow rate. It is the characteristic feature of the heat transfer augmentation by radiation that the augmentation becomes more effective when the fluid temperature is higher. Also this heat transfer augmentation is more effective when the fluid flow rate is small. For example the heat transfer augmentation effect at 75% flow rate is larger than that at 100% flow rate. This is due to the fact that the thermal resistance by radiation between the radiation plates and the heat transfer tubes is independent of the thermal resistance by convection. So when the fluid flow rate decreases, the convective heat transfer coefficient decreases, and relatively the heat transfer augmentation by radiation becomes more effective.

Thermal insulation characteristics

The inner duct in the intermediate heat exchanger is the boundary wall between the primary helium in the helical coil zone and the low temperature helium in the outer annular flow path. In order to affirm the performance of the thermal insulation structure, the temperature distribution in the inner duct was measured. The radial temperature distribution in the inner duct is shown in Figure 8. The solid line is the results calculated from the thermal conductivity of the thermal insulation material in helium gas which is gained from the basic research of the project. As shown in Figure 8, the calculated result is in good agreement with the experimental results. It is affirmed that concerning the thermal insulation structure of the inner duct, the short pass or convection of the helium gas in the thermal insulation material has not occurred and the fabrication of the thermal insulation structure was sufficient.

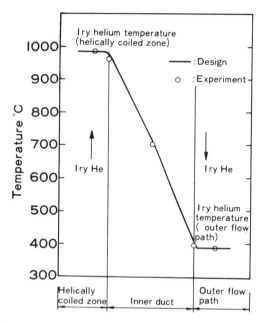

Figure 8. Radial temperature distribution in inner duct of He-He intermediate heat exchanger

Photo 2. Releasing operation of helically coiled tube bundle of He-He intermediate heat exchanger

Extraction and inspection of helical coil tube bundle

In 1978 the helical coil tube bundle was extracted from the shell for inspection of the internals. By the inspection of the internal structures, it was confirmed that all of the components were in good condition the same as the initial fabrication. The helical coil tube bundle which was released by the truck crane is shown in Photo 2.

THEORETICAL ANALYSIS

The convection and radiation heat transfer characteristics of the heat exchanger, in which the radiation plates are inserted between the tube bundle, are analyzed by Mori et al [7]. Here, applying the above mentioned analysis we studied the heat transfer characteristics in a helical coil tube bundle. The flow model used in this analysis is shown in Figure 9. The following assumptions are introduced in the analysis:

(1) The fluid is a non-radiating gas and tubes and radiation plates are gray bodies.
(2) Heat transfer by radiation between tubes and radiation plates is approximated in a one dimensional model and the temperature changes in the primary fluid flow direction.
(3) Heat transfer is considered at one layer of the tube bundle and the surrounding two radiation plates, and the heat loss is transferred through the radiation plates outward.

Following the above conditions, the energy balance equations of the primary helium (suffix: $\dot{m}_1$), the secondary helium (m_2) the outer surface of the tube (i=1), the radiation plates (i=R) and the inner surface of the tube (i=2) are given as follows:

$$(\rho u C_P)_{m1} S_{m1} \frac{dT_{m1}}{dx} = \sum_{i=1,R} A_i h_i (T_{m1} - T_i) \quad (2)$$

$$(\rho u C_P)_{m2} S_{m2} \frac{dT_{m2}}{dx} = A_2 h_2 (T_2 - T_{m2}) \quad (3)$$

$$q_{R\,out\,i} = -q_i + q_{R\,in\,i} + q_{ci} \quad (i=1, R) \quad (4)$$

$$\frac{2\pi\lambda}{d_1 \ln(d_1/d_2)} (T_2 - T_1) + A_2 h_2 (T_2 - T_{m2}) = 0 \quad (5)$$

boundary conditions:

$$\left. \begin{array}{l} x = 0 \; : \; T_{m2} = T_{m2\,in} \\ x = x_0 \; : \; T_{m1} = T_{m1\,n} \end{array} \right\} \quad (6)$$

where T_{mi}, T_i, S_{mi}, A_i are the temperature of the fluid m_i, temperature of wall i, free flow area for the fluid m_i and the heating surface of the tubes respectively. Also q_{ci}, $q_{R\,out\,i}$, $q_{R\,in\,i}$ are convection heat flux ($q_{ci} = h_i (T_{m1} - T_i)$), radiation heat flux leaving the wall i and radiation heat flux from the surrounding walls respectively. q_1 is the heat flux transferred through tube wall and defined as: $q_1 = h_2 A_2 (T_2 - T_{m2})/A_1$. q_R is the heat flux for the heat loss through the radiation plates and calculated from the temperature difference between the primary helium in the helically coiled tube bundle and that in the outer annular

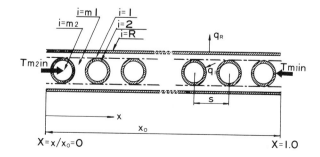

Figure 9. Flow model of helically coiled tube bundle

channel provided between the inner duct and the inner surface of the shell.

The q_{Routi} and q_{Rini} are expressed as follows by definition:

$$q_{Rout\,i} = \varepsilon_i \sigma T_i^4 + (1 - \varepsilon_i) q_{Rin\,i} \tag{7}$$

$$q_{Rin\,i} = \sum_{j=1,R} q_{Rout\,j} F_{i-j} \tag{8}$$

where F_{i-j} is the shape factor from wall i to wall j, and is given as follows:

$$\left. \begin{aligned} F_{1-R} &= 1 - 2 \left\{ \sqrt{(S/d_1)^2 - 1} + \sin^{-1}(d_1/s) - S/d_1 \right\} / \pi \\ F_{R-1} &= F_{1-R}/\bar{A} \end{aligned} \right\} \tag{9}$$

By Eqs. (7) and (8), q_{Routi} is expressed as follows:

$$\begin{aligned} q_{Rout\,i} &= \sigma T_i^4 + (1 - \varepsilon_i)(q_i - q_{ci})/\varepsilon_i \\ &= \varepsilon_i \sigma T_i^4 + (1 - \varepsilon_i) \sum_j q_{Rout\,j} F_{i-j} \quad (i = 1, R) \end{aligned} \tag{10}$$

Substituting Eq. (10) in Eq. (14), the following equation is obtained:

$$\sigma T_i^4 + \frac{q_i - q_{ci}}{\varepsilon_i} - \sum_{j=1,R} \left[\sigma T_j^4 + \frac{1 - \varepsilon_j}{\varepsilon_j}(q_j - q_{cj}) \right] F_{i-j} = 0 \tag{11}$$

$$(i = 1, R)$$

Eqs. (2), (3), (11), (5) and (6) are rearranged with the non-dimensional parameters, Eq. (17), as follows:

$$\frac{d\bar{T}_{m1}}{dX} = \frac{A_1}{S_{m1}} \frac{N_{u1}}{R_{e1} P_r} \left[\bar{T}_{m1} - \bar{T}_1 + \bar{H}_R (\bar{T}_{m1} - \bar{T}_R) \right] \tag{12}$$

$$\frac{d\bar{T}_{m2}}{dX} = \frac{A_2}{S_{m2}} \frac{N_{u2}}{R_{e2} P_r} (\bar{T}_2 - \bar{T}_{m2}) \tag{13}$$

$$M \varepsilon_i T_i^4 - \frac{\bar{H}_i}{\bar{A}_i}(\bar{T}_{mi} - \bar{T}_i) + Q_i$$

$$- \varepsilon_i \sum_{j=1,R} \left[M \bar{T}_j^4 + \left(\frac{1}{\varepsilon_j} - 1 \right) \left\{ Q_j - \frac{\bar{H}_j}{\bar{A}_j}(\bar{T}_{mj} - \bar{T}_j) \right\} \right] F_{i-j} = 0 \tag{14}$$

$$(i = 1, R)$$

$$\bar{H}_2(\bar{T}_2 - \bar{T}_{m2}) + \lambda^* (\bar{T}_2 - \bar{T}_1) = 0 \tag{15}$$

boundary conditions:

$$\left. \begin{aligned} X &= 0 : \bar{T}_{m2} = 1.0 \\ X &= 1 : \bar{T}_{m1} = \bar{T}_{m1\,in} \end{aligned} \right\} \tag{16}$$

$$\left. \begin{aligned} \bar{T}_i &= \frac{T_i}{T_{m2\,in}}, \quad M = \frac{\sigma T_{m2\,in}^3}{h_1}, \quad \lambda^* = \frac{2\lambda}{h_1 d_1} \frac{1}{\ln(d_1/d_2)} \\ Q_1 &= \bar{H}_2(\bar{T}_2 - \bar{T}_{m2}) \;, \quad Q_R = q_R/h_1 T_{m2\,in} \end{aligned} \right\} \tag{17}$$

As for the convective heat transfer coefficient for the cross flow tube bundles and for the radiation plates, the experimental equations (18) and (19) were applied with the correction of Prandtl number. These two equations were obtained for the crossflow outside of tubes for air by Mori et al [7]. And for the convective heat transfer coefficient of the inner surface of helically coiled tube, Mori and Nakayama's equation [8] was used. The coefficient f in Eq. (18) is the correction factor for the eccentricity of the tube from the center between the radiation

plates, and is obtained from the experimental results of the basic research with the use of inspection results of the released tube bundle.

$$N_{u1} = 0.050 R_{e1}^{0.8} f \quad (P_r = 0.71) \tag{18}$$

$$N_{uR} = 0.233 R_{e1}^{0.54} (H/d_1 - 1)^{-0.47} (S/d_1)^{0.5} \tag{19}$$

$$(P_r = 0.71)$$

$$N_{u2} = 0.0228 R_{e2}^{5/6} P_r^{0.4} (d_2/D)^{1/12} \cdot$$

$$\left[1.0 + 0.0732 / \{ R_{e2}(d_2/D)^{2.5} \}^{1/6} \right] \tag{20}$$

By Eqs. (12) - (16), $\bar{T}_{m1}$, $\bar{T}_{m2}$, $\bar{T}_1$, $\bar{T}_2$ and $\bar{T}_R$ are obtained by numerical iteration method. The finite difference in X direction and convergent condition of numerical calculation was chosen to be 0.02 and 0.5×10^{-4} respectively. The physical properties of helium were assumed to be dependent upon temperature. The analytical results of temperature distribution in the axial direction of the helical coil tube bundle are shown by solid lines in Figure 7. It is shown that calculation results of the temperature of primary helium, secondary helium, outer surface of tube and radiation plate coincide with the experimental results. The overall heat transfer coefficient for the analytical results obtained by the above mentioned method was in good agreement with the experimental results with 3% error. It is affirmed that this analytical method expresses accurately the heat transfer characteristics of the high temperature heat exchanger provided with the radiation plates.

CONCLUSIONS

Following are the summary of our studies and test results on the 1.5 MWt He-He intermediate heat exchanger:

(1) When the inlet temperature of the primary and secondary helium was 1000°C and 325°C respectively, and helium flow rate was 100% ($G_1 = 0.510$ kg/s, $G_2 = 0.494$ kg/s), the heat of 1.53 MWt was transferred and the outlet secondary helium temperature of 925°C was attained as designed.

(2) Extracting the helical coil tube bundle from the shell and inspecting the internal structures, it is affirmed that the internal structure has been in good condition as same as the initial fabrication and the structural design has been sufficient.

(3) It is affirmed that the heat transfer characteristics of the helical coil tube bundle provided with the radiation plate is improved at the high temperature zone and the overall heat transfer coefficient is increased. Moreover the radiation plate functions effectively and safely at temperature of 1000°C.

(4) The heat transfer characteristics of the high temperature heat exchanger provided with the radiation plate are expressed accurately by the simultaneous analysis of radiation and convection in one dimensional model.

Starting on the basic research in 1973, we have been devoted to the research and development of this project. We expect that our experiences up to now will be effectively utilized for the fulfilment of the overall project, but also our efforts and endeavors will contribute to the extended application of nuclear heat energy in the world.

ACKNOWLEDGEMENT

The authors would like to express sincere gratitude to Dr. Yasuo Mori, Professor at the Tokyo Institute of Technology, for the exhaustive and painstaking guidance which he has so generously extended to the research group throughout the entire course of the research and development activities.

REFERENCES

1. Nakanishi, T., NUCLEX 78, A4 (1978).
2. Itoh, M., IHI Engineering Review, Vol. 12, No. 2, 1-12 (1979).
3. Mori, Y., ASME Paper 80-HT-39 (1980).
4. Itoh, M., Priprint at IAEA Specialists' Meeting (1979).
5. Negishi, M., et al., Annual Meeting Jap. Soc. Mech. Eng. (in Japanese), No. 750-16, 97-110 (1975).
6. Nakada, T., Watanabe, K., et al., IHI Engineering Review (in Japanese), Vol. 15, No. 1, 33-46 (1975).
7. Mori, Y., Watanabe, K., et al., ASME Paper 76-HT-3 (1976).
8. Mori, Y., Nakayama, W., Int. J. Heat Mass Transfer, 10, 681-695 (1967).

High Temperature Heat Exchange: Nuclear Process Heat Applications

D. L. VRABLE
General Atomic, San Diego, California, U.S.A.

ABSTRACT

The high-temperature gas-cooled reactor (HTGR) is
unique among nuclear systems in its ability to
serve as a heat source up to 1000°C, high enough
for many industrial uses of energy, including
steelmaking, coal liquefaction and gasification,
and in the future, large scale thermochemical production of hydrogen. The HTGR process heat applications have been studied at General Atomic for
many years in programs sponsored by the Department
of Energy (DOE) and process heat user industries
(steel, chemical, and gas).

The first generation of HTGR was designed to take
full advantage of modern steam plant technology, as
exemplified by the Peach Bottom and Fort St. Vrain
reactors. The study of advanced HTGR process heat
variants [also termed very high temperature reactors (VHTR)] is aimed at further exploiting its
unique high-temperature capability.

The unique element of the HTGR system is the high-
temperature operation and the need for heat exchanger equipment to transfer nuclear heat from the
reactor to the process application. This paper
discusses the potential applications of the HTGR in
both synthetic fuel production and nuclear steel
making and presents the design considerations for
the high-temperature heat exchanger equipment.

INTRODUCTION

Energy consumption in the industrialized nations
has spiraled upward during the past decades. As
Table 1 shows, petroleum and natural gas meet more
than 75% of the U. S. energy demands (Ref. 1). However, petroleum and natural gas comprise only 7.9 %
of the estimated total recoverable U.S. fossil fuel
resources.

As Table 2 shows, coal comprises more than 75% of
the estimated total remaining recoverable U.S.
fossil fuel. However, coal accounts for only 18%
of the U.S. energy consumption.

TABLE 1
U.S. ENERGY CONSUMPTION
(1979)

Resources	J x 10^{18}	%
Petroleum	38.8	46
Natural gas	26.1	31
Coal	15.2	18
Hydropower	3.4	4
Nuclear	0.8	1
	84.3	100

TABLE 2
ESTIMATED TOTAL REMAINING RECOVERABLE RESOURCES IN THE U.S. (1979)

Resources	J x 10^{18}	%
Coal	36,300	77.3
Shale oil	6,700	14.4
Crude oil	2,100	4.5
Natural gas	1,600	3.4
Natural gas liquids	200	0.4
	46,900	100.0

Natural gas, which provides more than 30% of the
U.S. energy demands, cannot be expected to maintain
its hold over such a large share of U.S. energy
production. Limited natural gas availability will
require using other energy resources. Since oil is
the predominant fuel used in the U.S., it seems a
logical alternative to assume most of the burden.
However, domestic resources are inadequate to meet
the increased demand, and the U.S. urgently needs to
move away from its heavy reliance on foreign oil.
Coal and nuclear power remain among the proven energy sources to meet U.S. energy requirements. Coal,
the most abundant remaining fossil fuel in the
U.S., forms the basis for the emerging synfuel
programs.

Japan faces the same dilemma of increasing its
energy consumption and relying more heavily on
imported oil. The steel industry, one of the
largest Japanese energy consumers, uses more than
20% of the national energy. The Japanese steel
industry ranks first in coal, gas, and electricity
consumption and third in oil consumption.

Thus, a steel industry switch to alternative energy
sources would be particularly meaningful. Japan
has directed effort toward designing and developing
an experimental 50-MW(t) VHTR with a helium outlet
temperature of 1000°C (1832°F). Both governmental
and industrial suppliers are participating in this
national project. The Japan Atomic Energy Research
Institute (JAERI) is scheduled to start construction on the 50-MW(t) VHTR in 1984.
The energy future of both the U.S. and Japan will
depend on improved energy efficiency, conservation, and development of new energy sources.

HIGH-TEMPERATURE HTGR APPLICATIONS

The HTGR offers a unique heat source for either
power generation or process heat production, since
its operating temperature is significantly higher
than that of other nuclear reactor types. A key
feature of the HTGR is the nuclear reactor core,
which uses helium as the primary coolant, has
ceramic-coated fuel particles containing uranium

and thorium, and employs graphite as the moderator and structural material. The helium heat transfer medium used as the primary coolant is chemically inert and remains in a gaseous phase under all possible operating conditions. The entire primary coolant system of the HTGR is contained in a prestressed concrete reactor vessel (PCRV), which provides the necessary biological shielding and pressure containment. The aforementioned elements of the HTGR offer unprecedented safety characteristics.

The first generation of HTGRs was designed to take full advantage of modern steam plant technology, as exemplified by the Peach Bottom and Fort St. Vrain reactors (Ref. 2). The study of advanced HTGRs is aimed at further exploitation of the unique high-temperature capability made possible by the all-graphite core, inert gas coolant, and ceramic-coated fuel particles.

Process Heat Plant Concept

The high-temperature heat available from the HTGR makes it suitable for many process applications. The high-grade heat can be used to produce reducing gas, hydrogen, and synfuel, using coal, lignite, residual oil, or oil shale as the carbon source. The HTGR can also serve as the heat source for thermochemical water-splitting processes to produce hydrogen without carbon, and it may be useful in large-scale energy transport and storage systems for industrial or utility applications.

The HTGR process heat plant is envisaged as a nuclear-chemical process whose product is hydrogen (or a mixture of hydrogen and carbon monoxide) generated by steam reforming of a light hydrocarbon mixture. The reactor energy is transported to an externally located process plant by an intermediate heat transport loop. The intermediate loop provides an additional boundary between the nuclear heat source and the process, thereby enhancing plant safety and offering considerable flexibility for alternate applications.

References 3 and 4 reported a process heat plant based on a reactor core thermal rating of 842 MW(t). This value corresponds to the HTGR steam cycle constructed and operating at Fort St. Vrain for the Public Service Company of Colorado. Figure 1 shows an isometric view of the nuclear heat system for the HTGR-process heat (HTGR-PH) plant. The system consists of two loops, each embodying an intermediate heat exchanger (IHX) and primary system circulator. Figure 2 shows a loop diagram

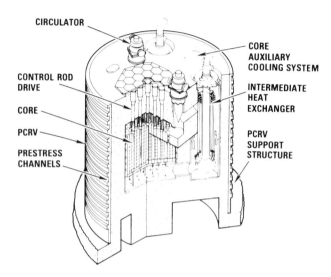

Figure 1 Nuclear heat source arrangement for 842-MW(t) HTGR-PH plant

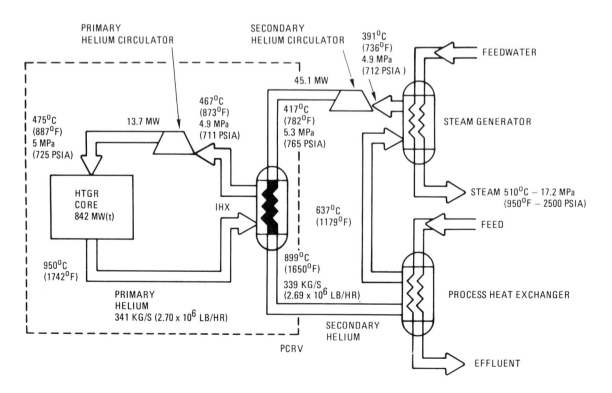

Figure 2 Cycle diagram for 950°C HTGR-PH plant

160

for the HTGR-PH plant. This cycle is based on a reactor outlet temperature of 950°C (1742°F), but studies are under way for an 850°C (1562°F) system. Figure 2 shows that the two secondary loops are supplied with energy from the IHXs and that each loop consists of a reformer, steam generator, and helium circulator connected in series. The HTGR-PH plant has several high-temperature heat exchangers, but this paper concentrates on various aspects of the most critical of these, namely the helium-to-helium IHX.

Integration of Nuclear Process Heat with Coal Liquefaction Processes

The vast U.S. coal resources will produce synthetic fuel via fundamentally two processes: (1) coal liquefaction and (2) coal gasification. The most readily marketable synthetic fuel would directly substitute for current liquid fuels (e.g., diesel fuel, jet fuel, or gasoline). These fuels can be made from coal, basically by adding hydrogen.

Production of liquid fuel from coal has two routes: (1) direct and (2) indirect. The direct route liquefies coal, then upgrades the basic product by hydrotreating. The Solvent Refined Coal (SRC II), Exxon Donor Solvent (EDS) and H-Coal processes are under development. The indirect route gasifies coal to produce a mixture of hydrogen, carbon monoxide, and methane, then uses this gas to synthesize, or hydrogenate, one or more liquid fuels.

Either liquefaction route can be used with the HTGR heat sources to produce synfuel from coal. The HTGR is integrated into the process to supply the utilities (steam and electricity) and process heat (reforming). Figure 3 illustrates integrating the HTGR-PH with a direct liquefaction process.

Table 3 compares the product yields with and without a nuclear heat source. Basically, adding the nuclear reactor can decrease the coal requirements by 33% per process or increase the product output by 50% for the same coal requirements.

Of the two routes for liquefying coal, only the indirect route is commercially used. This process of breaking down, then rebuilding hydrocarbons is the least efficient coal liquefaction route. Direct liquefaction, or hydrogenation, will probably become the major source of liquid fuels from coal.

Integration of Nuclear Process Heat With Steelmaking

Steelmaking would use nuclear energy as process heat to manufacture reducing agents and to supply heat at reaction temperature.

In the current refining process, iron ore, coke, and limestone are charged into a blast furnace. Then air blown into the furnace under high pressure partially burns coke, producing CO, which with

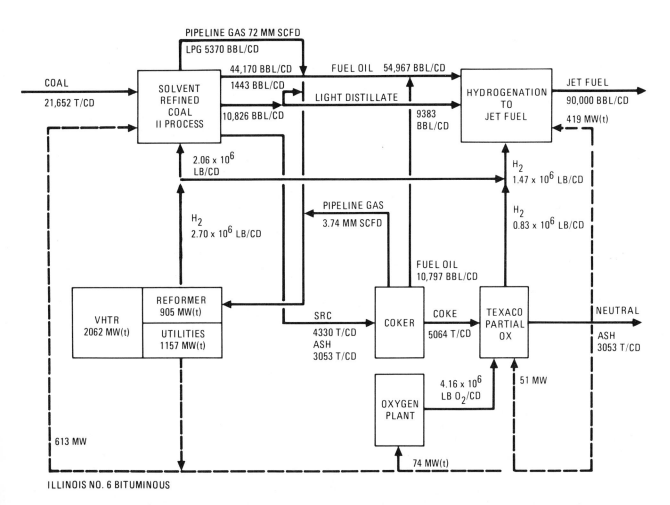

Figure 3 Coal to jet fuel, using VHTR reforming for H_2 (prime) utilities

TABLE 3
COMPARISON OF NUCLEAR AND NON-NUCLEAR COAL
LIQUEFACTION PROCESSES

Process	Indirect		Direct	
	Conventional Lurgi-Fischer-Tropsch	Nuclear Lurgi-Fischer-Tropsch nuclear reforming	Conventional SRC-II Conventional	Nuclear SRC-II nuclear
Coal feed, tons/day	46,400	340,800	32,210	21,700
Nuclear heat source Reforming,[a] MW(t) Steam, MW(t)	-- --	775 2,300	-- --	905 1,155
Product output bbl/day tons/yr	90,000 4.4×10^6	90,000 4.4×10^6	90,000 4.4×10^6	90,000 4.4×10^6
Thermal efficiency, %	42	48	59	67
Product/coal ratio, bbl/ton	1.9	2.9	2.8	4.2
Ratio of heat in product to heat in coal	0.42	0.64	0.59	0.95

[a]Includes steam production for reformer.

residual coke, reduces the iron ore to molten pig iron. In this process, coke is both reducing agent and heat source. Nuclear energy steelmaking would use direct reduction. Iron ore, while remaining solid, is reduced into sponge iron by a reducing gas (hydrogen or a mixture of hydrogen and carbon monoxide). The HTGR heat produces a reducing gas by steam-reforming a light hydrocarbon, such as natural gas (CH_4), naphtha, or liquefied coal. The reaction

$$CH_4 + H_2O + heat \rightarrow CO + 3H_2$$

occurs best at high temperatures and requires a considerable amount of heat. Hot reactor helium would meet both these requirements. The hot carbon monoxide-hydrogen mixture produced by the reforming reaction will reduce iron oxide to sponge iron. Thus, the heat needed to reduce the oxide is indirectly supplied by the hot helium from the HTGR (Ref. 5).

The directly reduced sponge iron can then be refined to steel in an electric furnace; electric power would be produced from a portion of the HTGR heat. Thus, the nuclear reactor provides all the energy needed to reduce the iron oxide and to make the steel. The hydrocarbon is only the reducing agent and not combusted for process heat. Figure 4 illustrates integrating the HTGR into a system to directly reduce iron ore.

HIGH-TEMPERATURE HEAT EXCHANGER

Both the Japanese and the U.S. process heat programs use the high-temperature helium-to-helium

IHX. This heat exchanger transfers high-grade thermal energy [up to 1000°C (1832°F] from the reactor core to the end process.

Although applications for high-temperature energy comprise a much broader spectrum than the helium-to-helium IHX can handle, this heat exchanger represents a key element in successful nuclear-process-heat programs. The IHX design will require major programs in materials development, heat exchanger design, and fabrication techniques.

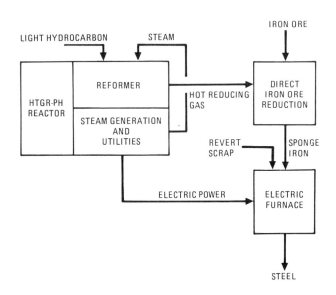

Figure 4 HTGR for direct reduction of iron ore

Heat Exchanger Design Considerations

1. The heat exchanger designs should use
tubular construction and conventional fabrication
techniques for state-of-the-art reliability.

2. Only counterflow and multipass cross-counter-
flow designs are thermodynamically practical. The
counterflow approach yields more favorable heat
exchanger proportions for vertical PCRV cavity
installations, lower metal temperature gradients,
and lower potential flow-induced vibration.

3. Compact surface geometries are desirable,
because very large heat transfer surface areas are
required in the gas-to-gas heat exchangers, which
must transfer heat at reactor transfer rates.

4. Since IHX materials (Inconel 617, Hastelloy X,
etc.) are costly, the heat exchangers, which are
important plant capital equipment items, should be
designed for long life.

5. The heat exchangers must be designed in accor-
dance with the appropriate provisions of the Ameri-
can Society of Mechanical Engineers (ASME) Boiler
and Pressure Vessel Code.

Additional heat exchanger design constraints are
created by the integrated HTGR plant arrangements,
where the PCRV acts as the heat exchanger containment and support. While these arrangements remove
the need for large vessels (typical of industrial
heat exchangers) and for shell-side fluid couplings
(connecting the heat exchangers to the primary
system), additional restrictions must be
accommodated:

a. Orientation. Vertical heat exchangers help
minimize both PCRV diameter and prestressing
complexity.

b. Diameter. Heat exchanger diameters should be
minimized to avoid directly impacting PCRV and
reactor containment building diameters, which are
major plant cost considerations.

c. Height. PCRV (and thus the reactor containment
building) height should not be governed by the heat
exchangers.

d. Accessibility. The restricted accessibility of
PCRV-mounted heat exchanger installations influences the heat exchanger mechanical designs in the
areas of in-service inspection (ISI), tube plugging
capability, and maintenance.

The IHX is the interface between the primary and
secondary helium circuits and therefore must be
leak-tight. For the process heat plant with a
reactor outlet temperature of 950°C (1742°F), the
maximum metal temperature in the IHX is 927°C
(1700°F); the equivalent value for a reactor outlet
temperature of 850°C (1562°F) is 835°C (1535°F).
Thus, IHX metal temperatures of this order mandate
using superalloys in at least the hot end of the
unit. If the IHX design is to use existing materials, pressure must be balanced to meet the low
allowable long-term stress levels at the temperatures mentioned above. HTGR-PH pressure is
balanced by maintaining the secondary helium pressure only slightly above the primary helium
pressure; thus, secondary helium inleakage to the
primary system will result if leaks develop in the
IHX.

International efforts are under way to design
helium-to-helium IHXs for nuclear process heat
plant applications; both straight-tube and helical
geometries are being evaluated (Refs. 6 through
10).

Conceptual IHX mechanical designs IHX (Refs. 11,
12) are detailed enough to establish sizing and
thermal performance, major structural details and
load paths, mechanical provisions for assembly,
ISI, maintenance, component installation and removal, and cost. Figure 5 illustrates the straight-
tube counterflow design, based on Table 4 design
information.

The initial design concept adopted the highly
effective modular arrays to facilitate fabricating

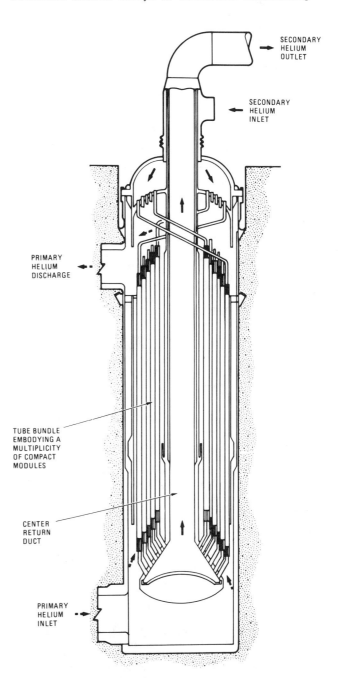

Figure 5 Overall view of modular, straight-tube
IHX concept for HTGR-PH plant

163

TABLE 4
INTERMEDIATE HEAT EXCHANGER DESIGN DATA

Plant rating, MW(t)   842
Exchangers per plant   2
Loop rating, MW(t)   421

Fluid Circuit	Secondary Helium	Primary Helium
Fluid routing	Tube	Shell
Flow per unit, kg/s	170.7	170.7
Inlet temp, °C (°F)	419 (787)	946 (1735)
Inlet pressure, MPa (psia)	5.24 (760)	4.95 (718)
Outlet temp, °C (°F)	899 (1650)	467 (872)
Pressure loss, MPa (psid)	0.055 (8.0)	0.046 (6.6)

Effectiveness	0.91
Log mean temperature difference, °C (°F)	52.2 (94)
Thermal conductance per unit, MW/°C (Btu/h-°F)	8.1 (15.4 x 10^6)
Heat duty per unit, MW(t)	428
Flow configuration	Counterflow
Type of construction	Straight tube
Assembly type	Modular
Tube o.d., mm (in.)	11.1 (0.4375)
Wall thickness, mm (in.)	1.5/1.09 (0.06/0.043)
Tube pitch	1.4
Surface compactness, m^2/m^3	167
Surface area per unit, m^2	10,488
Tubes per unit	32,512
Modules per unit	256
Effective tube length, m (ft)	9.2 (30.2)
Assembly diameter	4.27 (14)
Assembly height, m (ft)	16.15 (53)
Assembly weight, tonnes (ton)	218 (240)
Pressure boundary ΔP, MPa (psid)	0.29 (43) pressure balanced
Max metal temp, °C (°F)	927 (1700)
Thermal density, MW/m^3	6.70
Thermal flux, W/cm^2	4.03
Tube material	Inconel 617,2-1/4 Cr-1 Mo
ISI/repair level	Module
Assembly location	Factory
Transporatiaon mode	Barge/rail
ASME Code Class	Section III

and handling and to help promote good flow distribution. A monolithic tube field is the optimum heat exchanger packaging arrangement under development for the PCRV diameter. The IHX concept employs conical tubesheets. The headering approach based on tubesheet configurations. Since these configurations occupy less space than the hexagonal tube bundle envelope, the modular tubes must be bent locally to penetrate the desired tubesheets. This compact headering facilitates shellside fluid entry and exit, results in heat exchanger frontal area use approaching that of homogeneous tube fields, and eliminates the need for antibypass seals between modules. However, it obviously complicates module fabrication. Studies are continuing to identify more attractive alternatives.

The inlet and outlet secondary helium connections are located at the top of the unit. The secondary helium is routed downward inside the module tubes for counterflow operation. A drum header at the bottom of the unit collects the secondary helium leaving the modules, and a central return duct returns it to the top. Lead tubes connect each

module to the top of the main support plate and to the drum header. The main support plate also acts as a tubesheet. The upper set of IHX lead tubes (cold end of the unit) are coiled for elasticity to accommodate module differential growth.

Material Selection Considerations

Figure 6 shows approximate operating boundaries for various heat exchanger materials. Material temperature limits greatly constrain the high-temperature heat exchangers. Essentially, only stainless steels and the nickel- and cobalt-base superalloys can be used as metallic materials above 650°C (1200°F). The heat-resistant cast alloys have specialized applications. Figure 6 shows that the specific strength (strength/density) of metallic materials decreases very rapidly at elevated temperatures. The operating islands shown in Fig. 6 are very approximate; for a particular application, the actual properties of the candidate materials must be minutely examined over the appropriate temperature range in the operating environment. Many high-temperature systems use superalloys, and the design and fabrication of superalloy heat exchangers is a well-developed technology. Nickel- and cobalt-base wrought superalloys, such as Inconel 600, Incoloy 800H, Inconel 807, Hastelloy X, Inconel 617, and Haynes 188, have been used for elevated-temperature service. Figure 6 shows the upper limit for practical metallic heat exchanger design as 955°C (1750°F). This value represents an upper limit for using the wrought superalloys, which can readily be formed into a variety of heat exchanger surface geometries and construction types.

In the petrochemical industry, centrifugally cast tubes of 25 Cr-20 Ni steel, such as HK40, IN-519, and Manaurite 36C, have operated successfully in hundreds of plants for many years with tube wall temperatures above 982°C (1800°F). The centrifugally cast tubes in modern reformers are about 100 mm (4 in.) in diameter and have wall thicknesses of 20 mm (0.8 in.). While these cast tubes exhibit excellent high-temperature capability, they cannot be formed in the small-diameter size (11 mm) with wall thicknesses of about 1 mm, necessary for the compact surface geometry required in the process heat IHX. For the high-temperature IHX, wrought materials must suffice, and candidate alloys include Inconel 617, Hastelloy X, and Incoloy 800H.

Conclusions

As the HTGR-PH plant design progress beyond the conceptual design stage, design effort will intensify on high-temperature heat exchangers. For such long-term programs, the heat exchanger designer has an obvious responsibility to monitor international developments and to factor technology advancements into the design.

This paper outlines the IHX concept using existing technology and currently available and demonstrated materials. Advanced technologies could be incorporated by using new IHX materials (to permit higher system operating temperatures) and by adopting advanced recuperator and IHX surface geometries (to reduce cost, reduce size, and improve ISI capability and maintenance).

Both Japan and the U.S. nuclear process heat programs use the high-temperature heat exchanger.

164

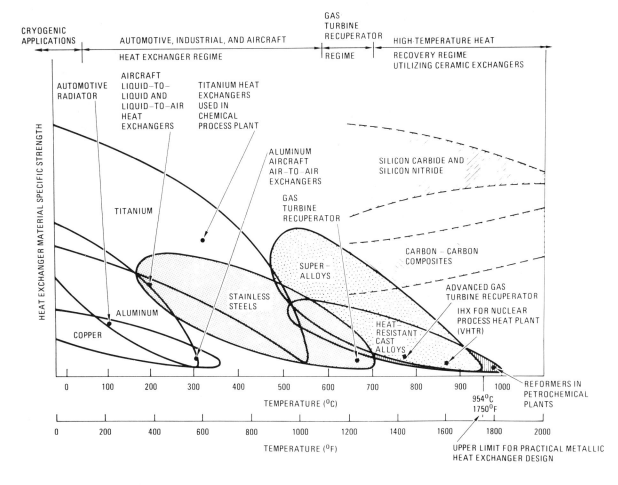

CRYOGENIC APPLICATIONS

AUTOMOTIVE, INDUSTRIAL, AND AIRCRAFT

HEAT EXCHANGER REGIME

GAS TURBINE RECUPERATOR REGIME

HIGH-TEMPERATURE HEAT RECOVERY REGIME UTILIZING CERAMIC EXCHANGERS

AUTOMOTIVE RADIATOR

AIRCRAFT LIQUID-TO-LIQUID AND LIQUID-TO-AIR HEAT EXCHANGERS

TITANIUM HEAT EXCHANGERS USED IN CHEMICAL PROCESS PLANT

ALUMINUM AIRCRAFT AIR-TO-AIR EXCHANGERS

GAS TURBINE RECUPERATOR

SILICON CARBIDE AND SILICON NITRIDE

TITANIUM

SUPER-ALLOYS

CARBON – CARBON COMPOSITES

ALUMINUM

STAINLESS STEELS

COPPER

HEAT RESISTANT CAST ALLOYS

ADVANCED GAS TURBINE RECUPERATOR

IHX FOR NUCLEAR PROCESS HEAT PLANT (VHTR)

REFORMERS IN PETROCHEMICAL PLANTS

HEAT EXCHANGER MATERIAL SPECIFIC STRENGTH

TEMPERATURE (°C)

0   100   200   300   400   500   600   700   800   900   1000

954°C 1750°F

TEMPERATURE (°F)

0   200   400   600   800   1000   1200   1400   1600   1800   2000

UPPER LIMIT FOR PRACTICAL METALLIC HEAT EXCHANGER DESIGN

Figure 6 Approximate operating boundaries for various heat exchanger materials

These countries should encourage complementary heat exchanger design and development programs. Joint success in the developing of this high-temperature heat exchanger is critical to deploy nuclear process heat programs. The current energy crisis could be substantially alleviated by introducing alternative energy sources (nuclear energy) with improved conversion efficiencies.

ACKNOWLEDGMENTS

The author would like to thank the management of General Atomic Company for permission to publish this paper. Work on the HTGR-PH plant heat exchangers has been supported by the DOE under Contract DE-AT03-76SF71061.

REFERENCES

1. Quade, R. N., D. L. Vrable, and L. Green, Jr., "Production of Liquid Fuels with an HTGR," Proceedings of 2nd Miami International Conference on Alternative Energy Sources, December 10, 1979, Miami, Florida.
2. Landis, J. W., "The GGA High-Temperature Gas-Cooled Reactor: A General Discussion," J. Brit. Nucl. Energy Soc. 12, No. 4, 367-385 (1973).
3. Vrable, D. L., R. N. Quade, and J. D. Stanley, "Design of an HTGR for High-Temperature Process Heat Applications," ASME Paper No. 79-JPGC NE-2, 1979.
4. Vrable, D. L, and R. N. Quade, "Design of the HTGR for Process Heat Applications," 15th Intersociety Energy Conversion Engineering Conference, August 18-22, 1980, Seattle, Washington.
5. Blickwede, D. J., "The Use of Nuclear Energy in Steel Making - Prospects and Problems," Nucl. News 17, No. 13, 65-68 (1974).
6. Nakanishi, T., "Development of High-Temperature Heat Exhanger on Heat Utilization from VHTR," NUCLEX 78, October 3-7, 1978, Basel, Switzerland, Paper No. A4/17.
7. Crambes, M., "Intermediate Heat Exchanger for HTR Process Heat Application," IAEA Specialists Meeting on Process Heat Applications Technology, November 27-29, 1979, Jülich, FRG.
8. Itoh, M., "Design, Construction and Operation Experience of the Helium-Helium intermediate Heat Exchanger," IAEA Specialists Meeting on Process Heat Applications Technology, November 27-29, 1979, Jülich, FRG.
9. Neimeyer, W., "Heat Exchanging Components for Coal Gasification - He/He Intermediate Heat Exchanger (IHX) and Steam Reformer," IAEA Specialists Meeting on Process Heat Applications Technology, November 27-29, 1979, Jülich, FRG.
10. Walravens, M. J., et al., "Preliminary Orientation and Parametric Survey for Very High Temperature Reactors with Intermediate Heat Exchangers," IAEA International Symposium on Gas-Cooled Reactors," October 1975, Jülich, FRG.

11. Mori, Y., "Performance of Heat Exchangers on HTGR Application," Joint ASME/AICHE National Heat Transfer Conference, July 27-30, 1980, Orlando, Florida, ASME Paper No. 80-HT-39.
12. McDonald, C. F., et al., "Heat Exchanger Design Consideration for HTGR Plants," Joint ASME/AICHE National Heat Transfer Conference, July 27-30, 1980, Orlando, Florida, ASME Paper No. 80-HT-62.

High Temperature Heat Transfer Application to Nuclear Power Safety Reflooding and Core Uncovery Phenomena

BILL K. H. SUN

Nuclear Power Safety and Analysis Department, Electric Power
Research Institute, 3412 Hillview Avenue, Palo Alto, California 94303,
U.S.A.

ABSTRACT

The perspective on fluid flow and heat transfer in
the safety aspects of light water reactor systems
is discussed. The discussion is focused on the
reflooding and the core uncovery phenomena which
may lead to high temperatures in the reactor core
in the event of accident conditions. During the
processes of reflooding and core uncovery, the
heatup of the reactor fuel cladding is shown to be
governed by the thermal-hydraulic mechanisms in
the flow channel and the two-phase levels in the
core. The two-phase mixture level and the quench
front serve as interfaces at which flow regime
transition occurs and sharp temperature gradient
arises. The detailed flow regimes and heat
transfer mechanisms are described.

I. INTRODUCTION

High temperature heat transfer does not exist in
normal steady-state operations of nuclear power
plants. The heat transfer modes for normal
reactor operations are forced flow nucleate
boiling for the boiling water reactors (BWRs) and
turbulent flow convection for the pressurized
water reactors (PWRs). High temperature conditions can occur when there are disturbances in the
plant system such that the coolant flow can no
longer efficiently transport the energy from the
reactor core. One disturbance that has received
worldwide attention concerning the safety of the
nuclear power plants is the loss-of-coolant
accident (LOCA).

The loss-of-coolant accident can be generally
categorized into the large-break LOCA and small-
break LOCA. The large break is hypothesized by a
double-ended break of a recirculation line. The
water inventory in the reactor is rapidly depleted
during the blowdown process. The Emergency Core
Cooling Systems (ECCS) is designed to activate at
the end of blowdown to initiate the refill and
then the reflooding phases. The reflooding process eventually cools down the core, quenches the
fuel rods, and terminates the LOCA transient.

The small break LOCA, such as the one that occured
at the Three Mile Island (TMI) Unit 2 incident in
1979, can be caused first by a loss of heat sink
in the steam generator [1,2]. This leads to the
overpressure of the reactor system and the opening
of the relief valve. If a mechanical malfunction
causes the valve to fail to close, water inventory
will continue to loose through the valve, resulting in a small break LOCA. If the inlet water
which may be introduced through the feed system or
the ECCS does not replenish the loss-of-coolant
due to the boil off, the core can eventually be
uncovered. As happened in the TMI-2 incident, the

core uncovery leads to very high temperatures for
the fuel rods.

The objective of this paper is to describe the
mechanisms of fluid flow and heat transfer in the
reflooding and the core uncovery process with particular emphasis on the heat transfer boundaries;
the quenching front and the two-phase mixture
level. While some representative result will be
described, this paper does not intend to be an
exhaustic review of accomplishments in these
areas.

II. HEAT TRANSFER IN LIGHT WATER REACTORS

The heat transfer in a normal reactor system, like
the other power generation systems, can be simply
considered as the balance of heat source from the
fuel with the heat sink of the primary fluid. In
PWRs, as shown in Fig. 1, the heat sink for the
primary side water is the steam generator where
the enthalpy of the hot primary side water is
transferred through the heat exchanger walls to
generate steam from the secondary side water. The
steam is in turn transported to the turbine generator to produce power and returned to the steam
generator via a condenser. In BWRs, steam is generated directly from boiling of water in the core
and transported to the turbines for power
generations.

The ultimate heat sink in reactor steady-state
operations is the ambient atmosphere and water
supply. However, in the loss-of-coolant situations, either small break or large break, it is
generally assumed that the heat exchangers and

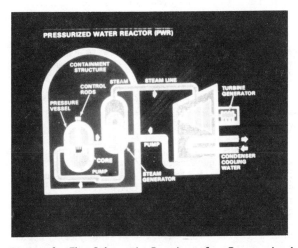

Figure 1 The Schematic Drawing of a Pressurized
Water Reactor System Under Normal Operating
Conditions

the condensers are incapacitated. The reactor then only releases its heat through the break flow to the containment.

In the following discussions, the phenomena of reflooding and core uncovery will be described in this conventional heat transfer approach, i.e., to investigate the source, the sink, and the transport mechanisms.

III. REFLOODING

Reflooding is considered as the third phase after the initiation of the LOCA transient. The first phase is rapid blowdown and loss of water inventory in the system. The second phase is the ECCS refill of the reactor vessel. The description of a typical PWR vessel is shown in Fig. 2.

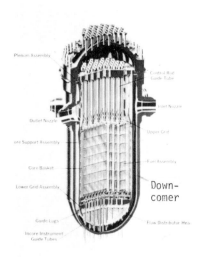

Figure 2 The Description of a Typical PWR Reactor Vessel

Reflooding heat transfer starts when the water reaches the bottom of the core whose cladding temperature is on the order of 500-1000°C (a typical PWR fuel rod bundle is shown in Fig. 3). Since the system pressure during this phase is about two to three times of the atmospheric pressure, the cladding temperature is higher than the Liedenfrost temperature or the minimum film boiling temperature. Because of this high surface temperature, the water can not rewet the surface instantly. The heat transfer process for water to rewet the high temperature surface is called rewetting or quenching [3,4,5,6,7].

As the quenching front progresses upward, the cladding surface undergoes a successive heat transfer regimes following inversely the conventional boiling curve; from film boiling along a dry surface, through transition boiling to nucleate boiling. All these occur within a narrow region whose dimension is on the order of one centimeter [7]. The velocity of the quench front depends on the axial conduction along the fuel rods as well as the surface heat transfer both in the dry region ahead of the quench front and the wet region behind the quench front [6,7]. The

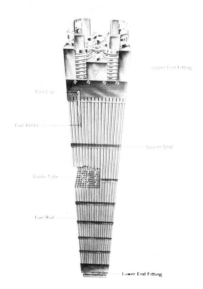

Figure 3 A Typical PWR Fuel Rod Bundle

transient heat transfer coefficient in the wet region can be several orders of magnitude larger than that in the dry region. Thus, the quench front serves as an interface that separates two distinctive heat transfer regions. In the reflooding process, the quench front is a moving heat transfer boundary, above which high temperature heat transfer prevails. With the understanding that the quench front is a reference during reflooding, we shall consider next the thermal and hydraulic conditions in the core with respect to the quench front.

As subcooled water is fed into the bottom of the core, the single-phase laminar or turbulent flow is first heated up by the decay heat generated from the fuel rod. As the water adjacent to the wall reaches the saturation temperature, bubbles begin to form from heterogeneous nucleation though the bulk flow is still subcooled. This location is called net vapor generation point (NVG) [8]. The NVG point separates the single-phase flow region below from the two-phase flow region above. In the subcooled nucleate boiling region, the bubbles are populated in the boundary layers along the wall, and the flow can be characterized as bubbly flow.

As the flow moves further upward, the bulk fluid becomes saturated. In this region, the bubbles sizes are larger. Since the vapor flow rate increases at higher elevations, the void fraction increases as a result of bubble merger to form slugs. The two-phase flow regime is called slug flow or churn-turbulent flow. The heat transfer is in a saturated nucleate boiling state. Because the heat transfer mode of nucleate boiling is very efficient, the fuel cladding is kept near the saturation temperature of water from the bottom of the core up to the quench front.

For the cases of low inlet flow and subcooling, the vapor flow rates can be very high. The high vapor velocities in the core can lead to a transition from the churn-turbulent flow to an annular flow regime below the quench front (as shown in Fig. 4).

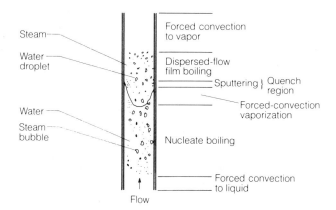

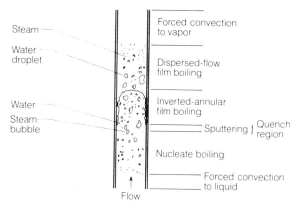

Figure 4 The Description of Flow and Heat Transfer Regimes in a Flow Channel for the Cases of Low Reflooding Rate and Low Subcooling

Figure 5 The Description of Flow and Heat Transfer Regimes in a Flow Channel for the Cases of High Reflooding Rates and High Subcooling

Above the quench front, the cladding surface is adjacent to a dispersed droplet flow regime with droplets generated from entrainment from the annular film, from bubble bursting through the pool, and from the sputtering phenomenon at the quench front. The heat transfer mechanism in the dispersed droplet flow region is often referred to as dispersed flow film boiling. The heat sink in this region is the droplets and the superheated steam, both of which absorb heat from the fuel cladding through radiation and convection. The droplet sizes are reduced as they are vaporized in the superheated steam environment. At further downstream of the flow the droplets may diminish and single-phase steam flow prevails [9,10].

For the cases of high reflooding rates and high inlet subcooling, the vapor flow rates are lower than the former cases and the bulk flow can be slightly subcooled even above the quench front. As shown in Fig.5, there exists a liquid core above the quench front. The high temperature cladding is separated from the liquid core by a thin vapor film. The regime is called inverted annular flow and the heat transfer mode is often called inverted-annular flow film boiling [4]. The heat sink in this region is the liquid core which absorbs heat from the cladding via the conduction mechanism across the thin vapor film. The liquid column eventually breaks up into droplets as a result of Helmholtz instability. At further downstream, the thermal and hydraulic conditions are similar to the cases of low reflooding rates.

In typical reflooding experiments, the heat transport due to forced-convection to steam and dispersed flow film boiling is on the same magnitude of the decay heat generation. As a result, the fuel cladding temperature can maintain at high temperatures (800°C-1200°C), as shown in Fig. 6, then drop sharply when the quench front approaches.

In a typical PWR reflooding process, the thermal and hydraulic regimes described previously are further complicated because (1) the reflooding feed flow is governed by the water head in the downcomer (see Fig. 2 for reactor geometries) and (2) the fuel rod geometry may deform as a result of cladding swelling and perforation. The

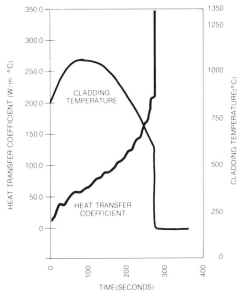

Figure 6 The History of Cladding Temperature and Surface Heat Transfer Coefficient in a Typical Reflooding Test from a Simulated PWR bundle (NRC/EPRI/W FLECHT-SEASET Reflooding Program)

condition (1) imposes a manometer-type oscillation on the fluid flow which is coupled to the quench and heat transfer process. Available experimental result have shown that the quench front velocity and the heat transfer downstream depend on the amplitude and the frequency of the oscillations [11,12]. The condition (2) may create zones in the core where the flow is partially blocked. The partial flow blockage has the overall effect of flow redirection to the unblocked zones. In the local blockage region, the blockage tends to act like a cooling fin. It can break up the droplets in the dispersed flow thereby desuperheats the steam and can initiate a local wet patch or quenching front [13].

IV. CORE UNCOVERY

During the small-break accident at the TMI-2 plant, the reactor core was partially uncovered because the make-up flows was insufficient to compensate for the mass loss due to boiling. (a schematic of the TMI-2 plant is shown in Fig. 7) The prolonged period of core uncovery for about one hour caused very high fuel rod temperatures which resulted in the damage of reactor core [2,14]. The core eventually recovered by cooling water. The core recovery transient is similar to the reflooding phenomenon. The only difference is that recovery transient refers to a high system pressure condition, while reflooding is at low pressure.

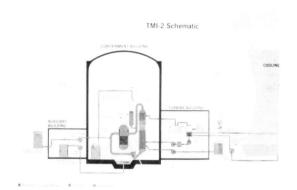

Figure 7   The Schematic of the Three Mile Island Unit 2 (TMI-2) Plant

The uncovery or boiling dry of a flow channel or a reactor core refers to the situation of gradual depletion of water inventory. The water inventory is often represented by the collapsed liquid level defined by neglecting the vapor bubbles. For the case of very small flow where the acceleration and viscous forces are negligible, the collapsed liquid height is equal to the hydrostatic head of water. To consider the heat transfer conditions during the uncovery of a core, a relevant parameter is the level of the two-phase mixture. The height of the two-phase mixture depends on the void fraction and the vapor flow rate below the level. It is generally marked as a sharp increase in the axial void distribution and in the surface temperature [15].

The two-phase mixture level represents a hydro-dynamically controlled dryout point for the heated surface. Similar to the quenching front in the low reflooding rate cases, the two-phase mixture level is a moving heat transfer boundary which separates two distinctively different heat trans-fer regimes. As shown in Fig. 8, the major dif-ferences between reflooding (Fig.4) and core uncovery is that, in the former case, the refer-ence is a conduction and thermal-hydraulic con-trolled quench front which progresses upward, and in the later case, the reference is a hydrodynami-cally controlled dryout front which retreats down-ward. Also, in reflooding, the turbulent sput-tering phenomenon at the quench front generates droplets which serve as heat sinks in the dis-persed flow film boiling region. While in the dryout process, sputtering does not exist.

The heat transfer mechanisms in the dry regions for the reflooding and the core uncovery condi-tions are basically similar. Except in a severe small break core recovery situation, such as happened in the TMI-2 accident, the fuel cladding temperature can reach so high that leads to chem-ical reactions between the zircaloy cladding and

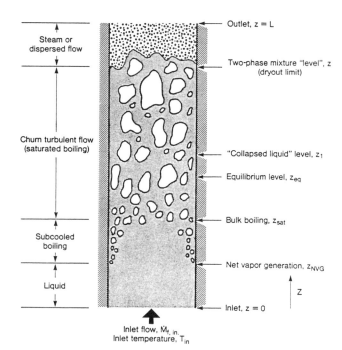

Figure 8   Schematic Diagram of a Vertical Flow Channel Uncovering or Boil Off

the steam, the oxidation process. The exothermic reaction process could produce a energy level much higher than the core decay power. As a result, the heat transfer modes of convection and radi-ation was insufficient to remove the heat and the fuel temperature could reach a level on the order of 2000°C.

The zircaloy and steam reaction produces hydrogen. The existence of hydrogen in the reactor system changes the nature of flow from the conventional single component two-phase flow to two-component two-phase flow. It couples the process of flow in the loop with mass transfer through absorption and extraction, and also influences the heat transfer process through the role of a noncondensible gas.

The heat transfer that cools the fuel rods above the two-phase level depends on the rate of vapor generation below the level. As the steam flows upward from the two-phase level, some of it is absorbed by the zircaloy cladding surface. The steam further penetrates into the cladding through the diffusion process while the surface is oxi-dized. Thus, the steam diffuses from the main flow through a concentration boundary layer to the cladding surface, then diffuses into the solid cladding material. The steam diffusion process is balanced by a counterflow diffusion of hydrogen which is generated inside the cladding material. The hydrogen diffuses from the cladding surface through the concentration boundary layer to the bulk flow.

The surface heat transfer is governed by the thermal boundary layer near the surface and is thus coupled to the mass diffusion concentration boundary layer which, in turn, is related to the zircaloy-steam reaction process inside the clad-ding. The cladding temperature is, therefore, dependent on the decay heat from the fuel, the heat generated by the chemical reaction, and the

surface heat transfer to the steam-hydrogen
mixture.

IV. CONCLUDING REMARKS

The worldwide interest in nuclear power safety, in
particular, the concern about LOCA reflooding and
core uncovery, provides a challenge to the heat
transfer technology community. There are still
phenomena and transport mechanisms of basic interest not fully understood. Examples are droplet
generation and entrainment, dispersed flow around
a blockage, inverted annular flow film boiling,
quenching with oscillatory flow, transient void
modeling, and coupling between convection and
oxidation, etc. The approaches of solving conventional heat exchanger problems are not sufficient
to resolve these concerns, largely because of the
following conditions:

(1) The reflooding and core uncovery heat
transfer are coupled with the transient locations
of the quench front add the two-phase mixture
level, respectively. The quench front and two-
phase mixture level, and the downstream heat
transfer modes, are in turn related to the
upstream thermal and hydraulic conditions.

(2) The core is a component which generates heat
for the nuclear plant system. The heat transfer
phenomena inside the core are coupled to the
thermal and hydraulic behaviors in the rest of the
system, for example, the momentum transport and
hydraulic resistance of the loop, the heat removal
capabilities in the heat sinks, and possible plant
operating disturbances.

The resolution of these concerns is not merely an
endeavor to develop analytical models for understanding. It also addresses the issues and
concerns of public safety. To achieve these purposes, a typical analytic effort involves not only
basic experiments and model development, but also
large-scale experiments and development and quali-
fication of system codes.

REFERENCES

1. Kemeny, J.G. et al., "Report of the
    President's Commission on the Accident at
    Three Mile Island," Washington, D.C., October
    1979.
2. Rogovin, M. and Frampton, G.T., "Three Mile
    Island, A Report to the Commissioners and to
    the Public," Vols. I and II, USNRC,
    Washington, D.C. Jan. 1980.
3. Yadigaroglu, G. et al, "Heat Transfer During
    the Reflooding Phase of the LOCA - State of
    the Art," EPRI Topical Report 248-1,
    September 1975 (also, Nuclear Safety,
    Vol. 19, 1978, pp. 20-36).
4. Arrieta, L. and Yadigaroglu, G.,"Analytical
    Model for Bottom Reflooding Heat Transfer in
    Light Water Reactors," EPRI Report NP-756,
    August 1978.
5. Yamanouchi, A., "Effect of Core Spray Cooling
    in Transient State After Loss-of-Coolant
    Accident," J. of Nuclear Science and Tech-
    nology, Vol. 5, 1968, p. 547.

6. Sun, K.H., Dix, G.E., and Tien, C.L.,
    "Cooling of a Very Hot Surface by a Falling
    Liquid Film," J. Heat Transfer, Vol. 96,
    1974, pp. 126-131.
7. Duffey, R.B. and Porthouse, D.T.C., "The
    Physics of Rewetting in Water Reactor Emer-
    gency Core Cooling," Nuclear Engineering and
    Design, Vol. 25, 1972, p. 212.
8. Saha, P. and Zuber, N., "Point of Net Vapor
    Generation and Vapor Void Fraction in Sub-
    cooled Boiling," Heat Transfer 1974, Pro-
    ceedings of 5th International Heat Transfer
    Conference, Tokyo, Vol. IV, paper B4.7, 1974,
    pp. 175-179.
9. Sun, K.H., Gonzalez, J.M., and Tien, C.L.,
    "Calculations of Combined Radiation and Con-
    vection Heat Transfer in Rod Bundles under
    Emergency Core Cooling Conditions," J. Heat
    Transfer, Trans. ASME, Vol. 98, 1976,
    pp. 414-420.
10. Lilly, G.P. et al., "PWR FLECHT Skewed
    Profile Low Flooding Rate Test Series Evalu-
    ation Report, " NRC/W/EPRI Cooperative R&D
    report, WCAP-9183, November 1977.
11. White, E.P. and Duffey, R.B., "A Study of the
    Unsteady Flow and Heat Transfer in the
    Reflooding of Water Reactor Cores," Annals of
    Nuclear Energy, Vol. 3, 1976, pp. 197-210.
12. Waring, J.P. and Hochreiter, L.E., "PWR
    FLECHT-SET Phase B1 Evaluation Report,"
    NRC/W/EPRI Cooperative R&D Report, WCAP-8583,
    August 1975.
13. Hochreiter, L.E. et al., "PWR FLECHT-SEASET
    21 Rod Bundle Flow Blockage Task, Task Plan
    Report," NRC/EPRI/W FLECT-SEASET Program
    Report No. 5, EPRI NP-1382, March 1980.
14. Ardron, K.H. and Cain, D.G., "Calculations of
    the Core Temperature Transient During the
    Early Phase of the Core Uncovering at TMI-2,"
    EPRI NSAC Report (to be published).
15. Sun, K.H., Duffey, R.B., and Peng, C.M., "The
    Prediction of Two-Phase Mixture Level and
    Hydrodynamically-Controlled Dryout Under Low
    Flow Conditions," EPRI Report NP-1359-SR,
    March 1980 (to appear in International J. of
    Multi-Phase Flow).

NOVEL HEAT TRANSFER TECHNIQUES FOR ENERGY UTILIZATION, HEAT STORAGE, AND RECOVERY

Heat Transfer Considerations in the Use of New Energy Resources

N. LIOR and G. F. JONES
Department of Mechanical Engineering and Applied Mechanics,
University of Pennsylvania, Philadelphia, Pennsylvania 19104, U.S.A.

H. OZOE
Department of Industrial and Mechanical Engineering,
Okayama University, Okayama, Japan

P. CHAO and S.W. CHURCHILL
Department of Chemical Engineering, University of Pennsylvania,
Philadelphia, Pennsylvania 19104, U.S.A.

ABSTRACT

The use of many of the new energy resources, especially those of low energy density (or flux), as well as the increased interest in efficient energy use, give new life to some well established areas of heat transfer, and introduce new advances and needs in other areas. This paper briefly reviews the state of the art, describes work in progress, and recommends new research directions in the energy-related aspects of natural convection in enclosures, conjugate fluid mechanics and heat transfer problems in fin-tube radiant energy absorbers, and in ranking heat transfer problems in passive solar heating and cooling, solar ponds, ocean thermal energy conversion, and thermal storage.

INTRODUCTION

The economical use of low density (or flux) energy resources (such as the sun) in particular, and of more conventional resources which are rapidly increasing in cost in general, and which includes also the need to use energy effectively by reducing waste or reclaiming energy from sources which constituted such waste in the past, poses new research challenges for established heat transfer fields and establishes the need to develop new fields. Such research needs also arise due to increasing restrictions on pollution during the whole energy conversion cycle.

The ultimate objective of the heat transfer research, apart from shedding light on hitherto unexplored areas, is to (1) provide quantitative design tools, (2) improve the heat transfer coefficients without undue increase in parasitic (say pumping) power or hardware cost penalties, so that economically justifiable energy can be provided for the small driving temperature differences inherent with many alternate energy resources in particular, and with low process irreversibilities in general, (3) to reduce heat transfer coefficients (improve insulation) under similar constraints, so that thermal losses are diminished, and (4) to allow the proper thermal conditions for energy conversion (typically combustion) which produces minimal pollution. Heat transfer enhancement methods and equipment which had limited or somewhat exotic use in the past, are being increasingly considered or used in such new applications. For example, heat pipes are used in heat recovery in the HVAC industry (cf. [1,2]) and in solar collectors (cf. [3,4]). The latter two use also vacuum insulation around the solar radiation absorber to minimize heat losses. Direct contact heat transfer methods, including submerged combustion, are getting renewed attention, and the use of fluidized beds is increasing, with coal conversion via fluidized bed combustion playing a prominent role.

This paper reviews several heat transfer problems which are of importance in solar collectors, thermal insulation, passive solar heating and cooling, solar ponds, ocean thermal energy conversion, and in thermal storage. Apart from a brief presentation of the state of the art, work in progress by the authors and others is described.

NATURAL CONVECTION IN ENCLOSURES

This topic has received added impetus during the last few years because of several energy-related applications, which include thermal insulation in general and that of solar collectors in particular, and heat circulation in spaces heated by passive solar systems.

Air spaces provide good thermal insulation, and their application in the building industry is known from ancient times. It became also clear that it would be desirable to minimize the natural convection in these spaces and thus to reduce the heat transfer between the enclosing walls to as close to pure conduction as possible. Much work has been done on this topic since the early experimental correlations by Mull and Reiher [5] in 1972. The state of the art was reviewed by Ostrach [6] and Catton [7] in 1978. The problem involves a set of coupled non-linear partial differential equations, and the solution, (when obtainable) is very sensitive to the boundary conditions. Consequently, the first solutions dealt with cases where one-dimensional or two-dimensional approximations were acceptable, such as for the space between infinite plates, or for infinite slots and annuli. The boundary conditions were also chosen to be as simple as possible - typically uniform and constant wall temperature or heat flux.

In order to reduce the natural convection, the Rayleigh Number needs to be made as small as possible, and that could be obtained either by reducing the thickness of the air layer or by partitioning it into smaller cells, such as by honeycomb (cf. Veinberg [8], Hollands [9, 10], Buchberg et al [11,12], Lior [13], Arnold et al [14], Cane et al [15], Smart et al [16]). The first method increases the conductive heat loss through the air. The second introduces conduction through the cell walls, interferes with incoming solar radiation when used in solar collectors, and increases re-radiation. In any case, the problem becomes clearly three-dimensional, with nonuniform boundary conditions which are impressed by the insolation and the configuration of the heat-removing absorber. Important inroads into both the fundamental understanding and engineering evaluation were made by Hollands, Raithby, Unny and their co-workers [9,10, 15-24] and by Catton, Edwards, Buchberg and their co-workers (reviewed in [7]). Two-dimensional analyses, accompanied by experiments,were performed by both groups.

The great interest in passive solar heating, where the house's structure is used for the collection, conversion and storage of solar energy, and where typically one of a room's surfaces is warmer than the others and heats the space by radiation and natural convection, has also added to the vigorous interest in natural convection. One of the potential problems of passive solar heating (or cooling) is the attainment of fairly uniform enclosure temperatures in space and time. To minimize electrical energy consumption and extent of mechanical equipment, it is necessary to understand how heat is convected naturally in such enclosures (cf. Bauman et al [25]),and the problem again is three-dimensional with nonuniform and nonlinear boundary conditions.

Following the early work by Aziz and Hellums [26], three-dimensional calculations became feasible only recently, mostly through the work of Ozoe, Churchill and their co-workers [27-33], Mallinson and de Vahl Davis [34], and Chan and Banerjee [35, 36], all of whom applied advanced numerical techniques for the solution of the three-dimensional Navier-Stokes, energy,and continuity equations with proper boundary conditions and with the Boussinesq approximation. All but Chan and Banerjee transform the equations to their vorticity form and introduce a solenoidal vector potential for the velocity vector. The resulting abridged equations are then solved by the ADI finite difference method. Chan and Banerjee retain the equations in terms of their primitive variables and use a "marker and cell" finite difference method for the solution.

Ozoe, Churchill and their co-workers have also verified the convective flow patterns and resulting heat fluxes by experiment, and have extended the work to include non-uniform boundary conditions, partial internal partitions, arbitrary inclination and rotation of the enclosure, and cylindrical annular enclosure geometries, in addition to the rectangular box-enclosures also treated by the other researchers.

There is common agreement in the analytical results, and verification by experiments, that real natural convection flows in enclosures are not two-dimensional, and that a velocity component parallel to the familiar roll-cell axis is also present. The flow thus resembles a double helix, with fluid particles moving along the direction of the principal axis of the roll cell, from the walls into the enclosure, up to a certain point, and then back towards the wall (Fig. 1).

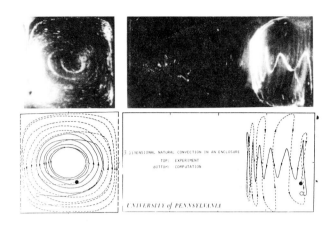

Fig. 1. 3-dimensional natural convection in a
horizontal enclosure [37].
Top: experiment, Ra = 15500, Pr = 4300
Bottom: computation, Re = 4000, Pr = 10

These flows are due both to the drag at the end walls, and to thermal gradients generated at these walls because of the diminished rate of circulation [34]. The significance of the three-dimensionality of the flow becomes even more pronounced when the enclosure is tilted, or when partial partitions are inserted. Past studies (cf. Wilkes and Churchill [38], Davis [39], Hart [40], Ozoe et al [41-43]) have shown that steady laminar natural convection in enclosures with a bottom temperature higher than the top, is first characterized by a train of roll-cells with their axis parallel to the short side of the enclosures, a fact which also served to justify the many two-dimensional analyses of the phenomenon. This is no longer correct when the box is tilted: inclination about its longer side causes the axis of the roll cells to become oblique to the side, and at some critical angle all the smaller roll cells form one large cell which has an axis perpendicular to the short side of the box; tilting the box along its shorter side gradually merges the parallel small roll cells into one large circulating cell, with the axis still parallel to the short side of the box [31]. As the box is tilted from the horizontal position, the Nusselt number is first seen to decrease gradually to a minimum which coincides with the transition from one convective pattern to another, reaches a maximum at a higher angle of inclination (~60° - 90°) and then diminishes monotonically as the angle is increased to 180°. Both the minimum and maximum of the Nusselt number occur at slightly higher values for boxes inclined about the long axis than for those inclined about the shorter axis, but the values of Nu are about the same in both cases and similar to those of Fig. 2 (which also gives results for other boundary conditions, as explained below).

Although quite evident, it is also noteworthy that 3-d calculations provide lower Nu values than 2-d ones, for the same case, principally because the 3-d calculations account for the slow-down of the circulation and redirection by the solid ends.

In an attempt to control natural convection in enclosures, and to improve the understanding of the phenomenon, similar experimental and analytical

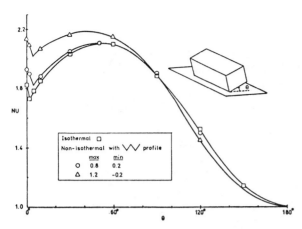

Fig. 2. Mean Nusselt number at various angles of
inclination, for a double sawtooth
distribution with a maximum temperature
at the central plane. Ra=6000, Pr=10 [33]

 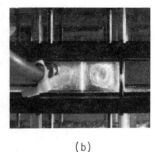

(a)   (b)

Fig. 3. Vacuum insulated natural convection
experimentation facility.
a. Top view
b. Flow visualized in cell

studies were performed by Ozoe, Churchill and
their co-workers on cylindrical annular enclosures, and on rectangular enclosures containing
partial baffles (cf. [32]). For vertical cylinders, the vertical wall drag was found to increase
the critical Rayleigh number for the onset of convection, and to decrease Nu, as compared to the
case of infinite parallel plates. This Nu number
difference becomes however negligible with the
onset of turbulent motion at Ra≈35,000. The
stable mode of circulation was found to consist
of a symmetrical set of equally sized roll cells
with their axes in the radial direction and in a
single horizontal plane. The motion of each fluid
particle in a roll cell is again a coaxial double
helix, confined to an inner or an outer region
roughly separated by a mid-circumferential plane.
Inclination of the cylinder's axis results in a
reduction in Nu and in change of flow from the
radial roll-cell mode to a circulating flow rising
along the hot plate and descending along the cold
one. At this point, Nu is at a minimum. For
further inclination, Nu climbs to a maximum at
about $\pi/3$ rad, and decreases to the conduction
state at $\phi = \pi$, corresponding to heating from
above.

The fact that the minimal Nu occurred at angles
of inclination at which the roll cells were lined
up with a longest or most tortuous axis, indicated
that the manipulation of roll-cell orientation by
such means as internal partial baffles may result
in the reduction of Nu. Recent analytical and
experimental studies by the same researchers,
for a 2x1x1 box, with a partial (0.5 long) fin
perpendicular to the long side in the enclosure
and attached to it, and for Ra=6000, show that
the fin produces a 30% reduction in Nu for all angles of inclination as compared to the fin-less
box. The general shape of the curve is similar
to that of Fig. 2. The results are now being
verified experimentally, in a vacuum insulated
test cell with internal dimensions of 2-1/2" x
1-1/4" x 1-1/4", which allows flow visualization
(Fig. 3) and measurement of heat flux and temperature on the high and low temperature surfaces.

It is well known that natural convection phenomena are quite sensitive to the boundary conditions. Recent studies by the same researchers

have addressed three such problems:

(1) a nonuniform temperature distribution on the
lower surface, using a sawtooth profile to stimulate such situations as the effect of cooling
pipes along the absorber of a flat-plate solar
collector [33].

None of these distributions, whether normal or
oblique to the side walls, influenced either the
number or the orientation of the roll cells for the
horizontal orientation. However, the distributions
did predetermine the direction of circulation,
a peak in temperature at the center of the enclosure producing upflow in the central plane
and vice versa. Typical streak lines are shown
in Figures 4 to 7.

With the non-uniform temperature distributions, a
slightly higher angle of inclination was required
to produce the transition from two roll cells
with their axis parallel to the short dimension
to a single roll cell up the hot plate.

Increasing the amplitude of the non-uniformity in
surface temperature increased the average Nusselt
number (Fig. 2) and the rate of circulation
significantly for angles of inclination less than
90 degrees but caused a slight decrease for
angles between 90 and 180 degrees. Inverting the
temperature distribution such that a minimum
occurred at the center of the enclosure, produced
a slightly lesser effect on Nu and the rate of
circulation. It is noteworthy that uniform temperature of the hot plate results in lower heat
transfer rates than any of the nonuniform temperature cases investigated.

(2) A 2x1x1 rectangular enclosure where half of
the 1x2 lower plate is heated, the other half is
insulated and the upper 1x2 plate is kept at a
lower temperature. The rest of the walls are
insulated. The experimental study, performed for
Ra=21,000 to 25,000, indicated that for a horizontal box a single roll cell is generated with
flow directed towards the corner next to the
heated part of the bottom plate. At small angles
of inclination around the short axis at the same
corner, this roll cell persists with the same
rotational direction, now flowing against gravity(!);

177

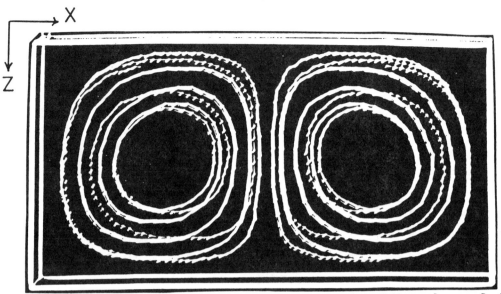

Fig. 4. Streaklines for uniform heating with a horizontal orientation of the enclosure. Ra=6,000, Pr=10 [33]

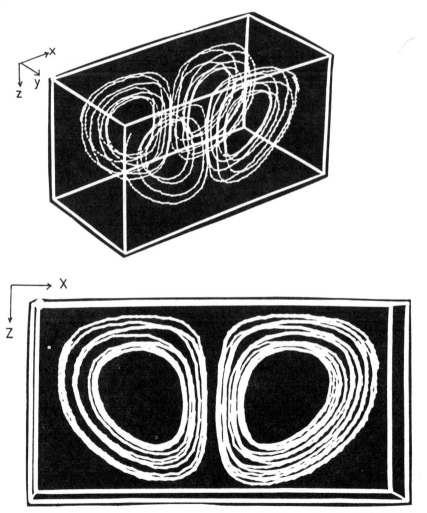

Fig. 5. Perspective and front view of streaklines for a double sawtooth distribution with maximum temperature at the central plane and a horizontal orientation of the enclosure. Ra=6,000, Pr=10 [33]

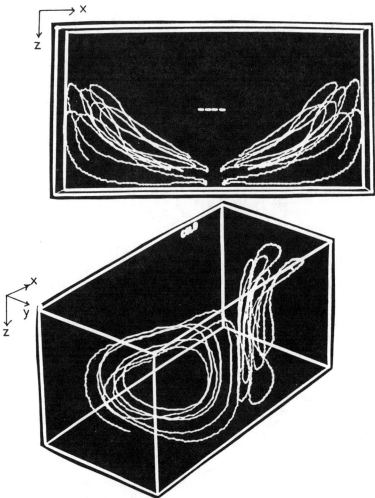

Fig. 6. Top and perspective view of streaklines for a double sawtooth distribution with maximum temperature at the central plane and a 2 degree inclination of the enclosure about the long (x) axis. Ra=6,000, Pr=10 [33]

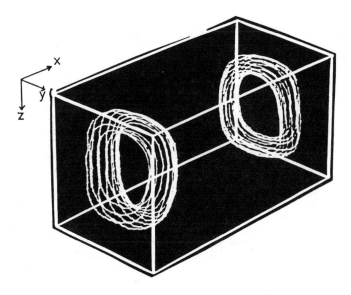

Fig. 7. Perspective view of the streaklines for a double sawtooth distribution with a maximum temperature at the central plane and a 60 degree inclination of the enclosure. Ra=6,000, Pr=10 [33]

as the angle is increased, two roll cells are formed, and as it is increased further (to 35π/180 rad), a single cell is formed again, flowing with the direction of gravity.

(3) Previous 2-d analytical studies and experiments by Chu and Churchill [44,45] of the effect of localized heating in square channels analyzed the effect of the position and width of a strip heater installed in the vertical wall of a horizontally-long channel, on heat transfer and circulation, for Ra $\leq 10^5$, Pr = 0.7.

Only minimal work has been done on the effect of radiation on natural convection in enclosures. Since the media typically considered (mostly air) are transparent in such thickness, the radiation effects only the bounding surfaces. Edwards and Sun [46] determined analytically that the existence of radiation increases the critical Ra number because it tends to smooth out any surface temperature perturbations. Normally, radiation is at present considered to be calculable independently of the natural convection phenomena, and the resulting convective and radiative heat fluxes are simply added to find the total heat flux, but more work is needed to determine the existence or extent of any coupling effects.

All of the above studies have dealt with the laminar regime. Attempts to analyze turbulent natural convection in enclosures have just started recently (Hjertager and Magnussen [47], Fraikin et al [48]),by using numerical methods. The latter researchers apply the k-ε turbulence model [49] with the time-averaged conservation equations of mass, momentum, and energy to the horizontal box with one vertical wall at a temperature higher than the other, and solve the problem by an ADI finite difference method. Reasonable agreement was obtained with some limited experimental data, and significantly more work is needed for a satisfactory solution of this class of problems.

Even if an economical way is found to suppress natural convection altogether, the air in the enclosure will still transfer heat by conduction. It became obvious as early as 1911 [50] that the evacuation of the air from the space around the solar energy absorber will reduce heat losses and improve collector efficiency. These type of collectors were successfully built(by Owens-Illinois, Corning, Sunmaster, Sanyo, Phillips, Philco Italiana and General Electric), attaining efficiencies approximately twice as high as comparable non-evacuated collectors, and some performance analyses were done (cf. [4, 51-53]). It is noteworthy that the pressure needs to be reduced to lower than 0.1 mmHg if the conductivity of the air is to be reduced to 1% of its value at atmospheric pressure (Fig. 8). In the absence of air, the heat transfer problem is principally one of radiative exchange, with some conduction through the glass window.

HEAT TRANSFER AND FLOW DISTRIBUTION WITHIN RADIANT-CONVECTIVE FINNED-TUBE MANIFOLD ASSEMBLIES

The present state-of-the-art of thermal and fluid flow analysis for radiant-convective finned-tube manifold assemblies such as those typically used

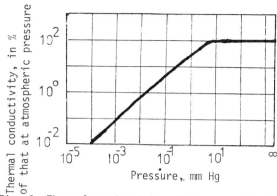

Fig. 8. Thermal conductivity of air as a function of pressure

for flat plate solar heat collectors includes many assumptions that simplify the governing equations sufficiently so that an algebraic closed-form solution for the thermal performance is obtained. The most notable of these includes a "lumped" overall thermal loss coefficient, one dimensional fin and tube temperature distributions, neglection of buoyancy effects within the coolant, thermal symmetry at the tube or fin centerlines and approximate radiant interchange mechanisms to name a few (cf. Duffie and Beckman [54]). Such assumptions may provide accurate design-quality results in some cases while in many others, where nonstandard finned-tube geometries are considered, their application is questionable. More importantly, the assumptions preclude a detailed understanding of the heat transfer and fluid flow processes.

In order to obtain this understanding and to predict overall finned-tube thermal and fluid performance, a conjugated analytical model of the thermal and fluid-flow processes which occur within a radiant-convective finned-tube manifold assembly was developed by two of the authors (Jones and Lior). The formulation yields a system of nonlinear algebraic and second-order nonlinear partial differential equations that are solved simultaneousoy with a system of linear second-order partial integro-differential and linear integral equations. Because of the complexity of the governing equations, numerical solutions are obtained. Contributions include the quantification of the effects of: 1) combined convection and simultaneous flow development within the riser tube fluid; 2) riser tube fluid maldistribution; 3) semi-gray diffuse fin and tube radiant interchange; and 4) direct and diffuse high temperature-source incident radiation at finned-tube surface, on the finned-tube manifold assembly thermal and hydrodynamic performance.

The governing equations are solved numerically by employing finite differences. All integrals are approximated by sixth-order Simpson's rules with the end correction terms approximated by fourth-order first difference expressions.
The resulting system of algebraic equations are solved by a novel combination of Newton-Raphson, Gauss-Seidel iteration and matrix decomposition numerical techniques.

Solutions for several limiting cases have been obtained to date. Specifically, for the case of isothermal riser tubes for both parallel and reverse hydordynamically developed flow, the flow within the riser tubes was seen to become more nonuniform as the ratio of riser tube to manifold diameter, riser tube spacing and the number of riser tubes increased and as the riser tube length decreased. Also in general the riser tube flow maldistribution is less and the overall dual-manifold system pressure drop is greater for parallel versus reverse flow (Jones and Lior [54]).

Preliminary heat transfer results for a simplified one-dimensional fluid temperature distribution case shows that riser tube flow maldistribution (Fig. 9) causes about a 7% nonuniformity in finned-tube temperature distribution for the case investigated (Fig. 10). In addition, a 70% circumferential variation in tube-wall temperature is noted. State-of-the-art thermal performance models cannot make these predictions and would provide erroneous results should accuracy be required. The flow and heat transfer problem inside the individual riser tubes of the collector's absorber involves the development of the momentum and thermal boundary layers from the tube's entrance, as well as that of buoyancy due to the heating of the tube walls. This complex problem was solved for the horizontal tube by Yao [56] and for the vertical tube by Lawrence & Chato [57] and Collins [58], but never for nonuniform tube-wall thermal boundary conditions as it occurs in the real situation. Jones and Lior have solved the full problem for flow and heat transfer in a tube of arbitrary inclination and wall boundary conditions, and with flow development and buoyancy effects, by addressing two limiting cases: (1) the horizontal tube with flow development but with buoyancy effects excluded, which provides the lowest thermal performance and (2) the vertical tube with flow development and buoyancy included, which provides the highest thermal performance. It should be noted that buoyant effects influence thermal performance in two ways: they improve internal convection and also reduce flow maldistribution among the manifolded risers when the main circulation is against gravity. The influence of flow development, buoyancy, and wall boundary conditions is shown in Fig. 11. It is noteworthy from Fig. 11 that Nu for the linearly increasing wall temperature case is 28%-40% larger than for the constant temperature case (both with buoyancy effects included).

The solution of the complete conjugate problem, i.e., (1) the fin-tube, with (2) the flow at heat transfer in the tube, is in progress by an iterative method.

PASSIVE SOLAR HEATING AND COOLING

Passive solar heating and cooling is the method by which the building itself is used as a solar energy collector (for heating) or heat dissipator (for cooling), with minimal utilization of mechanical devices to circulate fluids. Thermal storage is also incorporated as an integral part of the building, such as in heavy walls, floors, etc. Heat into or out of the heated space is transferred by conduction through the walls, by radiation to and from walls and through windows, and by convection from the walls. Inherently with this method, the interior room air circulation is by natural convection, while the

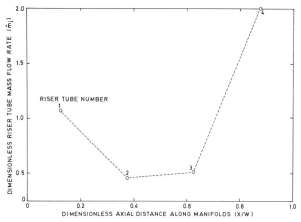

x = dimensional coordinate along manifolds
w = overall manifold width
$\hat{m}_j = \dot{m}_j/(\dot{m}_T/n)$ m_j = actual riser tube flow rate
$\dot{m}_T$ = total dual-manifold   n = number of riser system flow rate   tubes

Fig. 9. Dimensionless riser tube flow rate for case of four-riser tube dual-manifold system

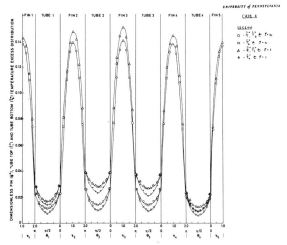

η = Dimensional axial distance along fin
θ = Angular location on tube circumference

Fig. 10. Fin-tube absorber temperature distribution, for the flow distribution of Fig.9

exterior is also influenced by wind. Proper natural ventilation of buildings, even without attempting solar heating, is being increasingly recognized as an important element of energy conservation (cf. [61].

The initial work in this field (Balcomb et al [62]) utilized a conventional electric analog model for the heat transfer, employing constant average values for heat transfer coefficients, and found acceptable interior temperature variations when adequate thermal storage was applied. It is obvious that more knowledge of natural convection in room- or building-type geometries, especially in the multiple-room case, is needed. Although the convection is typically turbulent (Ra > 10[10]), subsequent analyses [63,25] only dealt with the laminar case, utilizing two-dimensional finite difference methods.

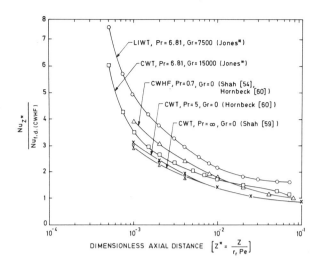

$$z^* = \frac{z}{r_r \, Pe}$$

$$Pe = Re \, Pr$$

$$Re = \frac{2\bar{w} \, r_r}{\nu} \quad , \quad \bar{w} = \text{mean axial velocity}$$

z = axial location

r_r = inside radius of tube

$$Nu_{fd(CWHF)} = 48/11, \quad Gr = \frac{g\beta r_r^3 (\bar{T}_w - T_{fi})}{\nu^2}$$

$$Nu_{z^*} = \frac{2h \, r_r}{K_f} \quad , \quad \bar{T}_w = \text{mean wall temperature}$$

T_{fi} = inlet fluid temperature

Legend

Symbol	Description	Source
o	linearly increasing wall temperature (from the bottom) Pr=6.81, Gr = 7500	Jones
△	CWHF, Pr = 7, Gr = 0	Shah [54] and Hornbeck [60]
□	CWT, Pr = 6.81, Gr = 15,000	Jones
X	CWT, Pr = 5, Gr = 0	Hornbeck [60]
X̱	CWT, Pr = ∞, Gr = 0	Shah [59]

Fig. 11. Simultaneous flow development within vertical tubes having uniform entrance velocity profile for large Pr fluids for various thermal boundary conditions for pure forced,or combined,convection

The latter study also incorporated experiments in a small two-dimensional enclosure with water as the fluid. Significantly more work is needed in these areas to understand and characterize the heat transfer phenomena in realistic configurations.

Passive solar cooling can be obtained by radiation from the surface that needs to be cooled to the sky (which under clear air conditions can have an infrared temperature around 30°F below ambient); by circulation of cooler night-air through the building, or by evaporation of water. For example, an ambient temperature surface can dissipate between 10 Btu/hr ft² and 20 Btu/hr ft²

by radiant exchange with the natural sky. This method is now being explored for practical applications (cf. Hay and Yellott [64, 65], Niles [66], Givoni [67, 68], Agha and Lior [69], Clark and Allen [70]). Work is underway to evaluate the radiant properties of the sky [71], to identify or develop materials which are spectrally selective for minimal absorption of radiation and maximal emission [72-74], and to understand the combined radiative-convective heat transfer problem. In the latter, wind has a strong influence on nullifying any cooling effect obtained by radiation, and thus wind-screens are being considered (cf. [74, 76]). The problem of combined free and forced convection, with spectrally dependent radiation through a non-uniform atmosphere, still remains to be solved for this application.

SOLAR PONDS

Salt-Gradient Ponds

This saline water pond (for early work see Kalecsinsky [77], Tabor [78], Tabor and Matz [79]), ranging in depth from a fraction of a meter to several meters, is exposed to the sun which heats its absorbing bottom. It is installed with an internal vertical salt gradient, with the highest concentration at the bottom and the lowest at the top, so that a density gradient is created to counteract the density gradient arising from the fact that the bottom is warmer than the top. Internal convection is thereby minimized, and so are thus the heat losses from the pond to the ambient air. The body of the pond also serves as storage for the thermal energy collected. Experimental ponds have reached bottom layer temperatures of slightly below boiling, while the surface layer is at ambient or slightly lower (due to evaporation or night sky radiation). A pond can even freeze in winter and yet supply heat from its storage layer. The warm water is drawn from the storage layer, and water is returned to the pond, by devices which minimize mixing. Alternatively, the heat can be withdrawn by an in-pond heat exchanger.

This type of solar pond introduces a large number of extremely interesting heat transfer problems: propagation of solar radiation through the water; thermohaline stability and convection under the influence of radiation and a free surface (which may be evaporating); free surface phenomena such as radiation reflectance, evaporation, surface tension effects, influence of wind and waves; influence of rain or ice; bottom and perimeter heat loss; and the thermal capacity of the pond and the ground surrounding it. Reviews of the state of the art (Nielsen in[80], Sargent and Neeper [81])and existing analyses (cf. Weinberger [82], Rabl and Nielsen [83], Hull [84]),indicate that the heat and mass transfer analysis of the ponds, including naturally the associated fluid-mechanics aspects, are in their infancy. Furthermore, these analyses have not even tapped as yet some existing fundamental information and research methods related to thermohaline stability and convection, whichwere generated through continuing interest in oceanography and limnology (cf. Turner and Stommel [85], Veronis [86, 87], Nield [88], Turner [89], Shirtcliffe [90], Huppert and Linden [91]). Recent studies on free surface phenomena associated with warm water exposed to the ambient (cf. [92-94]), and especially the research by Viskanta and his co-workers [95-99] which included the study of absorption of solar radiation in fresh water, and wind effects, are important first steps in the understanding of the solar pond problem, but much more work needs to be done to characterize all of the aspects of the problem.

Shallow Solar Ponds

The pond here is only a few cms. deep, and a transparent (usually plastic) cover is in contact with the top surface of the water to prevent evaporation (Clark and Dickinson in [80]). An insulating air space is usually created above the pond by adding a second transparent cover. Although the heat transfer problem is significantly simpler than that of the salt-gradient pond, the existing level of analysis (lumped system energy balance) does not take into account either radiative transfer in the liquid, nor questions of stability and natural convection which, if applied, may lead to a better understanding and design of these solar collectors.

OCEAN-THERMAL ENERGY CONVERSION (OTEC)

This method uses the temperature difference between the warmer surface water of the ocean (about 80°F) and the colder deep water (about 40°F at 3000 ft) to drive a power cycle and generate electricity. The energy resource, compared to direct use of solar energy, is continuously available, day and night. The small available temperatures require, however, large hardware and heat exchangers. Two principal types of cycles are under development: a closed one and an open one. In the closed cycle, the power cycle fluid is ammonia or a Freon, circulated in a closed loop, where the warm seawater is pumped through one heat exchanger to evaporate the working fluid, and the deep, cold water is circulated through the power cycle's condenser. In the open (Claude and Boucherot [100]) cycle, the warm seawater is deaerated and flashed in a chamber, in which a part of it is thus converted to steam. The steam passes through a turbine and condenses in a closed or open condenser. In the first case, the fresh-water condensate can be utilized, and in the second, the condensate mixes with the cold seawater and is discharged subsequently into the ocean. One variant of the open cycle concept under evaluation is the steam-lift approach [101], where the warm surface water is arranged to evaporate in such a way that it entrains with the vapor a large quantity of seawater into the condenser positioned at a higher elevation. From there, the water's potential energy is used to drive a hydraulic (instead of a steam) turbine. The lift is generated either by foaming induced by adding surfactant [102] or by generating a mist.

In the closed cycle concept, a significant number of projects is underway to study the evaporation and condensation characteristics of ammonia (and to some extent, Freon), with an effort specifically oriented to enhance the heat transfer coefficients and to design more compact and economical heat exchangers (cf. [103]). Of very important impact on these heat exchangers is biofouling, by seawater, which needs to be considered in conjunction with the heat transfer question.

In the open cycle concept, the major insufficiently known heat/mass transfer questions are: (1) noncondensable gas desorption in flashing seawater, and (2) flash evaporator performance, both at ambient temperatures. To improve heat transfer coefficients, and reduce cost and biofouling, direct contact condensers (vapor on cold seawater), and free surface type flash evaporators are being considered.

The flash evaporators considered are either of the open channel type, such as used in the multi-stage flash evaporation process of water desalination, or, in an attempt believed to enhance evaporation rates, they could be of the turbulent falling film or jet types. Although some (albeit incomplete) knowledge exists about flash evaporation of free-surface horizontal streams of seawater at temperatures of 125-250°F (cf. Lior and Greif [104], Fujii et al [105]), and of falling films (Jansen and Owzarski [106], Van der Mast et al [107]), practically no information is available for flash evaporation of seawater at around 80°F. Extrapolating from the existing higher temperature data, it is likely that the mode of evaporation changes gradually from the bubble-nucleation type at higher temperatures, to surface evaporation at the lower temperatures, thus entirely changing (and most likely slowing down) the heat transfer process. This is most likely caused by the fact that the hydrostatic head plays an increasingly important role in depressing bubble nucleation as the saturation temperature (and hence pressure) go down. The evaporation problem is complicated further by the need to desorb the evaporating liquid from noncondensable gases, so they do not get to the condenser in any significant quantity. The Schmidt number of the dissolved gas is, however, about 50 times larger than the Pr number. Consequently, a unit designed for the needed amount of evaporation may desorb only a small fraction of the dissolved gas, a fact which needs to be both analyzed further, and taken into account in the design of the evaporator.

Condensation is accompanied by the severe problem of noncondensables, which even in very small concentrations reduce condensation rates significantly, especially in direct contact condensers which have potentially high heat transfer coefficients. Increasing the steam flow rate across the condensing surface will improve condensation by removing gas from the interface, and is desirable if it can be attained without undue pumping power loss. Similarly to the case of flash evaporation, the coupled heat and mass transfer two-component problem of condensation has a data base (cf. [108-118]), but not for the conditions applicable to open cycle OTEC. Mostly under government sponsorship, work is presently underway in the U.S., Japan and Europe [103] to study some of these fundamental problems along with the overall study and design of OTEC pilot plants.

THERMAL STORAGE

Economical energy storage offers one of the most important ways to improve the efficiency of energy use, particularly when both the energy source and the load are transient. The storage of heat (or coolness) is a major method of energy storage, and can be employed through sensible heat, phase change, or thermochemical processes (cf.

Lior et al [119], Dickinson and Cheremisinoff [80]), all of which introduce challenging heat transfer problems.

Sensible Heat Storage

Here the thermal storage material is heated by the working fluid, gaining temperature and thermal energy. The storage material is a solid or a liquid, and does not change phase. Solids such as rock, sand, iron, and salts have been considered, and typical liquids are water, oil, or some other special organic liquids. The major disadvantage of this system is the fact that relatively large temperature swings are needed for a practical amount of storage material, which is somewhat detrimental to the end user.

The analysis of sensible heat storage units incorporating solid storage media is typically based on conventional transient heat conduction methods, such as employed with regenerators (cf. [54, 120-123]) and usually employ a one-dimensional solution, constant fluid-to-solid heat transfer coefficients, and constant properties. Analytical solutions with such restrictions were obtained by Yang and Lee [124], and a finite-difference solution of the one-dimensional fluid problem, but for two-dimensional conduction in the solid, was obtained by Schmidt and Szego for rectangular and cylindrical geometries and for the single fluid and two-fluid (one heated and the other simultaneously cooled, each flowing at a different side of a thermal storage solid)[125-127]. The transient temperature distributions and energy storage capacity were shown for different combinations of the governing parameters: the Fourier, Biot and Graetz Numbers. Improvements in the analysis can be obtained by accounting for the entry-length problem, by allowing properties' variation with temperature, and by performing a three-dimensional analysis, when warranted by the fluid or configuration contemplated.

With liquids as storage media, or with granular solids where the working fluid causes their thermal stratification, the question of natural convection in the storage system becomes of great interest. Such convection is complicated by the fact that it is not only influenced by the boundary conditions and sometimes by a two-phase medium, but also by the forced introduction and withdrawal of the fluid to and from the storage enclosure. Stratification allows the separation of the warmer portions of the system from the cooler discharged ones, so that working fluid can always be extracted at the highest possible temperatures. A detailed analysis of this problem is in progress by Lin and Sha [128] who have developed a three-dimensional transient thermohydrodynamic program (COMMIX-SA), utilizing the semi-implicit ICE finite difference method. The program allows for arbitrary in- or out-flows, and arbitrary internal tank geometries, to include such elements as baffles for the control of flow and natural convection. These are represented as non-linear velocity-dependent resistance-force functions which are introduced into the momentum equations. The solution provides both the transient temperature and flow fields in the tank, and has allowed the assessment of the influence of tank and internal baffle geometry, as well as of the mode and

geometry of fluid intake and discharge, on the stratification and the transient discharged fluid temperature.

Another application of sensible heat storage is through the utilization of a "steam accumulator", which was fairly common in Europe [129]. While several versions of the system are in existence, it essentially utilizes steam to heat water by direct contact condensation (or may just mix heated water with stored water), and steam is discharged from the accumulator when needed, by slight pressure reduction and flashing. One important advantage of this system may be that it can use the same fluid for the storage and for the needed process (such as power cycle). This type of storage is now under analysis and development by one of the authors (cf. Lior et al [130]), in a solar powered Rankine cycle cooling system.

Phase-Change Storage

Storing and recovering heat through the phase change process is advantageous due to the fact that the latent heat of most materials is much higher than their specific heat, thus requiring a much smaller amount of storage material. Furthermore, the thermal storage process occurs at a constant temperature, which is typically desirable for the needed process (during the discharge period). The heat transfer and design aspects for latent heat storage are much more complex than those in the sensible heat case. The problem is nonlinear and has analytical solutions for only a few, typically unrealistic, cases. For that reason, energy integral or numerical methods are usually employed. The problem is further complicated due to uncertainty in: (a) interfacial thermal resistance between container walls and fluid; if the fluid does not wet the container, it may collect in the form of drops, reducing markedly the interfacial heat transfer area, (b) interface resistance between walls and solids, (c) volume change with change of phase; upon shrinkage, void cavities arise in the frozen melt, and impede the heat transfer process significantly. Void formation was found to also be dependent on the rate of cooling; smaller (but more numerous) cavities being produced for faster rates of cooling). The configuration of the voids may also introduce complex new phenomena of internal radiation and natural convection, (d) scarcity of information on the properties of many phase-change storage candidate materials.

It is beyond the scope of this paper to review the general literature on heat transfer in phase-change problems, and only the highlights of new work directly related to phase change storage will be presented. Shamsundar and Sparrow [131] employed an approximate series solution for the freezing cylinder problem, and later developed an enthalpy model in which the interface does not need to be considered in the calculation, and the problem is thus transformed into one of nonlinear heat conduction without change of phase. The model is solved for the same configuration by a finite-difference method, but for constant boundary conditions, which do not reflect the actual situation in a pcm (phase-change material) heat exchanger, where the conjugate problem with the working fluid(s) needs to be solved. Yang and Lee [124] found an analytical solution to such a problem but they neglected the thermal resistance of the melting solid or the freezing liquid, which could have a significant effect on the process. A more realistic problem was solved by Lior et al [119], for the geometry shown in Fig. 12. A one-dimensional constant-property flow model was developed. There, $T_e(x)$ is the bulk average fluid temperature at any x. There are N/2 slabs of phase change material (N half-slabs) each 2d thick, Y wide, and separated by a distance 2G. The thermal resistance of the containment material as well as density variations during fusion/melting were neglected.

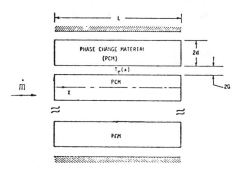

Fig. 12. Geometry for phase change storage

The slab was assumed to initially contain all solid for the charging process and all liquid for the discharging process at the melting temperature (T_M). Consider the discharging process. At t=0, cold fluid to be heated begins its flow over the slab, and the slab as it gives off heat begins to freeze. The downstream portions of the slab constantly see a higher temperature fluid since the fluid is heated as it flows along the slab. Consequently, the downstream portions freeze at a slower rate. Thus, a freezing front profile is established within the phase change material. As time progresses, more and more of the material is frozen and the front moves downstream. The variables are defined by

$$\theta = T_M - T/T_M - T_i, \quad \theta_e = T_M - T_e/T_M - T_i, \quad \eta = z/d$$

z* = position of interface

$\eta^* = z^*/d$, $\xi = x/L$ and, $\tau = t/(\rho \lambda d^2/k(T_M - T_i))$

The parameters are defined by the Stefan number, S, Biot number, B, where

$$S = \frac{C(T_M - T_i)}{\lambda}, \quad B = \frac{hd}{k}, \text{ respectively}$$

and

$$\gamma = L/V/\rho \lambda d^2/k(T_M - T_i), \quad H = \frac{YhL}{\dot{m}C_f}$$

The parameters λ and H are the ratio of the characteristic flow time to the characteristic freezing time, and the energy source strength per unit non-dimensional temperature difference, respectively. Precisely, the same equations govern the charging process so long as λ is considered to have a negative value and the material properties of the liquid are used.

The equations are solved by an energy integral technique under two limiting conditions. The first condition is that the Stefan number is zero. The error incurred by such an assumption is graphically displayed in Goodman [132] for the case of convective cooling-heating by a constant temperature fluid. There, the results indicate

that this assumption leads to a maximum error of 10-20% in the prediction of the interface motion for the conditions assumed in the analysis. The second limiting condition is that γ is equal to zero. Since the characteristic flow times are of the order of seconds or minutes and the freezing times are of the order of hours, this should be a reasonable condition. The integral equations are solved via a finite difference formulation. Fig. 13 shows the phase front position $\eta^*(\xi,\tau)$

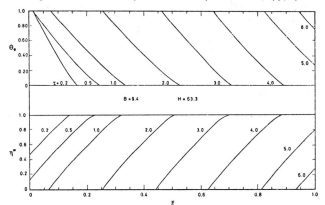

Fig. 13. Temperature and phase boundary location vs. axial position (all nondimensional) [119]

and the temperature profiles $\theta_e(\xi,\tau)$ for typical values of B and H. The similar propagation of the phase boundary proposed by Matveev [133] is evident, but his assumption of a relatively short phase boundary development is not borne out. An instantaneous, spatially-averaged heat transfer coefficient U, based on the characteristic terminal temperature difference $(T_M - T_i)$ is plotted (in the form of some type of Nusselt number) versus τ in Fig. 14.

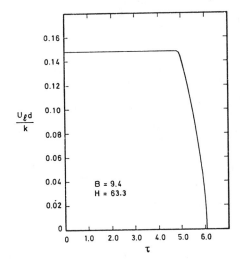

Fig. 14. Spatially-averaged overall Nusselt number vs. dimensionless time [119]

An essentially constant value of Ud/k=0.147 (for B=9.4, H=63.3) is observed until the phase boundary reaches the exit plane ($\tau \approx 4.7$) at which time the performance begins to seriously deteriorate due to the decreasing area involved in the heat exchange. This model was used to design the storage system for a solar power plant. It is noteworthy that the storage-cum-exchanger may need to provide preheat of the water, boiling, and superheating of the steam. Shamsundar and Sparrow [134] extended the use of their enthalpy model to solve multidimensional solidification in the presence of a growing shrinkage cavity, which is wholly located at the top of the freezing medium. Results were obtained for the transient temperature profiles, interface and cavity shapes, heat fluxes, and freezing rates. This work, as well as the quantification of the important effects of natural convection on melting performed by Sparrow and his co-workers [135-138] and Viskanta and his co-workers [139-141], help de-idealize the heat transfer problems associated with phase change storage systems and bring us closer to their better understanding and design.

REFERENCES

1. Deyoe, D. P., ASHRAE J., 36-38 (April 1973).

2. Basiulis, A., and Plost, M., ASME paper 74-WA/HT-48 (1975).

3. Mahdjuri, F., Proc. Solar Energy and Conservation Symp., Miami Beach FL, 11-13 Dec. 1978.

4. Ortabasi, U. and Fehlner, F. P., Solar Energy 24, 477-489 (1979).

5. Mull, W. and Reiher, H., Beih. Z. Gesundh,-Ing. 28, 1-28 (1930).

6. Ostrach, S., "Natural Convection in Enclosures." chapter 12, Advances in Heat Transfer, 8 (J.P. Hartnett and T. F. Irvine, Jr., eds.) Academic Press (1972)

7. Catton, I., Proc. 6th Int. Heat Transfer Conf., Toronto, Canada, 6, 13-31 (1978).

8. Veinberg, V. B., Optics in Equipment for the Utilization of Solar Energy, State Publishing House of Defense Industry, Moscow (English translation), 1959.

9. Hollands, K.G.T., Solar Energy 9, 159-164 (1965).

10. Hollands, K.G.T., Trans. ASME J. Heat Transfer, 95C, 439-444 (1973).

11. Buchberg, H., Lalude, O. and Edwards, D. K., Solar Energy 13, 193-221 (1971).

12. Buchberg, H., Catton, I., and Edwards, D. K., Trans. ASME J. Heat Transfer, 98C, 182-188, (1976).

13. Lior, N. and Saunders, A. P., Solar Collector Performance Studies, Report NSF/RANN/GI27976/ TR73/1, University of Pennsylvania, 1974.

14. Cane, R.L.D. et al, Trans. ASME J. Heat Transfer 99C, 86-91 (1977).

15. Arnold, J. N. Edwards, D. K. and Catton, I., Trans ASME J. Heat Transfer 99C, 120-122 (1977).

16. Smart, D. R., Hollands, K.G.T. and Raithby, G.D., Trans. ASME J. Heat Transfer, 102C, 75-80 (1980).

17. Hollands, K.G.T. and Konicek, L., Int. J. Heat Mass Transfer 16, 1467-1476 (1973).

18. Hollands, K.G.T., Raithby, G.D., Konicek, L., Int. J. Heat Mass Transfer 18, 879-884 (1975).

19. Hollands, K.G.T. et al, Trans ASME 98C, 189-193 (1976).

20. Raithby, G. D., Hollands, K.G.T., and Unny, T.E., Trans ASME J. Heat Transfer, 99C, 287-293 (1977).

21. El Sherbiny, S. M., Hollands, K.G.T. and Raithby, G. D., Trans ASME J. Heat Transfer 100C, 410-415 (1978).

22. Ruth, D. W., Hollands, K.G.T., and Raithby, G.D., J. Fluid Mechanics, 96, 461-479 (1980).

23. Ruth, D. W., Raithby, G. D. and Hollands, K.G.T., J. Fluid Mechanics 96, 481-492 (1980).

24. El Sherbiny, S. M., Raithby, G. D., and Hollands, K.G.T., ASME Paper 80-HT-67 (1980).

25. Bauman, F., Gadgil, A., Kammerud, R., and Greif, R., ASME Paper 80-HT-66 (1980).

26. Aziz, K. and Hellumns, J. D., Phys. Fluids, 10, 314-324 (1967).

27. Ozoe, H., et al, Trans. ASME J. Heat Transfer, 98C, 202-207 (1976).

28. Ozoe, H., Sayama, H. and Churchill, S. W., Int. J. Heat Mass Transfer 20, 123-129 (1977).

29. Ozoe, H., Sayama, H. and Churchill, S.W., Int. J. Heat Mass Transfer 20, 131-139 (1977).

30. Ozoe, H., et al, Proc. PACHEC 77, 1, 24-31 (1977).

31. Ozoe, H., et al, Proc. 6th Int. Heat Transfer Conference, 2, 293-298 (1978).

32. Ozoe, H., Okamoto, T. and Churchill, S. W., Heat Transfer-Japanese Research, 8, 82-93 (1979).

33. Chao, P., Ozoe, H. and Churchill, S. W., accepted for publication in Chem. Eng. Comm. (1980).

34. Mallinson, G. D. and de Vahl Davis, G., J. Fluid Mechanics, 83, 1-31 (1977).

35. Chan, A.M.C. and Banerjee, S., Trans. ASME J. Heat Transfer 101C, 114-119 (1979).

36. Chan, A.M.C. and Banerjee, S., Trans. ASME J. Heat Transfer, 101C, 233-237 (1979).

37. Ozoe, H., Sato, N. and Churchill, S.W., Int. Chem. Engrg. 19, 454-462 (1979).

38. Wilkes, J. O., and Churchill, S. W., AIChE J., 12, 161-166 (1966).

39. Davis, S. H., J. Fluid Mechanics 30, 465-478 (1967).

40. Hart, J., J. Fluid Mechanics, 47, 547-576 (1971).

41. Ozoe, H., Sayama, H., and Churchill, S.W., Int. J. Heat Mass Transfer 17, 401-406 (1974).

42. Ozoe, H., et al, Int. J. Heat Mass Transfer 17, 1209-1217 (1974).

43. Ozoe, H., Sayama, H. and Churchill, S.W., Int. J. Heat Mass Transfer, 18, 1425- 1431 (1975).

44. Chu, H.H.-S. and Churchill, S. W., Trans. ASME J. Heat Transfer 98C, 194-201 (1976).

45. Chu, H.H.-S. and Churchill, S. W., Computers and Chem. Eng., 1, 103-108 (1977).

46. Edwards, D. K. and Sun, W. M., Int. J. Heat Mass Transfer, 14, 15-18 (1971).

47. Hjertager, B. H. and Magnussen, B. F., in Heat Transfer and Turbulent Buoyant Convection, 2, 429-442, D. B. Spalding and N. Afgan, eds., Hemisphere, Washington, D. C. 1977.

48. Fraikin, M. P., Portier, T. J., and Fraikin, C. J., ASME Paper 80-HT-68 (1980).

49. Launder, B. E. and Spalding, D. B., Computer Methods in Appl. Mech. and Engrg, 3, 269-289 (1974).

50. Emmett, W.L.R., U.S. Patent No. 980,505, January 3, 1911.

51. Speyer, E., Trans. ASME J. Engrg. for Power, 87, 270 (1965).

52. Roberts, G. T., Solar Energy 22, 137-140 (1979).

53. Felske, J. D., Solar Energy 22, 567-570 (1979).

54. Duffie, J. A. and Beckman, W. A., Solar Energy Thermal Processes, John Wiley, New York (1974).

55. Jones, G. F. and Lior, N., Proc. Annual Mtg. AS/ISES, 2.1, Denver, CO., 362-372 (1978).

56. Yao, L. S., Trans. ASME J. Heat Transfer, 100C, 212-219 (1978).

57. Lawrence, W. T., and Chato, J. C., Trans. ASME J. Heat Transfer, 88C, 214-222 (1966).

58. Collins, M. W., Sixth Int. Heat Transfer Conf. Toronto, Canada, 1, 25-30 (1978).

59. Shah, R. K., Proc. Nat'l Heat Mass Transfer Conference, Bombay, India, 1 (1975).

60. Hornbeck, R. W., ASME Paper 65-WA/HT-36 (1965).

61. Milbank, N. O., Energy and Buildings 1, 141-145 (1977).

62. Balcomb, J. D., Hedstrom, J. C., and McFarland, R. D., Solar Energy, 19, 277-282 (1977).

63. Akbari, H. and Borgers, T. R., Solar Energy 22, 165-174 (1979).

64. Hay, H. R., and Yellott, J. I., Trans. ASHRAE 75, 165 (1969).

65. Hay, H. R., and Yellott, J. I., Trans. ASHRAE 75, 178 (1969).

66. Niles, P.W.B., Solar Energy 18, 413 (1976).

67. Givoni, B., Man, Climate and Architecture, Applied Science Publishers, London, 1976.

68. Givoni, B., Energy and Buildings 1, 141-145 (1977).

69. Agha, M. F., and Lior, N., Proc. Annual Mtg. of AS/ISES, Denver, CO, 2.2, 102-105 (1978).

70. Clark, G., and Allen, C. P., Proc. 14th IECEC, 1, 269-276 (1979).

71. Berdahl, P. and Martin, M., Proc. 2nd National Passive Solar Conference, AS/ISES 2.2, 684-686 (1978).

72. Catalanotti, S. et al, Solar Energy, 17 83 (1975).

73. Berdahl, P., University of California Berkeley Lawrence Laboratory Report LBL-9735, August 1979.

74. Sakkal, F., Martin, M., and Berdahl, P., University of California, Berkeley Lawrence Laboratory Report LBL-9697, August 1979.

75. Grenier, Ph., Revue de Physique Applique, 14, 87 (1979).

76. Addeo, A., et al., Solar Energy 24, 93-98 (1980).

77. Kalecsinsky, A. V., Ann. Phys. IV, 7, 408 (1902).

78. Tabor, H., Solar Energy 7, 189 (1963).

79. Tabor, H. and Matz, R., Solar Energy 9, 177-182 (1965).

80. Dickinson, W. C. and Cheremisinoff, P.N., Solar Energy Technology Handbook, Part A, Marcel Dekker, New York (1980).

81. Sargent, S. L., and Neeper, D. L., Proc. AS/ISES Annual Meeting, Phoenix, AZ, 3.1 395-399 (1980).

82. Weinberger, H., Solar Energy 8, 45-50 (1964)

83. Rabl, A. and Nielsen, C. E., Solar Energy 17, 1-12 (1975).

84. Hull, J., Solar Energy 25, 33-40 (1980).

85. Turner, J. S. and Stommel, H., Proc. Nat. Acad. Sci., 52, 49-53 (1964).

86. Veronis, G., J. Marine Research 23, 1-17 (1965).

87. Veronis, G., J. Fluid Mechanics 34 315-336 (1968).

88. Nield, D. A., J. Fluid Mechanics 29, 545-558 (1967).

89. Turner, J. S., J. Fluid Mechanics 33, 183-200 (1968).

90. Shirtcliffe, T.G.L., J. Fluid Mechanics, 35, 677-688 (1969).

91. Huppert, H. E. and Linden, P. F., J. Fluid Mechanics, 95, 431-464 (1979).

92. Ou, J. -W., Tinney, E. R. and Yang, W. J., in Flow Studies in Air and Water Pollution, ASME, New York, 9-23 (1973).

93. Katsaros, K. B., et al, J. Fluid Mechanics 83, 311-335 (1977).

94. Coantic, M. F., An Introduction to Turbulence in Geophysics, NATO Technical Report AGARD AG-232, 1978,

95. Viskanta, R. and Toor, J. S., Water Resources Research 8, 595-608 (1972).

96. Viskanta, R., and Toor, J. S., J. of Geophysical Research 78, 3538-3551 (1973).

97. Snider, D. M. and Viskanta, R., Trans ASME J. Heat Transfer, 97, 35-40 (1975).

98. Behnia, M. and Viskanta, R., Int. J. Heat Mass Transfer 22, 611-623 (1979).

99. Behnia, M. and Viskanta, R. in ASME Publication HTD-Vol. 8 Natural Convection in Enclosures, Torrance, K. E. and Catton, I., eds. 17-26 (1980).

100. Claude, G. and Boucherot, P., "Sur l'utilization de l'Energie Thermique de Mer",Proc. French Academie des Sciences (CRAS), 4 June 1928.

101. Beck, E. J., Science, 189, 293-294 (1975).

102. Zener, C., and Fetkovich, J., Science 189, 294 (1975).

103. Dugger, G. L., ed., Proc. Ocean Thermal Energy Conference, USDOE Conf-790631, Washington, DC, June 19-22, 1979.

104. Lior, N. and Greif, R., Proc. 5th Int. Symp. on Fresh Water from the Sea, 2, 95-106 (1976).

105. Fujii, T. et al, Heat Transfer-Japanese Research 5, 84-93 (1976).

106. Jansen, G. and Owzarski,   U.S. Dept. of Interior, O.S.W. Res. Devel. Progress Report 693 (1971).

107. Van der Mast, V. C., Read, S. M. and Bromley, L. A., Desalination 18, 71-94 (1976).

108. Chun, K. R. and Seban, R., Trans. ASME. J. Heat Transfer, 93C, 391-396 (1971).

109. Dukler, A. E., Chem. Engrg. Progress Symp. Ser., 56, 30, 1-10 (1960).

110. Mills, A. F., and Chung, D. K., Int. J. Heat Mass Transfer, 16, 694-696 (1973).

111. Limberg, H., Int. J. Heat Mass Transfer 16, 1691-1702 (1973).

112. Hubbard, G. L., Mills, A. F. and Chung, D. K., Trans. ASME J. Heat Transfer 98C, 319-320, 1976).

113. Seban, R. A. and Faghri, A., Trans. ASME. J. Heat Transfer 98C, 315-318 (1976).

114. Brumfield, L. K., Houze, R. N. and Theofanous, T. G., Int. J. Heat Mass Transfer, 18, 1077-1081 (1975).

115. Brumfield, L. K. and Theofanous, T. G., Trans. ASME J. Heat Transfer, 98, 456-502 (1976).

116. Benedek, S., Int. J. Heat Mass Transfer, 19, 448-450 (1976).

117. Theofanous, T. G., Houze, R. N. and Brumfield, L. K., Int. J. Heat Mass Transfer 19, 613-624 (1976).

118. Bakay, A., and Jaszay, T., Paper EC10, 6th Int. Heat Transfer Conf., Toronto, Canada (1978).

119. Lior, N. et al., Paper No. 769107, Proc. 11th IECEC, 613-622 (1976). Note: Errata were issued subsequently and are available from the authors.

120. Lambertson, T. J., Trans ASME 80, 586 (1958).

121. Bahnke, G. D. and Howard,C. P., Trans ASME. J. Engrg. for Power, A86, 105 (1064).

122. London, A. L.,Lampsell, D. F. and McGowan, J.G. Trans ASME J. Engrg. for Power, A86, 127 (1964).

123. Willmott, A. J., Int. J. Heat Mass Transfer 7, 1291 (1964).

124. Yang, W. J. and Lee, C.P., ASME Paper 74-WA/HT-22 (1974).

125. Schmidt, F. W. and Szego, J., Trans. ASME J. Heat Transfer, 98C, 471-477 (1976).

126. Szego, J., and Schmidt, F. W. Trans ASME J Heat Transfer 100C, 148-154 (1978).

127. Schmidt, F. W. and Szego, J., Trans ASME J. Heat Transfer, 100C, 737-739 (1978).

128. Lin, E.I.H., and Sha, W. T., Proc. ISES Silver Jubilee Congress, Atlanta GA 1, 586-590 (1979).

129. Goldstern, W., Steam Storage Installations, Pergamon Press, Oxford (1970).

130. Lior, N., Yeh, H. and Zinnes, I., Proc. AS/ISES Annual Neeting, Phoenix, AZ,1 210-214 (1980).

131. Shamsundar, N. and Sparrow, E. M., Trans ASME J Heat Transfer 96C, 541-544 (1974).

132. Goodman, T. R.,Trans. ASME, 80, 335-342 (1958).

133. Matveev, V. M., Geliotekhnika 7, 5 (1971).

134. Shamsundar, N. and Sparrow, E. M., Trans ASME J Heat Transfer, 98C, 550-557 (1976).

135. Shamsundar, N. and Sparrow, E. M., Trans ASME J. Heat Transfer, 97C, 333-340 (1975).

136. Sparrow, E. M., Patankar, S. V. and Ramadhyani, S., Trans ASME J. Heat Transfer 99C, 520-526 (1977).

137. Sparrow, E. M., Ramadhyani, S. and Patankar, S. V., Trans ASME J. Heat Transfer, 100C, 395-402 (1978).

138. Sparrow, E. M., Ramsey, J. W. and Kemink, R. G., Trans ASME J. Heat Transfer 101C, 578-584 (1979).

139. Bathelt, A. G., Viskanta, R. and Leidenfrost, W.; Trans ASME J. Heat Transfer, 101C 453-458 (1979).

140. Bathelt, A. G., Viskanta, R. and Leidenfrost, W., Trans ASME J. Heat Transfer, 101C, 227-241 (1979).

141. Bathelt, A. G. and Viskanta, R., ASME Paper 80-HT-10, (1980).

Acknowledgment: The recent work by the University of Pennsylvania authors of this study was supported in part by the U.S. Dept. of Energy, Solar Energy R & D Branch.

Heat Transfer in Thermo Chemical Energy Conversion System Utilizing Aqueous Solution of Inorganic Salts

N. ISSHIKI
Tokyo Institute of Technology, Ohokayama, Tokyo, Japan

I. NIKAI
Ishikawajima-Harima Heavy Ind. Co., Yokohama, Japan

J. KAMOSHIDA
Shibaura Institute of Technology, Omiya, Japan

ABSTRACT

Heat transfer and thermal performance of thermo-chemical Energy Conversion System utilizing aqueous solution of salts such as LiCl and $CaCl_2$ are studied in this report. The system is called CDE (Concentration Difference Energy or Potential) engine system here, and it is thought to be a typical new energy conversion system for picking and storage of various alternative energies of low temperature difference. From these results, a very high absorption heat transfer coefficient has been obtained in the absorbing evaporator, and it is thought to be caused by boundary layer agitation by the sudden collapse of bubbles of injected steam. Futhermore, it becomes clear that a high overall heat transfer coefficient enables the overall thermal efficiency of the absorbing evaporator to be higher than 0.9.

INTRODUCTION

In order to collect, store and utilize all kinds of alternative energy resources in our surroundings such as solar energy, geothermal energy, industrial waste heat, or excess electricity, many chemical working media are going to be used through various reversible thermochemical phenomena such as absorption, adsorption, dilution, hydrogenation, humidification and dissociation etc. Among them, the phenomena which causes a change of pressure of a particular media have the possibility to be used in the generation of "power" using the expander as well as the generation of "heat". In our laboratory, the combination of water(steam) and aqueous solution of inorganic salts as LiCl and $CaCl_2$ has been selected as the working media thermo-chemical reaction, and the CDE (Concentration Difference Energy or Potential) engine using these media has been studied experimentally with success in operation. A general description of this system has been reported in several previous reports (1) - (5) and (11).
Here in this report, the fundamental studies on heat transfer, such as absorption and boiling of aqueous solutions, are reported after some description of a general idea of thermo-chemical reactions and a particular CDE system. An especially high heat transfer coefficient is shown in an absorbing evaporator, and it enables the overall temperature efficiency of a CDE engine system to be high enough.

STORAGE AND GENERATION OF HEAT AND POWER BY THERMO-CHEMICAL REACTION

There are many reversible thermochemical reactions such as absorption of vapour by aqueous solutions, gas adsorption by metals, thermal dissociation of chemicals, humidification of dessicants, etc., which can be used in the storage of heat and power. Most of them can be expressed in general in the following relationship.

$$A + B \rightleftharpoons C \pm Q \text{ (heat)} \quad \dots\dots\dots\dots(1)$$

Generally, A corresponds to gas or vapour to be absorbed or adsorbed, B to highly concentrated solution or dry absorbent or adsorbent, C to thinner solution or wet absorbent or adsorbent, and Q to heat generated in each reaction.
For example, in the case of absorption of steam by any solutions, Line A corresponds to the saturation vapour pressure of water and, Line B and C to equilibrium vapor pressures of thick and thin solutions respectively, as seen in Fig.1.
If we add heat Q_2 at C_2, the reaction $C_2 \rightarrow A_2 + B_2$ proceeds. If A's steam is absorbed at C_1, heat Q_1 is regenerated again through the reaction $A_1 + B_1 \rightarrow C_1$.
The saturation pressure diagram (p ~ T diagram) of various solutions of salts is shown on Fig.2. As seen there, if we concentrate any solution, the line is shifted further to the right and energy is stored in CDE potential.
In Fig.3, the process of generation of power is illustrated by the generalized p ~ T diagram identical with Fig.1.
As seen there, if we feed heat Q_1 generated from the reaction $A_1 + B_1 \rightarrow C_1$ into A_1', the vapour or gas is produced at A_1' with higher pressure than that of A_1, so that mechanical power can be taken out through any expander set between A_1' and A_1, feeding the new vapour of A_1 into the original reaction.
In general, any reversible thermo-chemical reaction has a possibility of storage and production of "power" as well as "heat" if the phenomenon contains pressure change in the vapour or gas on the p - T diagram.

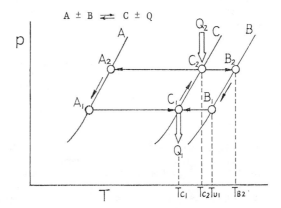

Fig.1. Reversible thermochemical reaction on p ~ T diagram

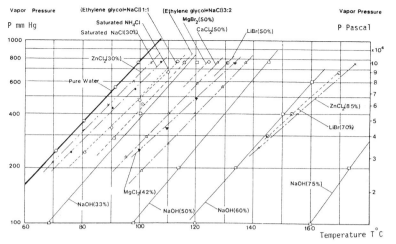

Fig.2. The saturation pressure diagram of aqueous solution of various salts

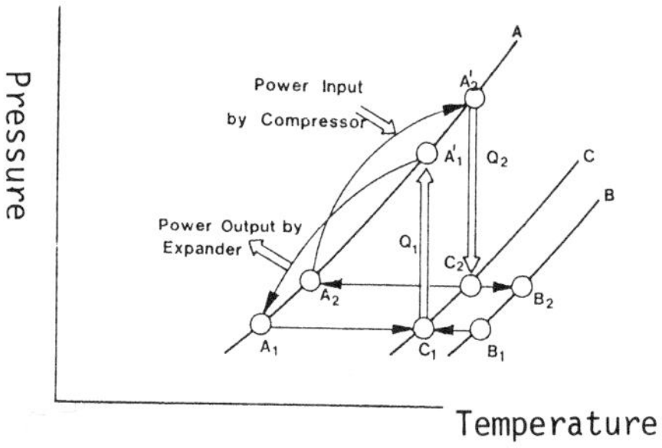

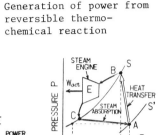

Fig.3. Generation of power from reversible thermo-chemical reaction

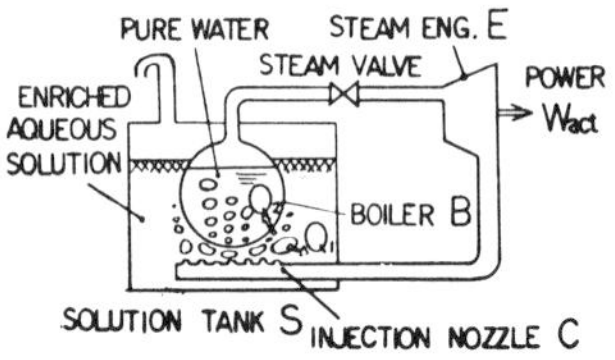

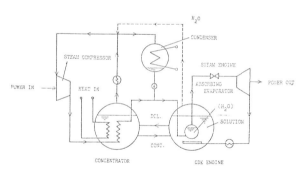

Fig.4. Basic CDE engine

Fig.5. p ~ T diagram basic CDE engine

CDE SYSTEM AND ENGINE USING AQUEOUS SOLUTION OF SALTS

Among the various reversible thermochemical reactions, we picked the phenomena of steam absorption into dense aqueous solution of inorganic salts and studied the CDE system and engine.

The CDE SYSTEM is one which collects, stores, transports, converts and utilizes thermal and mechanical alternative energy resources in surroundings such as solar, geothermal heat, waste heat from industry, wind, wave power, excess electricity, etc. using cyclic change of concentration of salt solvents in aqueous solutions, while the CDE engine is a device to liberate power in the process of concentration decrease of a solvent in dense solution by the absorption of steam into it.

The CDE engine consists of a pure steam boiler immersed in a thick aqueous solution of salts, and the exhaust steam from the steam engine is absorbed in the original solution liberating absorption heat. In Fig.4, a schematic construction of the basic CDE engine without concentrator is shown, and in Fig.5 the simplified p – T diagram of this engine corresponding to a special case of Fig.3 is shown. In Fig.5, the saturation pressure lines of thin and thick solutions are combined into one line of S', while the saturation pressure line of water is expressed in line S, and ΔT corresponds to the elevation of boiling point of the solution's compared to water's.

The flow diagram of the generalized closed cycle CDE engine system with concentrator system is shown in Fig.6, where the CDE engine side is just the same as in Fig.4, while the diluted solution is concentrated and enriched again at the concentrator side by heat input into the heater or by power input into the steam compressor. If we assume that the concentration m of the solution is m=1.0 at start and that steam is absorbed into the solution until the concentration becomes m=mf, the total theoretical amount of steam G, heat and power to be liberated by CDE engine can be calculated theoretically, knowing the change ΔT_0 and absorption heat L* as the functions of m, and assuming a Carnot cycle efficiency to the power generation.

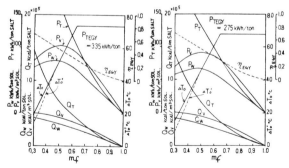

Fig.6. Generalized closed cycle CDE engine system with concentrator

Fig.7. ΔT_0, heat and power storage abilities of aqueous solutions of (a) LiCl (b) LiCl + CaCl$_2$ (3 : 2)

Also the heat and power storage ability Q_w, P_w per unit weight of solution at m=mf can be calculated, as well as the heat and power storage ability Q_v, P_v per unit volume of solution at m=mf. For example, on Fig.7, the ΔT_0 and power and heat storage abilities of aqueous solutions, of (a) LiCl (b) LiCl + CaCl$_2$ (3 : 2), are shown.

OVERALL EXPERIMENTS ON CDE ENGINES

1 KW class CDE engine

First, an experimental CDE engine set which is planned to generate 1 KW electric power was experimented overall.
A flow sheet of the engine is shown in Fig. 8. The solution used here was the aqueous solution of $CaCl_2$ + LiCl (2 : 3, weight ratio) mixture. This CDE engine ran for 1 hour without feeding water, and its maximum adiabatic power was 4 kw. The adiabatic power was evaluated by the steam flow rate throughout the engine and by the steam enthalpy drop between the evaporator outlet and turbine exhaust. As the efficiency of the electric generator was about 0.5 and that of the steam engine was 0.25 because of their small size, the maximum electric generation obtained was actually 0.6 kw. By feeding water to the evaporator, the engine can run about 4 hours without any heat input from the outside. In Fig. 9, one of the examples of temperature and power records for a typical water feeding test run is shown.
The calculated actual specific power P_W of the above solution was 19 kwh/ton of the final solution This value is about 2/3 times of an ordinary electric battery. and it will be elevated further by improving the efficiency of the absorbing evaporator, etc. In these experiments, the absorption heat transfer coefficient and overall heat transfer coefficient K of the absorbing boiler are also measured against the mean absorption steam velocity V=(steam flow)/(heat transfer area).

Concerning the overall heat transfer coefficient K (heat flux of boiling surface/temperature difference between boiling point of solution and steam temperature) of the absorption boiler, the measured values of K are plotted in Fig.10, where very high K values, as high as 1.0 kw/m²k (= 0.8 × 10⁴ kCal/m²h°C) are indicated.

A manned CDE car (tricycle)

The experimental CDE tricycle (manned CDE car No.1) is shown in Fig.11 which carries three small reciprocating steam engines with 2 double action cylinders of 20 mm diameter and 20 mm stroke. This car is driven for 20-25 minutes at the average speed of 10 km/h by the CDE energy of 22 liters of LiCl and $CaCl_2$ mixed aqueous solution and 6 liters of fresh water in absorbing boiler soaked in the solution kept at the temperature of 155°C to 135°C. Also the data of the overall heat transfer coefficient K of this CDE vehicle are plotted in Fig.12. Markedly, high K's as high as about 2.0 × 10⁴ W/m²K are obtained again.

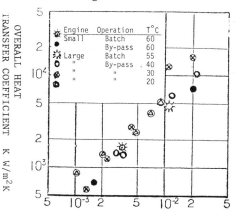

Fig.10. Overall heat transfer coefficient K of the absorbing evaporator of 1 KW CDE test engine

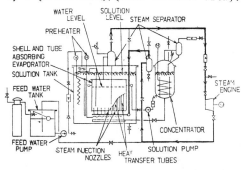

Fig.8. Flow sheet of 1 KW CDE test engine

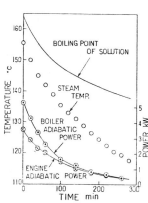

Fig.9.
Typical temperature and power record of the test engine of Fig.8

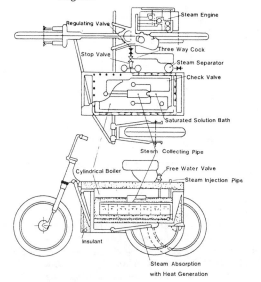

Fig.11. Manned CDE car No. 1

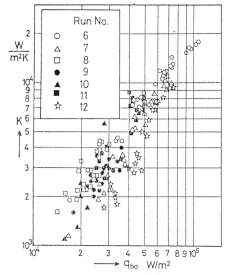

Fig.12. Overall heat transfer coefficient K of the evaporator of the manned CDE car No.1

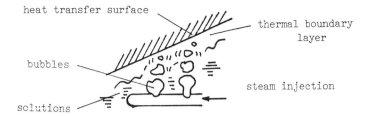

heat transfer surface

thermal boundary layer

bubbles

steam injection

solutions

Fig.13. Supposed agitation by sudden collapse of
injected bubbles by absorption

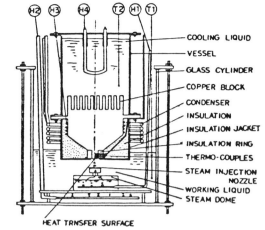

COOLING LIQUID
VESSEL
GLASS CYLINDER
COPPER BLOCK
CONDENSER
INSULATION
INSULATION JACKET
INSULATION RING
THERMO-COUPLES
STEAM INJECTION NOZZLE
WORKING LIQUID
STEAM DOME
HEAT TRANSFER SURFACE

Fig.14. Experimental apparatus for the
measurement and observation of
absorption heat transfer

HEAT TRANSFER IN THE ABSORBING BOILER

Considerations on absorbing heat transfer

As seen in Figs.10 and 12, a very high overall
heat transfer coefficient K is achieved in absorbing boilers, and the value of K increases as
the absorbing rate increases.

By consideration, the reason for this achievement
of such a high heat transfer rate is thought to
be that, as shown in the sketch of Fig.13, strong
heat and agitation are generated in the boundary
layer the heating surface accompanied with the
sudden collapse of the steam bubbles.

This phenomena seems to be very similar to the
condensation of steam on the cooled surface, and
also similar to the subcooled boiling in which a
bubble collapses instantaneously on the surface.

Experimental apparatus of absorbing heat transfer in aqueous solution

In Fig.14, the experimental apparatus of absorption of steam in the solution is shown.

Steam is generated in the small cell situated at
the bottom of the apparatus box by the heating of
electric heater H1, and steam bubbles are injected
into the working fluid (aqueous solution or pure
water) through the steam injection nozzle faced
upward.

The bubbles are absorbed in the working fluid near
the cooled test surface which is faced downward.
The cooled test surface is situated at the bottom
of a copper block which is cooled by cooling liquid flowing upward.

The diameter size of the cooled test surface is
20 mm, and it is situated 5 mm inside of the bottom surface of the insulation plate forming a shal-
low cavity which can trap bubbles. Additional
electric heaters H2, H3, H4 and a water cooled
condenser are installed for fine adjustment of the
system temperature, and several thermo-couples
are installed for measurement of temperature and
heat flux.

The total apparatus is soaked in a glass cylinder,
and the movement of bubbles can be observed from
the outside.

The steam injection nozzles are made by single or
several small holes drilled in various nozzle
plates. Three nozzle plates are prepared with 2
mmϕ nozzle × 1, 1ϕmm nozzle × 2, and 1 mm nozzle
× 25.

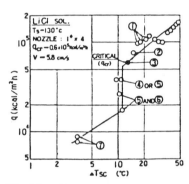

Fig.15. Measured heat flux q
versus ΔT_{SC} curve for
the case of multiple
nozzles

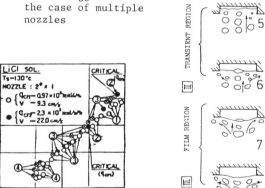

Fig.17. Measured heat
flux q against
ΔT_{SC} curve for the
case of single
nozzle

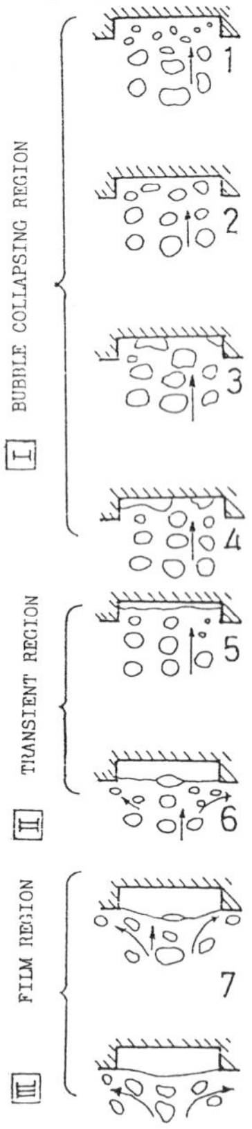

Fig.16.
Sketches of
collapsing bubbles
for the case of
multiple nozzles
corresponding to
Fig.15

Characteristics of absorption heat transfer with multiple nozzles

In Fig.15, typical data and characteristic curve
of heat flux q versus wall temperature difference
ΔT_{SC} for the case of multiple nozzles and LiCl
solution are shown. As shown there, the heat
transfer curve is divided into three regions which
are (I) high heat flux region at $\Delta T_{SC} > 13°C$ (II)
transient region and (III) low heat flux region
$\Delta T_{SC} < 13°C$.

By observation from the outside as shown, in Fig. 16, the region (I) corresponds to the bubble collapsing region where all bubbles are absorbed before reaching the cooled surface, as marked by descriptions 1, 2, and 3.

When steam film appears on the cooled surface, the transient region (II) begins, and heat flux falls down sharply passing conditions 4, 5 and 6. So, the maximum heat flux of the region (II) can be said to be a kind of critical heat flux q_{cr} here. If ΔT_{sc} is higher than its value corresponding to the point of q_{cr}, which is about 13°C at this case of Fig.15, the heat flux of absorption becomes very high, as high as the heat flux of direct condensing.

If ΔT_{sc} is decreased enough, the region (III) begins where steam film is formed on the cooling surface as shown in cases 7 and 8 of Fig. 15. In Figs. 15 and 17, V is the mean steam absorption velocity of the injected steam. Also, the value of T_s is the boiling point of the aqueous solution whose value is the criterion of concentration of the solutions.

Absorption heat transfer with a single nozzle

For the case of a single nozzle with LiCl solution, the data of $q \sim \Delta T_{sc}$ is shown in Fig.17. As seen there, the existence of three regions is not as clear as the multiple nozzle case, but at about $\Delta T_{sc} = 40°C$, the critical heat flux q_{cr} appears in the heat flux data, and the value of q here, film region are higher than that of the multi-nozzle cases in general. The sketches of the observation of this single nozzle case are shown in Fig.18.

General results of the absorption heat transfer

On the same line, several experiments of steam absorption in dense LiCl Solution, H_2O itself and Freon R-113 were carried out by the same apparatus. In Fig.19, some typical results of experimental data are shown in their mean lines, where the theoretical predictions by the Nusselt equation for the case of direct film condensation of steam are shown. By these results the following are known:

(1) For the case of steam injection into H_2O, the values of heat flux become very close to the values of the Nusselt equation of condensation.

(2) For the case of steam absorption into aqueous solution of LiCl, three regions appear in the heat flux curve, and the value of heat flux decreases by the increase of LiCl concentration.

(3) For the case of H_2O and LiCl solution, the values of heat flux for single nozzle cases are generally higher than those of the multi-nozzle cases.

(4) The value of absorption heat flux for R-113 is a little higher than that of the Nusselt equation of condensation for the same fluid.

By these results, it can be said that in actual absorbing boilers with vertical or horizontal tubes, the value of heat flux q should be high enough, because there exists no cavity which traps bubbles. So, the heat transfer coefficient of the absorption in an absorption boiler can be thought to be nearly the same as that of direct condensation of steam whose high value can explain the high overall heat transfer rate of absorption boilers as shown in Figs.10 and 12, although more detailed fundamental studies should be done on absorption heat transfer using open tubes or plates with induced circulation which take place in actual absorption boilers.

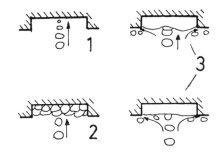

Fig.18. Sketches of collapsing bubbles for the case of single nozzle corresponding to Fig.17

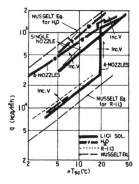

Fig.19. General results of absorption heat transfer

BOILING HEAT TRANSFER OF AQUEOUS SOLUTIONS OF SALTS

Necessity of getting data of boiling heat transfer of aqueous solutions of salts

In order to design the thermo-chemical energy conversion system using aqueous solutions of salts as the CDE engine system, the boiling heat transfer characteristics of solutions of salts should be known more clearly.

In the past, several studies of boiling heat transfers of aqueous solutions were carried on by Minchinko, (6), Tajima (7), Kozeki (8) on LiBr solutions, and by Yusufonva (9) on NaCl solutions, etc., where some decrease of boiling heat flux in salt solutions, and generation of forming and priming were reported.

Here, we have studied the boiling heat transfer characteristics of LiCl and $CaCl_2$ solutions by two apparatus, and have known some aspects of their characteristics in detail (10).

Experimental apparatus of boiling of aqueous solution of salts

In Figs. 20 and 21, the experimental apparatus is shown, where the test boiling surface of the former is the circular plate of 10 to 30 mm dia. facing upward, and that of the latter is the cylin-

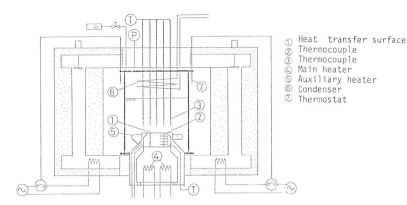

① Heat transfer surface
② Thermocouple
③ Thermocouple
④ Main heater
⑤ Auxiliary heater
⑥ Condenser
⑦ Thermostat

Fig.20. Experimental apparatus
for boiling heat transfer
of aqueous solutions with
flat plate heating surface

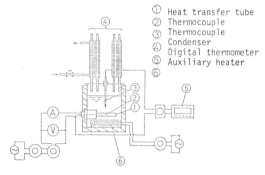

① Heat transfer tube
② Thermocouple
③ Thermocouple
④ Condenser
⑤ Digital thermometer
⑥ Auxiliary heater

Fig.21. Experimental apparatus
for boiling heat trans-
fer of aqueous solution
with horizontal cylindrical
heating surface

drical tube of 10 mm dia. in 100 mm length situated horizontally in the center.
Both boiling surfaces are heated electrically and
soaked in aqueous solutions of about 80 mm depth,
and heat flux data were taken after each surface
was carefully washed and polished by fixed fine
emery papers each time.

Experimental results

In Fig.22, the data of heat flux q of the experimental apparatus of Fig.20 are plotted against ΔT_S.
In Fig.23, those of the apparatus of Fig.21 are
shown for the case of LiCl solutions. It seems
that the results of both apparatus are quite different but the heat flux range of Fig.22 is far
higher than that of Fig.23, so that both data
connect in the low heat flux range.

By combining both results of Figs.22 and 23, a
general idea on the performance of boiling curves
of LiCl solutions seems to be such as shown in
Fig.24. As shown there, by increasing concentration m of the solvents, the boiling curve is
shifted to the right hand side accompanied with the
increase of ΔT_S for the generation of nucleate
boiling points (GNB), with the sharp decrease of
the heat flux of DNB points and with the slow decrease of the heat flux of burn out points (BO).
In Fig.25, the decreases of heat flux of BO points
and DNB points are plotted against the increase of
concentration m. And, in Fig.26 the increases of
ΔT_S of GNB points are plotted.

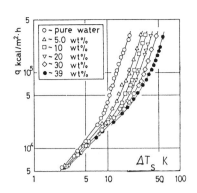

Fig.22. Experimental results
of the apparatus of
Fig.20

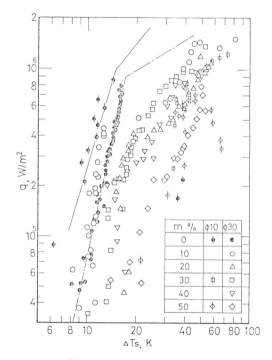

Fig.23. Experimental results
of the apparatus in
Fig.21

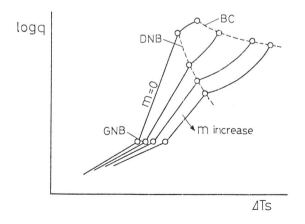

Fig.24. Generalized shapes of boiling
curve of aqueous LiCl solution

196

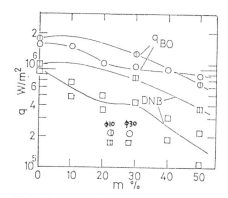

Fig.25. Variation of q_{BO} with increasing m

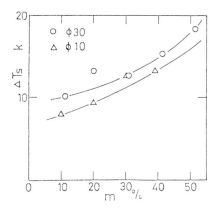

Fig.26. Variation of ΔT_S of DNB points with increasing m

By calculation, the shift of nucleate boiling heat fluxes between GNB points and DNB points are nearly identical to those predicted by the famous Nishikawa's nucleate boiling equation.

In our opinion the reason for the decrease in heat flux in general is thought to be due mainly to the decrease of surface tension in solutions.

By the above results, although the data is not enough, it can be said that, for the concentration of aqueous solutions of inorganic salts, we need a quite high super heat ΔT_S, maybe more than about $20 \sim 30°C$, to get a high enough boiling rate. That means that if the boiling point of solution is about 150°C, we need an outside heat source of a temperature higher than $170 \sim 180°C$.

So, in the future, augmentation techniques of boiling concentration of salt solutions should be studied, although there will exist strong contaminations and erosions.

OTHER ENGINEERING STUDIES ON CDE SYSTEM

Other several engineering studies on the CDE system have been carried out for near-future application of CDE to the energy storage and conversion of various alternative energies as follows: (see (1) to (5) and (11))

(1) The selections of the best solvent salts for the CDE system were studied. The candidate salts include NaOH, H_2SO_4, H_2NO_3, NaOH, $NaSO_4$, LiBr, $NgCl_2$, $ZnCl_2$ etc. But, for several engineering reasons.

the mixed solution of LiCl and $CaCl_2$ was selected as the best solvent salts for today.

(2) Selection of materials for the CDE system using the above solvents were studied carefully, and Titanium alloys and Copper-Nickel alloys were thought to be the best materials for this system.

(3) Vacuum concentration of solutions using a steam condensing injector was studied (10), using large size experimental apparatus of condensing injector, but from many experiments, it has become clear that, we need a driving steam of higher than 1.5 at a gauge pressure in order to achieve enough of a vacuum with steady operation.

(4) Several multi-stage bottoming cycles of the CDE engine system were theoretically studied which are aimed to utilize industrial waste heat for power generation.

By calculation, these multi-stage CDE systems are very good in energy efficiency and in storage capacity of thermal energy and power.

(5) Application of the CDE system for a domestic heating system or so called "inverse-stove", which utilizes the natural temperature difference between cold air and water in the earth's ground to heat up rooms without any fuel or electricity is now under study.

(6) A small four-wheel test automobile of CDE system is now under construction.

CONCLUSION

In this report, the principle of the thermo-chemical reaction is explained at first, and then as a typical example of utilization of such a reaction the CDE (Concentration Difference Energy or Potential) system and engine utilizing aqueous solution of inorganic Salts are introduced.

By conducting several experiments on the CDE engine system, the high heat transfer ability of absorbing boilers is remarked. Then, a fundamental study on heat transfer in absorption of steam into solutions is carried on using a experimental heat transfer surface faced downward receiving steam injection from the bottom. By this experiment, it has been clear that it is possible for the heat flux of absorption to be nearly equal to that of the direct condensing case as long as steam film is not formed on the cooled surface. Then, the heat transfer study of the boiling of aqueous solution of salts is carried out in order to know design characteristics of boiling concentration (enrichment) of aqueous solutions for energy storage and conversion.

By the experiment for the case of LiCl solutions, the boiling heat flux is known to decrease gradually with an increase of concentration of solvents. By the above several studies, many useful data and characteristics have been cleared, and they will be hopefully used for the better design, the utilization of the CDE system, and the utilization of the other thermo-chemical energy storage and conversion in the future.

REFERENCES

1. N. Isshiki, et al., "Energy conversion and storage by CDE engine and system "12th IECEC, 779180 Washington D.C. 1977-Aug.

2. N. Isshiki, "Study on the Concentration difference Energy System" J. Non-Equilibrium Thermodynamics, 2-2, 1977.

3. N. Isshiki, et al., "Heat transfer of Absorbing Evaporator in CDE Engine System" 6th International Heat Transfer Conference, Tronto, 1978-Aug.

4. N. Isshiki, et al., "Development of CDE system and Engine" 14th IECEC 799430 Boston 1979-Aug.

5. J. Kamoshida, et al., "Experimental Investigation of Manned Vehicle Utilized CDE Engine; Proc. of 2nd Miami Int. Conf. on Alternative Energy Resources" (1979-Dec.)

6. F.P. Minchenko et al., "Problems of Heat Transfer and Hydraulics of two phase Media" (Pergamon Press)

7. Tajima, et al., "Study on Boiling Heat Transfer of solutions" Reito (in Japanese) 53-607 (1978) 381 and 49-562 (1974) 687

8. Kozeki; Preprints of JSME Meeting No.790-16 (1979-Oct.) 292

9. V.D. Yusufova et al., Desalination 26 (1978) 175

10. J. Kamosida, et al., "Study on Boiling Heat Transfer of Aqueous Solution of Salts" Proc. 17th Japan Heat Transf. Conf. (1980-May. Kanazawa) 181

11. N. Isshiki "Storage and Generation of Power and Heat By Aqueous Solution of Salts" Int. Seminar on Thermochemical Energy Storage" (Stockholm) (1980-Jan.) 301-326

HEAT TRANSFER IN NONCONVENTIONAL ENERGY (POWER AND PROPULSION) SYSTEMS

Heat Transfer Problems in Some Advanced Power Systems

R. VISKANTA
School of Mechanical Engineering, Purdue University,
West Lafayette, Indiana, U.S.A.

ABSTRACT

Design studies and developmental experiments of some advanced power systems have revealed a host of challenging and sometimes extremely difficult heat transfer problems. These problems may be either of a fundamental or of design nature. Specifically, some heat transfer problems related to the design of solar thermal, coal-fired MHD, and controlled nuclear fusion electric power generating plants are reviewed. Only problems unique to these power generation concepts are discussed; those needing research attention are identified.

INTRODUCTION

The purpose of this paper is to acquaint the heat transfer community with some of the problems related to the development of advanced power generation systems. Specifically, attention is focused on solar thermal, open-cycle coal-fired MHD, and controlled nuclear fusion electric power generation plants. A full discussion of heat transfer in these systems is impossible as it presents too broad a topic to be adequately covered in this brief paper. Therefore only the current state-of-the-art understanding of the more fundamental heat transfer problems unique to the non-conventional parts of the plants is reviewed. The three novel power generation concepts will be discussed separately.

The author hastens to add that this paper is not intended to serve as a comprehensive, definitive, or exhaustive discussion of the forementioned problems. Nor is there sufficient space and time to review the array of special problems which ensue in each system. In discussing the problems, emphasis is placed on recent developments in the United States because of the author's greater familiarity with them, but recognition is given to similar developments under way in other countries, including Japan. References are given but they are merely illustrative and thus primarily useful only as starting points of further investigations. Opinions are freely offered, but hopefully they are clearly identified as such.

SOLAR-THERMAL RECEIVER TECHNOLOGY

The solar-thermal receiver is a key component of high efficiency systems used for electric power generation, for industrial manufacturing applications (such as fuel and chemicals production), and for industrial process heat. The function of this component is to capture concentrated solar energy and to transfer it to a working fluid within the receiver with minimum losses. The purpose of this section of the paper is to discuss some of the heat transfer problems related to the development of solar thermal central receivers for electric power generation demonstration plants and some other advanced concepts.

The two types of solar thermal receiver concepts pursued are the external-type and the cavity-type (Figure 1). The external receiver has been chosen for the Barstow, California pilot plant [1]. It consists of a cylindrical geometry and is constructed from 24 panels, each of which contains 70 tubes. A fraction of the incident solar radiation (reflected from heliostats surrounding the tower) is lost and the remainder is transferred to the working fluid which enters as water at 288°C and 1600 psia and leaves as superheated steam at 516°C and 1578 psia [2]. There is no steam drum in this design and the receiver is basically a once through boiler. The cavity (internal) receiver, with either multiple or single apertures, is an attractive alternative means of solar energy absorption and continues to be pursued [3, 4].

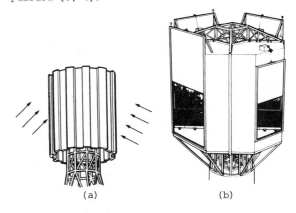

Figure 1 Solar central thermal receiver concepts for electric power generation: a) external receiver and b) cavity receiver.

External Receivers

An accurate knowledge of external receiver efficiency is crucial in the selection of one receiver over another for a given application and for the prediction of energy production costs when the receiver is incorporated into a power plant. The determination of receiver efficiency requires the prediction of simultaneous convective and radiation heat losses [2, 4]. However, even if it is assumed that the radiative heat losses can be calculated with sufficient accuracy and that there is no interaction between convection and radiation, the size and expected environments of typical external receivers (see Table 1 for representative conditions) place them in the combined (forced and free) convection regime. In the absence of wind heat transfer will occur by free convection, but even in the presence of wind buoyancy is expected to play a significant role in determining the convective heat losses by altering the flow field around the receiver (Figure 2a). Heat transfer data is lacking and the existing rudimentary predictive methods are as yet untested for calculating the convective losses with confidence.

Table 1   Representative Central Receiver
          Paramenters [2]

Reynolds Number	10^6 - 10^7
Grashof Number	10^{13} - 10^{14}
Dimensions	10 - 30 m
Surface Temperature	500 - 1200°C
Solar Flux	0.2 - 1.5 MW/m^2

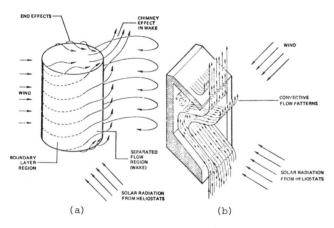

Figure 2   Models of physical process for
           external (a) and cavity type (b)
           central receivers, from Ref. 4.

Turbulent free convection heat transfer from a vertical isothermal plate for large Grashof numbers and high temperatures, characteristic of a receiver, have been predicted by Siebers [5] using the STAN-5 computer code and compared (Figure 3) with available correlations and extrapolations. The results indicate that the use of the McAdams empirical correlation to predict the free convective heat loss from various central receivers may result in an underestimation of heat loss by factors ranging from 1.5 to 2.0. The findings of the study lead one to conclude that extrapolating low Grashof number data to high Grashof numbers and high temperatures may result in significant errors in predicting heat transfer and that property variations across the boundary layer probably must be considered in any analytical modeling. Some preliminary forced and free convection heat transfer data from vertical cylinders at high Reynolds and Rayleigh numbers have recently been obtained in a cryogenic wind tunnel [6]. Turbulent combined convection from a vertical cylinder or a plate when the

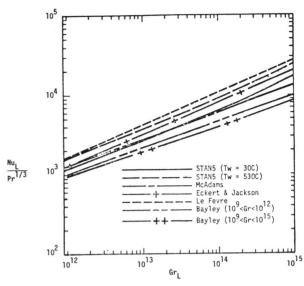

Figure 3   Average free convection heat transfer
           coefficients from a vertical plate
           in air at high Grashof numbers, from
           Ref. 5.

body force is perpendicular to the external flow direction (Figure 2a) does not appear to have been studied either analytically or experimentally. The analysis is complex because of the three dimensional nature of the problem and the uncertainty of the turbulence models, and experiments are difficult to perform and scale (see Table 1) not only because both the Grashof (Gr) and Reynolds (Re) numbers are large but also because Gr/Re^2 is of the order of unity. Combined forced and free convection heat transfer from a smooth vertical cylinder in cross flow has been estimated by combining the results for forced convection from a cylinder [7] with those for free convection from a vertical plate [8] using the procedure suggested in the literature [9] for parallel flows, e.g. $Nu_c^n = Nu_f^n + Nu_n^n$, where $2 \le n \le 4$ and Nu_c, Nu_f, and Nu_n are average Nusselt numbers for combined, forced and free convection, respectively. The results of Figure 4 in conjunction with Table 1 show that neither forced or free convection can be ignored in determining the heat loss from an external receiver. Of course, the receiver is not smooth because the tubes in the panels contribute a relative roughness of about $\epsilon/D = 60 \times 10^{-6}$. The available experimental data for forced convection must be extrapolated to proper roughness values, and there is no data on roughness effects in free convection flows at large Grashof numbers. The validity of the procedure used to estimate the convective heat transfer has not been verified experimentally, and it is not certain if the above relation is appropriate when the forced and free convection currents are perpendicular to each other. Sufficient boundary layer structure detail is needed for guidance in turbulence modeling, and local heat transfer data are required for checking two- and three-dimensional turbulent boundary layer heat transfer predictions. A combined experimental and analytical program has been proposed [2, 10] to obtain such data and to develop the modeling capability.

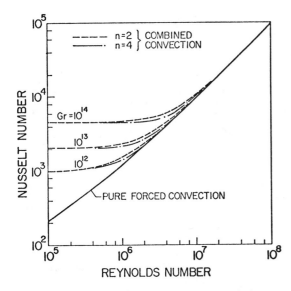

Figure 4 Air-side heat transfer coefficient
          for combined convection from a large,
          smooth vertical cylinder.

The occurrence of a critical boiling condition
and thermal-hydraulic instability have been
identified as two independently acting phenomena in the once-through-boiler for the Barstow
receiver [2, 11]. These phenomena are not
unfamiliar but they are not normally present in
conventional drum-type boilers. Because of the
timewise unsteady and circumferentially non-
uniform heating expected, there is concern that
these phenomena may contribute to the thermal
fatigue of the tubes. A boiling heat transfer
experiment employing an irradiated 17 m long
panel containing tubes of Incoloy 800 (1/2 inch
O.D.) is being performed to obtain needed data
[11].

Cavity Receivers

The combined heat losses from a cavity receiver (Figure 1b) are controlled by several
independent variables, namely: (a) the receiver geometry, orientation, baffling, and
wind shields; (b) the distribution of solar
radiation on the receiver surfaces; (c) heat
transfer in the cavity, working fluid, and
through the receiver walls; and (d) the wind
speed and direction, air temperature, and
turbulence intensity. Because of the size
(~ 10 m) and high temperature (~ 800°C), heat
transfer in the cavity irradiated through an
opening will be by turbulent combined convection and by radiation. The inflow and outflow
of air through the cavity opening is expected
to affect heat transfer within the cavity and
the convective heat losses through the opening.
There is no experimental data or predictive
models directly relevant to this problem. Combined convection and radiation within a closed
rectangular cavity has been studied analytically [12, 13] and some experimental velocity,
temperature and heat transfer results have
recently been reported for an open rectangular
cavity [14]. Simple models for estimating the
convective heat losses from an open cavity
receiver have recently been proposed [10, 15,
16], but they must be verified experimentally
if they are to be used with confidence.

Some problems relating to thermal design of
cavity-type receivers have been identified [10,
15, 17] and include: (a) turbulent combined
convection in open cavities, (b) radiation heat
transfer in open irradiated cavities, (c)
theoretical description of buoyancy influenced
turbulence, (d) interaction of convection and
radiation in nonisothermal wall cavities, and
(e) development of scaling procedures and relations for convective and radiative heat
losses from large receivers. The large size
of the receiver makes full scale experimentation difficult and costly. Analytical efforts
and supporting experiments are needed to
develop modeling capability to predict the
thermal behavior of cavity receivers. A computer code development combined with supporting experiments to test the code predictions
have been planned [4, 10]. Some specific
small and large scale experiments concerned
with flow visualization and obtaining velocity
and temperature distribution measurements in
cavities and apertures are presently operational. Parallel analytical modeling efforts
include the prediction of radiation transfer
and of 2-D natural and combined turbulent convection in cavities. Combined turbulent
natural convection and radiation heat transfer
in a receiver cavity is three-dimensional, and
therefore steady, two-dimensional models for
scaling heat transfer inside the cavity and
the convective losses through the opening are
not expected to be adequate.

Advanced Receivers

Three major stimuli for advanced high temperature receiver development are: (1) the goal
of achieving high system efficiency through
the utilization of high efficiency heat
engines for power generation, (2) the need for
industrial process heat, and (3) the energy
requirements of fuel and chemical production.
Temperatures in excess of 1100°C are needed in
some industrial processes because a substantial
fraction of the energy used is above this level
and also since some of the most attractive
processes currently being considered as part
of the fuel and chemical production program
require such high temperatures. However, the
requirement for increased receiver operating
temperature must be balanced against the increase in thermal (convective and radiative)
energy losses at these higher temperatures. For
a given collector system, the radiative losses
increase more rapidly than does the convective
losses for a given temperature increase. Hence,
the minimization of radiation losses is a key
factor in achieving increased system efficiency. The ability to attain high receiver
efficiency is also dependent on the quality of
the concentrator(s) [18].

The energy from the solar concentrator must be
absorbed by the working fluid within the con-
fines of the receiver cavity with a minimum of
temperature drop. Hence, heat transfer (in a
controlled manner) to the working fluid and
the fluid mechanics of the working fluid within
the receiver are other key factors which must
be taken into account in the design and development of advanced high temperature (from
about 1000°C to about 1700°C) receivers. For
example, the heat transfer components must be

capable of withstanding the stresses induced by
thermal gradients which in turn are due to the
non-uniform heating created by the concen-
trator(s). Also, the receiver must be capable
of operating during transient conditions such
as   a sudden cloud coverage of the sun.
Key receiver knowledge requirements (e.g. concerning energy losses, heat transfer, fluid
mechanics, mechanical design, and materials)
and current technology development efforts for
five classes of receiver applications: (1) high
temperature, (2) advanced Bryton, (3) Stirling
and (4) Rankine cycle engines, and (5) fuels
and chemicals have been identified [18]. However, specific heat transfer problems needing
solutions have not been discussed. For
example, the high temperature receivers considered include: (a) a honeycomb pressurized
matrix, (b) a coiled tube, (c) a cylindrical
ceramic, (d) a ceramic dome, and (e) sodium and
potassium heat pipe [19] receiver concepts.
One of the high temperature receiver concepts
uses heat exchanger (storage) modules filled
with a phase change material for both the
deposition of solar radiation and storage of
thermal energy. The solid-liquid phase
change heat transfer, including the effects of
natural convection and materials interaction,
are not completely understood and a design
data base is not available. Many challenging
heat transfer problems needing solution are
expected to surface as detailed design and
development of advanced receivers progresses.

COAL-FIRED MHD

Research in the field of MHD power generation
systems has progressed beyond the laboratory
stage and is now at the stage where the development of large scale test facilities is
feasible and operational systems are being
designed. Detailed design of the various
system components requires an accurate knowledge of heat transfer. The primary heat
transfer problems associated with the open-
cycle, coal-fired MHD power plant are in the
topping cycle which include the combustor,
MHD generator channel, diffuser, radiant
boiler, and the high temperature air preheater.
The steam bottoming cycle will rely primarily
on existing components for conventional coal-
fired power plants but significant technology
deficiencies [20, 21] exist here also. A
schematic diagram of a directly-fired MHD
steam power plant concept is shown in Figure 5.
Some of the thermal design problems in these
components have already been discussed [21, 22],
and therefore we will focus only on some of the
more fundamental heat transfer problems.

Combustor

Heat transfer in the combustor is tightly
coupled to the coal combustion process (which
is only partially understood [23]). The importance of residence time on the combustor
heat transfer has been discussed and it was
concluded that while current analytical models
are adequate for cycle optimization, refine-
ments are still required for optimal combustor
design. There is insufficient information on
the contribution of ash and slag particles
(from the mineral matter in coal) and of unburned

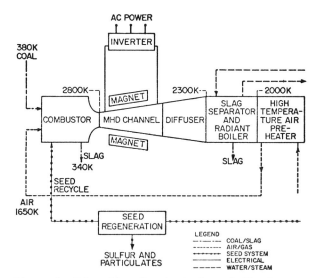

Figure 5   Schematic of a typical coal-fired
MHD/steam system, after Ref. 22.

coal particles on radiation heat transfer, which
is the major heat loss mechanism in the combustor. The formation of a frozen-molten
slag layer (the molten phase of the mineral
content of coal) provides some thermal protection to the walls. The molten slag flows along
the walls and is partly removed at the bottom
of the combustor. The thickness of the slag
and the degree of protection it provides to
the walls depends on the physical and transport
properties of the slag (both solid and liquid),
the heat flux at its surface, and the shear
forces. An additional complicating factor is
that the slag is not opaque but partly transparent to radiation, and its spectral radiation
characteristics are not known. The determination of radiation heat transfer at the wall
of combustor and net heat transfer through the
slag are crucial inputs to the estimates of
combustor wall heat loss rates (and hence of
the estimate of plasma temperatures). Knowledge
of heat transfer from combustor walls can also
be effectively used as a diagnostic tool to
optimize combustor configuration for direct
coal-fired MHD generators [24]. The development of a coal combustor for commercial, open-
cycle MHD power generation is greatly impeded
by the lack of any effective knowledge of coal
combustor scaling techniques [25]. Validated
scaling relationships for MHD cyclone coal
combustor are also lacking. Different MHD
combustor design concepts and methods for cooling the walls are discussed in the literature
[21, Chapter 10].

MHD Generator Channel and Diffuser

An accurate knowledge of heat transfer in MHD
generator channels and diffusers is required
for the proper selection of materials for the
walls, for designing a wall cooling system,
for optimal integration of the MHD topping
cycle with the steam bottoming cycle, and for
estimating the overall plant efficiency. Both
the channel and diffuser are subjected to
hostile thermal environments. Heat transfer to
the walls from the products of combustion containing ionized potassium seed and slag

particles is by convection and radiation. The coupling of the magnetic field with an electrically conducting fluid in a channel influences the fluid dynamics, and thus the wall shear and heat transfer [26]. The deposition of slag on the channel and diffuser walls produces further complications. Thus, many of the thermal problems encountered are unique to MHD power plants.

The plasma that flows in the channel in the presence of a magnetic field produces fluid dynamic phenomena that have a major effect on the convective heat transfer to the wall [22]. First, a properly aligned magnetic field retards the core flow and accelerates the flow in the boundary layer. Secondly, the coupling of the Lorentz force across the channel is nonuniform and leads to localized acceleration near the wall [27]. The higher velocity gradients near the wall result in an increase in wall shear and convective heat transfer. Thirdly, the coupling of the MHD initiated forces and secondary rotational flows produce unconventional, cellular flow patterns which could lead to a redistribution of the channel wall heat flux [22, 27]. The magnetic field reduces the heat flux to the channel wall [28] while the Joule heating increases the flux, the latter effect being of more importance than the former. Experimental results and analytical interpretations of the Joule heating phenomena in the wall boundary of a small channel have shown that the flux is strongly affected by the local current density [29].

The deposition of slag and seed material on the walls does not reduce the electrical system performance and the slag layer protects the electrode from arcing and excessive wear (see Figure 6). The advantages of operating the channel with a slag layer have been demonstrated [31, 32]. Thus, the wall heat transfer is tied to the thermodynamics and fluid mechanics of the slag layer adjacent to the wall. Since the slag layer is not opaque but semitransparent to radiation, the heat transfer through the slag and the boundary condition at the slag/plasma interface (Figure 7) are uncertain. Further, the apparent roughness of the slag layer due to non-uniform (wavy) thickness produced by the pressure and shear perturbations generated by the external gas stream must be considered [33]. Finally, the boundary conditions must be determined for different types of walls. Gaps and protruding electrodes (including the type of construction material used) will affect the heat transfer boundary conditions [22].

In most of the studies performed to date [20, 21, 28], the contribution of radiation to heat transfer from the plasma to the walls and on the electrodynamics and flow field in MHD channels has been ignored. This approximation is probably justifiable for small experimental channels where the opacity is small enough to make the radiation from the plasma a second order effect; however, this may not be the case for large base load channels because of their size. Heat transfer by convection is expected to exceed that by radiation at the front end of the channel where the boundary layer is thin. As the flow proceeds downstream and the boundary layer

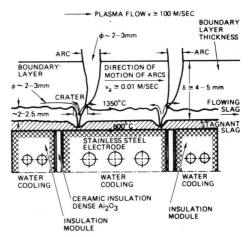

Figure 6 Arc-slag interaction on cold electrodes, from Ref. 30.

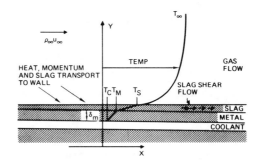

Figure 7 Model flow field near a slagging wall.

grows, convective heat flux decreases more rapidly than the radiative flux because the increase in opacity resulting from channel divergence will partly compensate for the drop in gas temperature. Thus, it is expected that at the back end of the channel radiative flux may be of the same order of magnitude or even higher than, the convective flux.

The effect of radiation from combustion gases to the walls of a large (enthalpy flow rate of 990 MW) MHD channel have been estimated [34]. The contributions from the major infrared radiating species and from ionized potassium seed have been accounted for. The wings of the potassium resonant doublet have been accounted for in the calculations. The investigation has found that the inclusion of radiation transfer is essential for the calculation of heat transfer in the channel, estimation of efficiency, and design of a large MHD power plant. The interaction of radiation with the boundary layer may also change the convective heat transfer. A similar analysis which included radiation by slag particles in the channel of a coal-fired MHD system has been reported [35]. Calculations performed for a base-load channel (see Figure 8) indicate that heat transfer by gas radiation almost equals that by convection for smooth walls and amounts to as much as 70% of the convective heat transfer for rough walls. It should be noted, however, that the analysis has neglected the effect of magnetic field on turbulence suppression, and therefore the convective heat

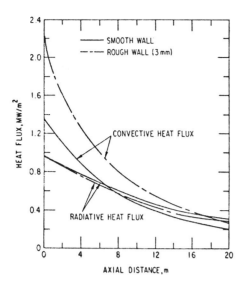

Figure 8 Variation of local heat flux along a
large MHD generator channel, from
Ref. 35.

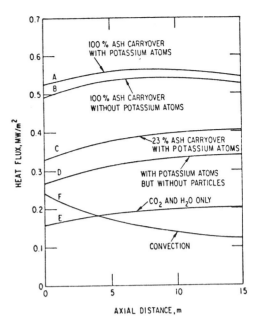

Figure 9 Effect of slag carryover and
potassium atoms on local heat transfer
along a large diffuser, from Ref. 37.

transfer to the generator wall may have been
overestimated. The relative importance of the
magnetic field and Joule heating effects on convective heat transfer has not been discussed and
their significance has not been ascertained for
large channels [35].

The flow in the diffuser and heat transfer to
the wall will depend on whether the magnetic
field is on or off. The Lorentz forces, which
influence the velocity field in the channel,
have to be taken into consideration in the
design of the diffuser. The performance of a
MHD diffuser with high blockage has been analyzed [36]; however, the contribution of gas
and particle radiation to wall heat transfer has
not been recognized. This may be justifiable
for small diffusers but, as in the case of the
MHD generator channel, the contribution of
radiation to the total wall heat transfer may
be significant. Heat transfer to the wall of
a subsonic MHD diffuser by convection and by
gas and particle radiation has been analyzed
[37]. The results (Figure 9) reveal that heat
transfer is sensitive to ash carryover into the
channel, slag particle size distribution, and
wall emissivity. Sensitivity studies [37]
demonstrate that slag particles play a dominant
role in promoting radiation, and the results also
exhibit the radiation shielding effect of the
particles through multiple scattering of the
short wave radiation emitted by potassium. As
expected, the radiation heat flux was also found
to be sensitive to the wall emissivity.

The interaction between the plasma and the slag
layer on the walls (Figure 7) of the channel
and diffuser has been neglected in determining
the heat transfer scaling relations for these
two components (as well as for the radiant
boiler [38, 39]). The thickness of the slag
layer is determined primarily by the thermophysical and optical properties of the slag,
shear at the interface, temperature conditions,
and heat input, and therefore its heat transfer
characteristics are expected to have a significant bearing on the heat loss from the walls

and on the plasma temperature and dynamics.
Therefore, it appears that the presence of the
slag layer needs to be accounted in more
realistic models of slagging components of the
coal-fired MHD plant. Simulated MHD tests have
recently shown that the slag has a destructive
effect on the electrodes [40], and therefore
it may be desirable to eliminate the layer
completely. This can be accomplished by increasing the temperature or changing the composition of the slag, so that its viscosity decreases until film (Rayleigh) instability causes
the layer to break up and vanish.

Radiant Boiler

In addition to providing for the necessary energy
transfer to the steam cycle, the radiant boiler
must reduce NO_x concentration to an acceptable
level, remove a large portion of the entrained
slag, and retain the potassium in the vapor
phase. Since radiation from a gas stream
containing slag and seed particles is the
principle mode of heat transfer, the thermal
design of the radiant boiler depends strongly
on the radiative properties of the gas-particle
stream and the heat transfer characteristics
of the slag film covering the vessel walls.

Radiative heat transfer in the boiler is
strongly influenced by the number density,
size distribution, and optical properties of
the slag droplets. The emissivity of isothermal gas-slag droplet clouds has been estimated by accounting for emission, absorption,
and non-isotropic scattering by the particles
[41], and radiation heat transfer in the
boiler has been calculated [42]. Unfortunately, there is great uncertainty in modeling the number density and size distribution
of slag particles, and in the optical properties of the slag and their variation with
temperature. Also, the one-dimensional and zonal
approximations of radiation transfer will be

206

suspect until more realistic multi-dimensional analyses become available and/or the models are verified experimentally.

In the analysis of radiation heat transfer in the radiant boiler [42] the interaction between the combustion gases and the slag layer on the wall has been neglected by assuming that the wall is isothermal and black. In reality, the temperature of the slag/combustion gas interface will be determined by combined conduction, convection, and radiation heat transfer. Since the slag introduces significant thermal resistance to heat transfer, the determination of this resistance is an important consideration in the design and scaling of radiant boilers [43]. The heat transfer rate through the slag is sensitive to the choice of the model (opaque or semitransparent to radiation) and as a consequence significantly different gas cooling rates can be obtained [43]. The results of Figure 10 show that for the thermal conditions and parameters considered, the conductive heat transfer ($k \, dT/dy$) predominates near the cooled wall ($y/y_s < 0.2$) whereas radiation (– F) is the dominant mode of heat transfer over the rest of the slag layer. The available results suggest that for reliable prediction of heat transfer through the slag not only physical and transport properties but also the optical properties of the slag must be known as a function of temperature and composition. Unfortunately, the spectral absorption and scattering coefficient data for coal ash slags are not available. Furthermore, the spectral radiative flux incident on the slag layer must be calculable to a sufficient accuracy.

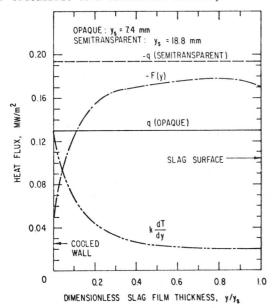

Figure 10 Distributions of local conductive, radiative and total heat flux across opaque and semitransparent coal-ash slag layers, from Ref. 43.

Air Preheater

The proposed open-cycle MHD power generation system shown in Figure 5 requires gas temperatures in the MHD channel of approximately 2800 K.

Since this temperature cannot be achieved by the combustion of coal and relatively cold air, it is therefore necessary to preheat the air to a temperature of about 1900 K. The proposed plan [22] is thus to preheat the air in two steps: (1) a low temperature (up to 1070 K) heat transfer, to be affected in heat exchangers constructed from high temperature alloys, and (2) a preheating of the air to about 1900 K in periodic ceramic regenerative heat exchangers. Since the energy source for these heat exchangers is the combustion products from the MHD channel, it is necessary to determine what effects the products of coal combustion (primarily the slag) will have on the heat exchanger performance. One anticipated effect is the deposition of particulates on the flow passage walls of the cored brick. Therefore, the determination of the particulate deposition rate is of primary importance in the design and operation of the air preheaters [22]. Mathematical models have been proposed [44] and tests have been performed [45, 46] to obtain the necessary data and develop modeling capability. A few test facilities for the development of high temperature air heaters are operational [47]. Experimental results [46] show a significant effect of particle diameter on the magnitude of increase of the pressure drop. The plugging of the gas flow passages by the coal slag is an important practical problem requiring a periodic slag removal. The effects of tube diameter, wall roughness, type of coal slag, particle size, and velocity on the deposition rate need to be established since all of these parameters may affect pressure drop and heat transfer characteristics within the air preheater.

NUCLEAR FUSION POWER SYSTEM CONCEPTS

The principal objective of the United States national fusion program is to develop the technology for the safe and economical production of electrical energy from fusion energy. The goal is to develop the engineering technology, fusion power reactor systems, and components for the demonstration of fusion power generation by the mid to late 1990's. In the United States, two major concepts are being pursued to produce practical fusion power [48]: (1) magnetic confinement (tokomak, mirror and theta pinch) and (2) laser-fusion. The general technological requirements for power by fusion have been discussed [49] and the unusual heat transfer problems encountered in the design of magnetically confined fusion reactor (MCFR) schemes has been reviewed by Hoffman et al. [50]. The heat transfer problems occurring in laser-induced controlled fusion reactors (LCFR) have been discussed by Frank et al. [51]. Unique heat transfer problems have been identified in the neutron moderator blanket, magnetic shield, plasma production, injection and removal system, and in the superconducting magnet components of a MCFR power plant. Because of the brevity of this survey only problems of a more general nature have been alluded to.

In order to obtain reasonably high thermal efficiencies for future fusion power plants employing magnetic confinement (Figure 11) it is necessary to operate with blanket temperatures in the range of 600 to 800°C (depending on the design of the plant). Concurrently, it is necessary to have as uniform a temperature distribution as possible throughout the blanket in order to minimize thermal stresses. The thermal-hydraulic characteristics of various possible coolants, including lithium, sodium, potassium, molten salt flibe (67% LiF and 33% BeF_2), boiling alkali metals (primarily potassium), helium, and water have been studied and are discussed in some detail within the context of general fusion reactor requirements [50, 52]. In addition, these references contain a comprehensive list of the many constraints and boundary conditions imposed by the MCFR environment and which the thermal hydraulic system designers must try to satisfy.

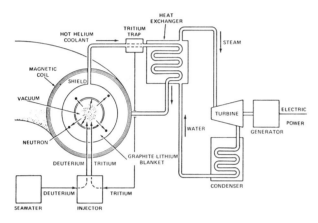

Figure 11 Schematic diagram of a magnetically confined (tokamak) fusion power plant, from Ref. 48.

Steady state magnetic fields can produce a number of MHD effects in a flowing conducting fluid [50]. For example, a magnetic field component perpendicular to the flow direction creates Hartmann (eddy) currents which can alter the velocity profile in the core as well as near the wall, and, depending on the strength and uniformity of the magnetic field, may cause additional retardation of the flow. The magnetic field may also induce flow separation in sharp bends, have profound effects on the flow development length, suppress turbulence and natural convection, affect bubble dynamics in liquid metals, and have other effects. The existence of these effects has been verified by experiments, but quantitative design information on fluid flow is incomplete [50]. It is well established, both theoretically [53] and experimentally [54], that in single phase flow the magnetic field inhibits convective heat transfer. Theoretical [55] and experimental [56] studies indicate that there is a significant effect on bubble growth and reduction of overall heat transfer in pool boiling under the influence of a horizontal magnetic field. And as of now heat transfer design information appears to be incomplete due to the lack of an experimental, and very often a theoretical, data base.

Detailed MCFR conceptual design studies [50, 52] have served in identifying important heat transfer problems which may be expected in the blanket, super-conducting magnet, and other components of a fusion reactor system (where design information is lacking). For example, some areas where additional research on magnetic effects (due to both uniform and nonuniform magnetic field components) appear to be needed are: (1) the study of the effects of a variable magnetic field along the flow direction on the hydraulic and thermal development lengths, on pressure gradients, and on local and average heat transfer coefficients; (2) the investigation of the effects of combined loop and Hartmann eddy currents and of very high magnetic fields on pressure gradients and local heat transfer; (3) a study of local pressure drop and heat transfer in straight ducts, bends, and turn-arounds under conditions in which magnetically induced separation is likely to occur; (4) experiments on pool and forced convection, nucleate and film boiling conditions of liquid metals under various orientations (with respect to the field of gravity) and magnetic fields should be carried out and the effect of magnetic field on critical heat flux be determined; and (5) natural convection circulation cooling of the blanket has been suggested [57], but magnetic forces tend to inhibit natural convection, - the effects have not been studied quantitatively and are not understood. For example, there is some experimental evidence [58] which indicate that strong magnetic fields (above a critical value) may completely suppress turbulence and thus affect heat transfer and pressure drop. Since experimental data is practically nonexistent for magnetically influenced turbulent flows of practical concern, there remains a need for a theoretical means of predicting heat transfer in conducting fluids under turbulent flow conditions in the presence of magnetic fields.

Detailed MCFR concept design studies performed [50, 52] have indicated that the use of lithium as a primary coolant of the blanket would result in a very high MHD pressure drop and thus require a high inlet coolant pressure. Therefore, the blanket would have to endure large stresses. Because of such difficulties alternative coolants such as helium and other inert gases, water and two-phase systems (helium with 30% volume fraction of lithium oxide dust as coolant plus a tritium breeder) have been proposed [50, 59]. Heat transfer problems with a helium cooled blanket concepts are highly design dependent, and simple flow paths result in poor heat transfer characteristics near the critical first wall region. More complicated (modular or segmented) blanket designs result in very complicated flow geometries and paths. There appears to be a significant uncertainty of heat transfer in coiled tubes and sharp bends, and the effects of flow separation and recirculation on heat transfer need to be investigated. Further research is needed to study the thermal-hydraulic behavior of the two-phase blanket coolant mentioned above when employed in ducts

with many turns and sharp bends. The collection
of solid particles in stagnation regions, the
plugging of ducts (clinker formation) and the
related hot spot problems would also need to be
investigated.

Heat transfer problems in components other than
the blanket or the first wall are no less
severe. For example, the dump for the ion and
neutral beams must be capable of absorbing an
incident flux of about 10 kW/cm^2. Use of
swirling flow and inserts in tubes has been
considered, but the large pressure drop in
small diameter tubes results in severe stress
problems. High velocity water flowing circumferentially to remove heat from the beam dump
by subcooled nucleate boiling has been suggested [60]. However, critical heat flux data
base for water in small diameter tubes, U-bends
and other shapes in the presence of extremely
large pressure drop (e.g. 90% of the inlet
pressure) and nonuniform circumferential heating is not available.

Heat Transfer in LCFR Concepts

The most challenging heat transfer problems encountered in the design of LCFR systems are
associated with the protection of the surfaces
of the cavity components from damage due to
x-rays and other charged particles. The most
hostile environment in laser fusion reactors
results from the fusion-pellet microexplosion
in a reactor cavity. Concepts for protecting
cavity component surfaces from impinging
charged particles include restorable protective layers, an application of magnetic
fields in order to deflect charged particles
away from the cavity walls [51], and of cavity
gas near the target to absorb the x-rays and
ionic debris (emanating from the microexplosion)
to protect the wall [61]. Some models for
evaluating the thermal response of the first
wall (or its protective liner), which is produced by x-rays and charged particles emanating
from the pellet and by reflected laser light,
have been developed [51, 62]. Other heat
transfer problems in LCFR concepts which do not
appear to have been seriously considered include the cooling of large mirrors and windows
that are exposed to intense short pulse laser
beams and possibly to radiation from a reactor
cavity. No doubt, other problems will surface as practical feasibility and preliminary
engineering design of conceptual LCFR's evolve
and technology is proven.

CONCLUDING REMARKS

In this review paper an attempt has been made
to introduce heat transfer researchers to some
interesting and complex heat transfer problems
identified to date in connection with development of demonstration plants for solar thermal
electric power generation, test facilities, and
plants for coal-fired MHD and also in the conceptual design of controlled nuclear fusion
reactor systems. Successful development and
design of full scale units require a much more
sophisticated data base. Many more challenging
thermal problems are expected to arise as the
development activities continue and detailed
designs evolve.

ACKNOWLEDGMENTS

The author wishes to acknowledge the editorial
assistance of Mr. Louis A. Diaz in the preparation of the paper.

REFERENCES

1. Schweinberg, R. N. and Leibowitz, L. P.,
   Sandia Laboratories Report SAND79-8073,
   pp. 30-50 (1979).
2. Siebers, D. L., Abrams, M. and Gallagher,
   R. J., ASME Paper No. 79-WA/HT-38 (1979).
3. Skinrood, A. C., in Ref. 1, pp. 21-29.
4. Gallagher, R. J., Abrams, M. and Kraabel,
   J. S., in Ref. 1, pp. 247-257.
5. Siebers, D. L., Sandia Laboratories
   Report SAND78-8276 (1978).
6. Clausing, A. M., et al., in Proceedings of
   the International Symposium on Solar
   Thermal Power and Energy Systems, Marseille,
   France, June 15-20, 1980 (in press).
7. Churchill, S. W. and Bernstein, M., Trans.
   ASME, Series C, J. Heat Transfer, 99, 300-
   305 (1977).
8. Churchill, S. W. and Chu, H. H. S., Int.
   J. Heat Mass Transfer, 18, 1323-1329
   (1975).
9. Churchill, S. W. and Usagi, R., AIChE
   J., 18, 1121-1128 (1972).
10. Gallagher, R. J., Abrams, M., and Kraabel,
    J. S., in Ref. 6.
11. Liebenberg, J., Sandia Laboratories
    Report SAND79-8073, pp. 185-199 (1979).
12. Larson, D. W. and Viskanta, R., J. Fluid
    Mech, 78, 65-88 (1976).
13. Larson, D. W., in International Conference
    on Numerical Methods in Thermal Problems,
    Swansea, England, July 2-6, 1979.
14. Penot, F., Revue Phys. Appl., 15, 207-
    212 (1980).
15. Clausing, A. M., in Ref. 6.
16. Schaut, E. "Gasgekühltes Sonnenturm-
    Kraftwerk Leistung 20 MWe," Dornier
    System Technischer Bericht, Friedrichshafen,
    West Germany (1980).
17. Yanagi, K., in Ref. 6.
18. Kudirka, A. A. and Leibowitz, L. P.,
    AIAA Paper No. 80-0252 (1980).
19. Kreeb, H. and Schaber, K., AIAA Paper No.
    AIAA-80-1507 (1980).
20. Heywood, J. B. and Womack, G. J., Editors,
    Open-Cycle MHD Power Generation, Pergamon
    Press, New York (1969).
21. Petrick, M. and Shumaytsky, B. Ya.,
    Editors, Open Cycle Magnetohydrodynamic
    Electrical Power Generation, A Joint
    USA/USSR Publication, Argonne National
    Laboratory, Argonne, IL (1978).
22. Postlethwaite, A. W. and Sluyter, M. M.,
    Mech. Eng., 100 (3), 32-39 (1978).
23. Sarofim, A. M., Howard, J. B., Padia, A.,
    and Kabayashi, H., in Sixth International
    Conference on Magnetohydrodynamic Power
    Generation, Washington, DC (1975), Vol.
    II, pp. 305-315.
24. Roy, G. D., ASME Paper No. 80-HT-125 (1980).
25. Wright, R. J., in 18th Symposium on
    Engineering Aspects of Magnetohydro-
    dynamics, Butte, Montana, June 18-20,
    1979, pp. K.6.8-K.6.14.

26. Romig, M. F., in _Advances in Heat Transfer_, T. F. Irvine, Jr. and J. P. Hartnett, Editors, Academic Press, New York (1964), Vol. l, pp. 267-354.

27. Oliver, D. A. and Maxwell, C. D., AIAA Paper No. 77-108 (1977).

28. Zaporowski, B. and Roszkiewicz, J., in Ref. 25, pp. B.6.1-B.6.6.

29. James, R. K. and Kruger, C. H., in Ref. 25, pp. E.4.1-E.4.6.

30. Bogdanska, M. et al., in Ref. 23.

31. Dicks, J. B., Tempelmeyer, K. E., Wu, Y. C. L. and Crawford, L. W., in _11th Intersociety Energy Conversion Conference Proceedings_, AIChE, New York (1976), Vol. II, pp. 1015-1019.

32. Stickler, D. B. and DeSaro, R., AIAA Paper No. 77-109 (1977).

33. Saric, W. S., Touryan, K. J., and Scott, M. R., J. Energy, $\underline{1}$, 108-114 (1977).

34. Biberman, L. M. et al., in Ref. 25, pp. B.5.1-B.5.6.

35. Im, K. H. and Ahluwalia, R. K., AIAA Paper No. 80-0250 (1980).

36. Doss, E. D., J. Energy, $\underline{1}$, 370-375 (1977).

37. Ahluwalia, R. K. and Im, H. K., AIAA Paper No. 80-0252 (1980).

38. Ahluwalia, R. K. and Kim, K. H., in _Proceedings of the 7th International Conference on MHD Power Generation_, Cambridge, MA (1980), pp. 187-194.

39. Im, K. H. and Ahluwalia, R. K., ibid, pp. 329-336.

40. Pober, R. L., Cannon, W. R., Bowen, H. K. and Louis, J. F., ibid, pp. 278-286.

41. Johnson, T. R. et al., Fourth U.S./U.S.S.R. Colloquium on MHD Power Generation, Washington, DC, October 4 and 5, 1978.

42. Sistino, A. J., ASME Paper No. 80-HT-44 (1980).

43. Chow, L. S. H., Viskanta, R. and Johnson, T. R., ASME Paper No. 78-WA/HT-21 (1980).

44. Im, K. H., Patten, J. and Johnson, T. R., in _18th Symposium on Engineering Aspects of Magnetohydrodynamics_, Butte, Montana, June 18-20, 1979, pp. C.2.1-C.2.7.

45. Townes, H. W., Reihman, T. C., Mozer, C. J. and Ameel, T. A., ibid, pp. C.7.1-C.7.7.

46. Townes, W. H., Reihman, T. C., Mozer, C. J. and Ameel, T. A., ibid, pp. C.8.1-C.8.5.

47. Saari, D. P. et al., in Ref. 38, pp. 371-378.

48. Williams, J. M., AIChE Symposium Series, $\underline{73}$ (168), 1-8 (1977).

49. Steiner, D., Nucl. Sci. Eng., $\underline{58}$, 107-165 (1975).

50. Hoffman, M. A., Werner, R. W., Carlson, G. A. and Cornish, D. N., in Ref. 40, pp. 9-44.

51. Frank, T. G., Bohachevsky, I. O., Booth, L. A. and Pendergrass, J. H., in Ref. 48, pp. 77-85.

52. Fraas, A. P., Oak Ridge National Laboratory Report ORNL-4999 (1975).

53. Gardner, E. C. and Lykoudis, P. S., J. Fluid Mech., $\underline{47}$, 737-764 (1971).

54. Brouillette, E. C. and Lykoudis, P. S., Phys. Fluids, $\underline{10}$, 995-1007 (1967).

55. Wagner, L. and Lykoudis, P. S., AIChE Symposium Series, $\underline{73}$ (164), 142-147 (1977).

56. Wagner, L. Y. and Lykoudis, P. S., in _Proceedings of the 8th Symposium on Engineering Problems of Fusion Research_, IEEE Publication No. 79CH1441-5 NPS, New York (1979), Vol. IV, pp. 2075-2077.

57. Giezszewski, P. J., Mikic, B. and Todreas, N. E., ASME Paper No. 80-HT-69 (1980).

58. Gardner, R. A. and Lykoudis, P. S., AIAA Paper No. 69-723 (1969).

59. Fillo, J. A. and Powell, J. R., in Ref. 48, pp. 45-61.

60. Moir, R. W. and Taylor, C. E., ASME Paper No. 80-HT-140 (1980).

61. Abdel-Khalik, S. I., Moses, G. A. and Peterson, R. R., ASME Paper No. 80-HT-64 (1980).

62. Abdel-Khalik, S. I. and Hunter, T. O., ASME Paper No. 77-HT-83 (1977).

High-Performance Mist Cooled Condensers for Geothermal Binary Cycle Plants

Y. MORI
Tokyo Institute of Technology, Tokyo, Japan

W. NAKAYAMA
Hitachi Ltd., Ibaraki, Japan

ABSTRACT

A Rankine cycle system, using organic fluids of low boiling point as the working fluid, has a great potential for generating electric power from liquid-dominated geothermal resources and waste heat of industrial plants. Its feasibility depends largely on the performance and cost of condensers. The mist cooled condenser has many attractive features such as a small parasitic power requirement, low capital cost, small consumption of cooling water and adaptability to ambient temperature change. This paper describes the experimental work undertaken by the authors' group to obtain the data required to optimize the design of mist cooled condensers. The research subjects include the finding of zones for effective mist cooling in a tube bundle, improvement of wettability on the outside surface and enhancement of condensation on the inside surface. Optimization is performed so as to minimize the volume of a condenser, the consumed power, the capital cost and the operation cost.

NOMENCLATURE

C_p = specific heat capacity, J/kg°C
d_i = inside diameter of tube, m
G_t = mass flow rate of working fluid per cross sectional area, kg/m^{2}s
h = depth of groove, m
L = length of tube, m
$\dot{m}$ = mist flow rate per cross sectional area, kg/m^{2}h
p = pitch of fins, m
T_i = temperature of working fluid inside of the tube, °C
T_w = temperature of tube wall, °C
V = volumetric flow rate, m^3/s
V_a = air velocity, m/s
y = vapor quality
α = heat transfer coefficient, W/m^2°C
ΔT = temperature difference between the tube wall and upstream air, °C
ΔT_c = temperature increase of cooling water, °C
ρ = density, kg/m^3

Subscripts

c = cooling water
l = liquid
v = vapor

INTRODUCTION

Most of the oil consumed in Japan is imported, and research and development of alternative energies is one of the most important projects. Among the alternative energies, geothermal energy is expected to be the most promising, national, reproductive and least polluting resource, the No. 3 position next to nuclear energy and coal, as Japan is totally located in the geothermal dominating zone. The total output of the geothermal power plant is about 250MW at present, mainly by use of geothermal steam, but above 40,000MW is expected to be generated from geothermal energy if geothermal steam and brine are totally developed and effectively utilized. Geophysical investigation made so far predicts plenty of geothermal energy is contained in geothermal brine than in steam. Therefore, efficient utilization of geothermal brine to generate electricity is highly desired. One of the technically and economically feasible ways for this purpose is to adapt a Rankine cycle, using an organic fluid of low boiling point as the working medium. The Rankine cycle, using geothermal brine is aptly called the binary cycle. A R&D program of binary cycle has been undertaken as one of the national projects in the development of geothermal energy. Two pilot plants of about 1MWe output were built and operated in 1977 through '78 under the sponsorship of the Sunshine project of the Ministry of International Trade and Industry and technically satisfactory results were obtained. It may be necessary to point out here the importance of the development of the binary cycle from a wider stand point of energy problems. The binary cycle is getting its importance also for waste heat recovery in industries such as steel or chemical plants. Energy conservation is one of the major items in the national energy projects, and it is scheduled to cut down about ten percent of the imported oil by energy conservations. In the energy consuming industries, waste heat recoveries at higher temperatures were studied and put into practice; however, to promote more energy conservation waste heat recovery in the middle or low temperature ranges has been attracting attention where the same cycles with the geothermal binary cycle are found to be economically feasible and attractive. In this respect the contents of research and the discussions of this paper should be understood from a wider scope of resource development and energy conservation. In the economical assessment of the binary cycles tested in Japan, it was confirmed that the in-plant power consumption is unexpectedly large compared with that in conventional steam plants. The main parts consuming power are the feed pumps of working fluid and the large cooling water pumps and fans in the cooling system due to the low temperature of the heat source and consequently low thermal efficiency and large exhaust heat. Therefore, it should be noted that the development of a high performance condenser which consumes less power is in high demand. Based on the experience obtained in the assessment of the operation results of the binary cycles, the following requirement should be satisfied in advanced condensers in the economical binary cycle which is expected to be built in the near future.

(1) compact size,
(2) low cost,
(3) small power consumption by fans and pumps,
(4) small consumption of cooling water,
(5) control of heat exchange performance adaptable to variation of ambient temperature.

The cooling tower system consisting of a surface condenser and a wet cooling tower has been considered as a conventional one, but has disadvantages associated with item (4) in addition to unsuitability to cold weather in winter causing environmental problems. On the other hand, the dry cooling system does not satisfy items (1) and (2). Besides, the cooling performance of both systems depends much on the ambient temperature. In consideration of the remarkable decrease of the binary cycle output partly caused by the increase of ambient temperature in summer compared with the conventional power plant and enormous seasonal demand of electric power in summer, a cooling system – which affords satisfactory cooling capacity in summer and in winter, provides the necessary performance without consuming much operation power or producing environmental problems – is highly required. A condenser cooled by mist carried with air flow could satisfy all the items mentioned above under an economically feasible condition when it is designed by well established data for optimal performance. The performance of the mist cooled condenser is closely connected with three basic heat transfer phenomenon, such as, forced convection and evaporation of water outside of the tube and condensation of working fluid inside of the tube, but a design method has not yet been fully established. Moreover, an accurate and reliable design is needed to guarantee the specified performance of the binary cycle, therefore, firstly, basic data of heat transfers should be studied, secondly, essential factors for economical design are selected, and finally, the design method for the optimal performance has to be established. In this paper, experimental researches of heat transfer performances on enhanced outside and inside tube surfaces are reported in order to give fundamental data for the optimized mist cooled condenser.

RESEARCH AND DEVELOPMENT OF MIST COOLED CONDENSER

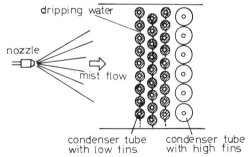

Fig.1　A new scheme of mist cooled condenser

In Fig. 1 one unit of mist cooled tube bundle is shown. Horizontal or tube bundles slightly inclined to the horizontal direction are set in a uniform air flow. Small water droplets ejected from nozzles are carried over with the air flow. The working medium of the binary cycle condenses

inside of the tube. Many of these tube bundles, fans and water pumps which eject water are the main components of the mist cooled condenser. The items of the research and development of the proposed mist cooled condenser are listed and explained below.

(1) Confirmation of zones of effective mist cooling

Droplets carried with the air flow are caught by the heat transfer tubes of upstream rows while tubes located at several rows downstream have low heat transfer coefficients since they are not advantaged by mist cooling. Tubes in the lower zone get wet receiving falling drops from the upper tubes, and are considered to have a heat transfer coefficient different from those around the upper tubes. These complicated distributions of heat transfer coefficients have been rarely reported so far, and are studied in detail in this paper.

(2) Improvement of heat transfer on the outside surface of a tube by extention of wetted area

On a smooth surface of a tube or on a smooth fin, the area getting wet extends over only a rather small fraction of the total surface. Droplets caught on the surface form a liquid film on the heat transfer surface, but the film separates from the surface with the main flow and rarely covers the rear surface of the tube, therefore, in order to effectively use the mist cooling effect, a study of surface configuration to extend the film to cover the rear surface has been conducted. Tiny fins are provided for this particular purpose by effective use of surface tension. The enhanced heat transfer performance caused by this means is described in detail in the following.

(3) Enhancement of condensation heat transfer on the inside surface of a tube

If the condensation heat transfer coefficient of organic fluid on the inside surface of a tube is not high enough, then the measure by use of adequate fins to enhance condensation performance cannot only make the condenser compact, but decrease the condensation temperature and pressure leading to an increase of the output power. However the provision of fins along the inner surface brings about disadvantages such as the pressure loss and the increase of tube material and fabrication cost. Therefore, the shape of fins should be determined while taking into account the advantages and disadvantages described above. The optimized fin shape thus investigated is that of the tiny fins of 0.1mm to 0.2mm height and pitch.

(4) Establishment of design process for optimized mist cooled condenser

By use of the results obtained from the fundamental investigations in items (1) to (3) described above, the specification for an optimized mist cooled condenser is to be given. In the optimization design, the volume of the condenser, the consumed powers to operate it, the capital cost and the operation cost are selected as the necessary independent parameters for optimization. The configuration of the outer surface structure of the tube, the pitch of the tube bundle, the air velocity and the water flow rate are determined so as to make each independent parameter selected above to have a minimum value. After careful and detailed basic investigations, several schemes emerge as candidates of the ultimate optimized structure of a mist cooled condenser. Fig. 1 shows one of them, where the three upstream bundles consist of low finned tubes with tiny fins along the surface as to extend the water film all along the surface by the action of surface tension and the high finned tubes are used in the far downstream row. This is based on the idea of heat load sharing among the tube bundles in seasons, in summer the three upstream bundles cooled by mist operate effectively to carry the major part of the heat load, while in the winter the downstream high finned tubes carry most of the heat load without injecting mist and introducing problems such as freezing or environmental pollution. Capacity and temperature condition of local geothermal resource or waste heat vary in a wide range and the objective of the mist cooled condenser for the plant to convert the heat into electricity is to fully satisfy various specified conditions to be highly accurate, and to prove the optimized performance. It is suggested firstly to determine the optimum construction of a module such as the one shown in Fig. 1 consisting of four row bundles, and secondly to construct a mist cooled condenser installing an adequate number of modules. In the following the experimental results associated with items (1) to (4) are described, and lastly the design procedure for optimized performance is briefly summarized.

HEAT TRANSFER FROM A MIST COOLED TUBE BUNDLE

1. Experimental devices

Heat transfer performances on the outside surface of tubes in a bundle were measured. The aluminum tubes of 25.4mm O.D. having various surface configurations were set horizontally in a 500mm rectangular flow duct. The tubes were electrically heated by heaters installed in the tubes. The tests were made with staggered arrangement of tubes in four rows. The air flow was changed from 1m/s to 3m/s. Mist was formed by nine atomizing nozzles set at 1.5m upstream of the tested tubes. The mist flow rate was controlled in the range of 50 to 400 kg/m^{2}h. Fig. 2 shows an experimental setup for the bundle pitch of 54mm and for this particular case the experimental data obtained by use of this setup are discussed. The temperature of the tube was measured by four thermocouples of 0.2mm dia-

meter inserted in the holes drilled in the tube wall. The wall temperature was varied from 20°C to 80°C.

2. Experimental results and discussions

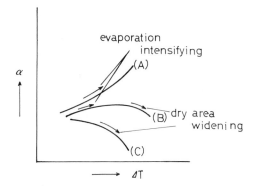

Fig.3 Physical interpretation of the curves of heat transfer cofficients on mist-cooled tube surfaces.

The general aspects of variation of the heat transfer coefficient on mist cooled tubes versus the tube wall temperature are as shown in Fig. 3, where ΔT is the temperature difference between the tube wall and the upstream air. The heat transfer coefficient on the ordinate is the mean value in the circumference of the tube. Curve A corresponds to the case when the tube surface is covered by water film and the heat transfer coefficient α increases with the increase of ΔT as evaporation is enhanced. Curve B represents the case when the supply of mist becomes insufficient as ΔT is increased to catch up the increasing evaporation of liquid film from the surface. Eventually α decreases with an increase of ΔT above a critical value of ΔT as dry surface extends. When most of the surface is dry even for small ΔT, the behavior of α shows a monotonous decrease with ΔT as indicated by curve C.

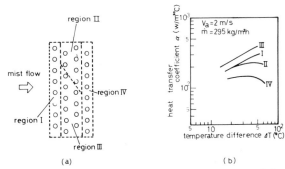

Fig.4   Division of the regions in a tube bundle (a) according to the dependence of their heat transfer coefficients on the temperature difference ΔT (b).

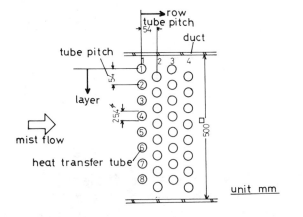

Fig.2 Arrangement of tubes in a test bundle.

The experimental results for the smooth tube bundle of four rows are shown in Fig. 4. The bundle is divided into four regions each having specific heat transfer performances. Mist is generated by use of nozzles of 0.25mm to 0.35mm in diameter and consists of droplets of about 44μm. The tubes of the first row, indicated by the region I, have the high performance which coincides with that of a single tube because the wet surface area is relatively large. The upper tubes of the second and third rows show the performance indicated by the region II and being similar to   case B in Fig. 3 as the supply rate of water is relatively small. This region extends to the front and lower zone with the increase of ΔT. The lower tubes of the second and third rows in the region III have the highest heat transfer performance enhanced by film evaporation of liquid dropping from above. Enhancement of forced convection also works here owing to the local high speed induced by a small gap between the tubes. The tubes in   region IV rarely get wet by mist. The influence of the increase of the tube pitch, which is the same in the flow and cross-flow directions, is found to bring the mist far downstream and make mist cooling effective to the downstream bundle. From these experimental results, it is found that installation of atomizing nozzles after every three rows is needed, or high finned tubes should be adopted in downstream rows to effectively utilize forced convective heat transfer.

<u>ENHANCEMENT OF HEAT TRANSFER ON THE OUTSIDE SURFACE OF A TUBE</u>

According to the previous papers [1], [2], [3], [4], [5], [6] heat transfer experiments in mist flows around circular or finned cylinders have been reported. The air speed ranges from 1m/s to about 40m/s, the tube temperature from ambient to 100°C, and the flow rate ratio of mist to air is 0.2 to 0.005. It is to be noted that the experiments made so far were conducted in the range of air speed rather higher than those preferably used in mist cooled condenser. The features of our research are as follows.
(1) In the mist cooled condenser, the air speed is not to be taken too high and is taken in the range of 1 to 3m/s.
(2) As reported so far, the rear zone of behind the separation point of the tube is not wet and not effective to mist cooling. To extend the wet zone for utilizing mist cooling enough, a new idea to be explained below has been introduced.
As pointed out by Simpson [5], the ratio of the wet area to the total area of finned tube gets smaller with the increase of fin height resulting in reduction of the mist cooling effect. In this research, tiny fins with circumferential trough in-between are provided along the tube surface to extend the wet area far over the tube rear surface, and to prevent the appearance of dry area so as to keep the high heat transfer coefficient by mist cooling along the whole tube surface. In the experiments for heat transfer enhancement, smooth tubes, tubes with tiny circumferential fins and high finned tubes with smooth base surface were tested. Also investigated were the finned tube with tiny crossing grooves of 0.1mm depth and 0.75 mm pitch on the fin surface. Tubes and fins are made of anti-corrosive aluminium.

1. Experiments and results

Experiments were made in a wind tunnel of 15cm X 15cm square cross section and a test tube was set in a horizontal way. Mist flow rate was measured by a small pipe of 19mm opening with its opening opposed to the air flow. The flow rate was actual-

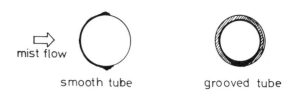

Fig.5 Spread of water film on a smooth tube and a grooved tube.

ly found from the weight of water collected through this pipe during a predetermined time interval. The dry-bulb and wet-bulb temperatures were measured at the exit of the sampling pipe after separating water. Visualization experiments around the tubes were made to observe the behavior of water film around the tube and the typical results are sketched in Fig. 5. The left picture shows the film around a smooth tube, the rear surface of which is dry as film separates and is thereby carried away with the main stream. On the other hand, the right picture shows that on the tiny-finned tube the liquid covers the whole tube surface. In Fig. 6, experimental results of heat transfer coefficients expressed by use of the base tube surface area are shown with the parameters of air speed and mist flow rate. The arrows show the value of ΔT obtained from visualization studies

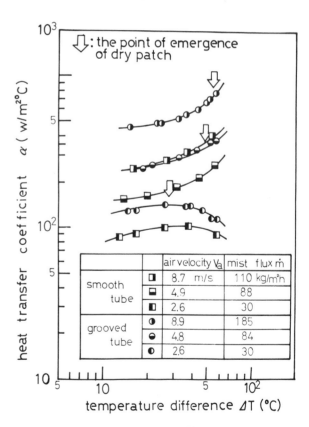

		air velocity V_a	mist flux $\dot{m}$
smooth tube	▯	8.7 m/s	110 kg/m²h
	▱	4.9	88
	▮	2.6	30
grooved tube	◑	8.9	185
	◒	4.8	84
	◐	2.6	30

Fig.6 Heat transfer coefficient on a smooth tube and a grooved tube placed in a mist flow.

when the dry zone appears on the rear surface of the grooved tube. For small mist flow rates, the dry zone appears with a small ΔT resulting in the deterioration of heat transfer. A similar effect of capillary extension is confirmed with the finned tube having tiny grooves or roughness on the fin surface. In comparison between smooth fins and roughened fins, in the case of the former, water is accumulated near the leading edge of the fin and the bottom of the tube and a wide dry zone was observed. In contrast, in the case of the roughened fin, tiny grooves on the fin surface have actions to extend water over a wider zone, and the tube with roughened fins is found to have about a 20% higher heat transfer coefficient than that with smooth fins. A comparison study between tubes with high fins and low fins of smooth surface was also made. Fins of 10mm and 5mm height were tested. The ratios of the wet area to the total surface area were calculated from the measured amount of heat transfer coefficient increase and confirmed by visualization studies. It was found from the comparison, that the tube with low fins has a higher ratio of wet area. The ratio decreases almost in a way inversely proportional to air speed. The decrease of the wet area ratio with the increase of ΔT is more remarkable for a high finned tube. Researches on further improvement of heat transfer performance by assisting extension of water film over a wider area of the surface are carried on to provide sufficient data for high performance mist cooled compact condensers.

AUGMENTATION OF IN-TUBE CONDENSATION

Reviewing the previous works [7] ~ [10], one finds that almost all of the previous attempts to enhance condensation by means of internal fins inevitably caused a higher pressure drop. To break up this link between heat transfer augmentation and increase in pressure drop is one of the objectives of the present research, namely, to develop a tube which yields a high condensation heat transfer coefficient without incurring a large pressure drop.

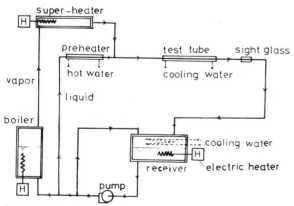

Fig.7 Schematic diagram of the in-tube condensation experimental apparatus

The schematic diagram of the experimental apparatus is shown in Fig. 7. Vapor is generated from the boiler, its temperature being controlled by the superheater, then undergoes mixing with the liquid before entering the test tube. The liquid to be mixed with vapor is extracted from the liquid line

after the circulation pump. Its temperature is controlled by the preheater. In this way, the flow rate and the quality of the working fluid are well controlled and set as predetermined values at the entrance of the test section. The remnant of vapor leaving the test condenser tube is condensed in the receiver. R-113 was used as the working fluid. The testing section is made of double tubes. The working fluid flows in the middle passage and the cooling water flows in the outer annular passage. The cooling water is circulated from the cooling water tank where the temperature is kept constant. The pressure of the working fluid in the test condenser tube is measured by the pressure transducers at the inlet and the outlet of the test tube. A saturation temperature at the average of these two pressures is defined as the temperature of the working fluid (T_i) in reduction of the experimental data. The tube wall temperature was measured by setting the thermocouples inside the tube wall. Small holes of 1mm in diameter and 1.5mm deep were bored, where the copper-constantan thermocouples having the wire diameter of 0.2mm were inserted and cemented by the epoxy resin. The average of the readings of the six thermocouples, located on the circumference of the middle section and two other points longitudinally separated, is defined as the wall temperature (T_w). Condensation heat transfer coefficient $\alpha(W/m^2°C)$ is computed by the following equation,

$$\alpha = \frac{C_p \rho V_c \Delta T_c}{\pi d_i L (T_i - T_w)}$$

where $V_c(m^3/s)$ is the volumetric flow rate of the cooling water, $\Delta T_c(°C)$ the temperature increase of the cooling water, $C_p(J/kg°C)$ the specific heat capacity of the water, d_i the inside diameter of the tube (=0.02m), L the length of the tube (=0.69 m) and $\rho(kg/m^3)$ is the density of the water. Flow rates of the liquid and the vapor were measured separately by turbine flow meters. The mass flux $G_t(kg/m^2s)$ and the quality y of the working fluid were computed by the following equations.

$$G_t = (\rho_v V_v + \rho_l V_l) / (\frac{\pi}{4} d_i^2)$$

$$y = \rho_v V_v / (\rho_v V_v + \rho_l V_l)$$

Here, ρ_l and ρ_v are the density of liquid and that of vapor, respectively. The pressure drop in the tested condenser tube was measured by the differential pressure transducers, with the accuracy of ±50Pa.

Table 1 Groove geometry of tested condenser tubes

tube No.	geometry of inner surface of tube	depth h(mm)	apex angle of crest φ	pitch p(mm)	screw angle	ratio of real to nominal surface area
1	smooth	-	-	-	-	1
2	groove A	0.2	90°	0.45	7°	1.4
3	groove B	0.4	30°	0.45	7°	2.3
4	groove C	0.4	90°	0.90	7°	1.4

The specifications of the four tested tubes are shown in Table 1. All have the innermost diameter

of 20mm and the working length of 690mm. The material is anti-corrosive alminium alloy. The grounds for selection of these dimensions are as follows.
(1) Enhancement of heat transfer without an appreciable increase in required tube material to erect internal fins.
(2) Easy manufacturing.
The depth of   groove A is set much smaller than the tube diameter (about 1/100).   Grooves B and C require more tube material and more manufacturing time than   groove A; however, they were taken up to see the effect of the groove geometry upon condensation. The experimental results are shown in Figs. 8 and 9. Fig. 8 shows the effect of the flow rate of the working fluid (G_t) upon the heat transfer coefficient α. The curves are drawn by combining the data points of nearly equal vapor qualities (y). Fig. 9 shows the effect of

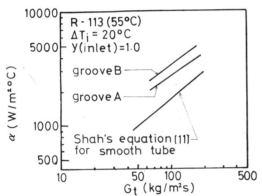

Fig. 8 Relation between heat transfer coefficient and mass flux of working fluid in horizontal tubes

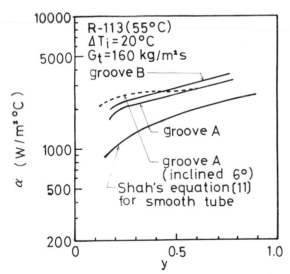

Fig. 9 Relation between heat transfer coefficient and vapor quality

the quality when the mass flux is kept constant at 160kg/m^{2}s. The important point revealed by the experiment is that the heat transfer coefficient becomes very low when the vapor quality is reduced to less tnan 0.2.

Table 2 Ratings of the various groove geometries

tube	$\left(\dfrac{\text{increase of heat transfer coefficient}}{\text{increase of pressure drop}}\right)$	$\left(\dfrac{\text{increase of heat transfer coefficient}}{\text{increase of material}}\right)$	productivity
smooth	1	1	o
groove A	1.5	1.5	o
groove B	1.7	1.5	x
groove C	1.5	1.3	o

The ratings of the four structures are summarized in Table 2.   Groove A has the following advantages over the other grooves.
(1) The ratio of the increase of heat transfer coefficient to that of material of fins is greater than that of   groove C.
(2) Manufacturing of   groove A is much easier than that of   groove B.
The effect of inclination of the tube on heat transfer has also been investigated. When an inclination of 6° is given, the heat transfer coefficient is almost constant in the quality range of 0.2~0.5 and is about 1.2 times as high as that of a horizontal tube in low qualities. This is likely because tne flow is in a regime of stratified flow and the inclination helps the drainage of condensate. The results of this research provide the basis for the future design of a condenser via the following two contributions. First, the performance of the groove tube is established quantitatively, giving a good prospect for reduction of the size of the condenser. Secondly, the data suggests that a further reduction of the condenser size can be achieved by providing condensate separators within a network of heat exchangers to maintain the local vapor quality above 0.2.

METHOD OF OPTIMIZATION OF CONDENSER DESIGN

The ultimate objective of the optimization is to minimize the contribution to the cost of generated power of a binary cycle plant. However, the factors affecting the condenser design and the electricity cost are too numerous and complicated, so that it is by no means easy to achieve the ultimate optimization from the outset. As a first step, the following approach is taken in this work. Confining attention to the factors concerned only with the technical aspects of the condenser design, the objective factors and the fixed factors are listed up as shown in Table 3. Among them, the cases 1~3 are selected as the most important ones and subsquent analysis of optimization were made for them. The BOX method [12] was used for the optimization analysis, which is one of the direct searching methods, and enables one to locate a minimum point in the space of nonlinear objective functions. A flow diagram of the program is shown in Fig. 10. Using this program, the condenser of a 10MW class binary cycle power generation plant was designed. An example of the cycle condition is shown in Fig. 11. Two cases have been studied until now;   one is a condenser built by conventional high-finned tubes, and the other is a condenser built by tubes having circumferential grooves on the outer surface and   groove A on the inner surface. When the minimization of the tube bundle volume is chosen as the objective of optimization, it is concluded that the use of high-performance tubes can reduce the volume almost half of that of the conventional structure. Analytical work in progress incorporates more attractive features such as a combined use of high-performance tubes and high-finned tubes and a suitable routing of the working fluid in a network of heat exchanger modules. A more detailed description of the results will be reported in the future.

CONCLUSION

The research program has been carried on to develop a mist cooled condenser which is a key component of the binary cycle power plant utilizing geothermal hot water as the heat source. The present paper reports the results heretofore obtained in this program. The following are the summary of the results.

(1) In order to guarantee sufficient accuracy in the performance estimation of a mist cooled condenser, it is essential to know how the heat transfer tubes are wetted by sprayed water. The experimental investigation has yielded detailed information on this problem. In a tube bundle, a relatively large quantity of water is captured by the front, the second and the third rows, whereas the fourth row receives only a small fraction of sprayed water, so that the enhancement of heat transfer is negligible there. The surface condition of a tube is another important factor in the design of a mist cooled condenser. On conventional bare tubes and finned tubes, only a limited part of the heat transfer surface gets wet, resulting in insufficient utilization of mist cooling enhancement.

(2) The important mode of heat transfer augmentation on the outer surface of a tube is evaporative heat transfer. While Yang and Clark [13] reported a negligible effect of evaporation on heat transfer from a mist cooled radiator for automobiles, it is concluded from the present study that the relative importance of evaporative heat transfer depends on water film formation. If the captured water forms localized thick films, a situation likely to exist in densely finned radiators, the resistance to heat flow is in the water film. On the other hand, heat transfer is controlled by evaporation when the water spreads over a wide area in the form of a thin film. Since the evaporative mode yields the highest augmentation, the extension of water film by means of small grooves on the fins was attempted. The experiment produced the results as expected; the heat transfer coefficient increases almost 20% from that on a smooth tube having smooth fins, a comparable figure to the

Table 3 Combination of objective and fixed factors

	case	1	2	3	4	5	6	7	8	9	10	11	12	13
objective of optimization	minimum volume of condenser	o			o	o								
	minimum pump and fan power		o			o	o							
	minimum consumption cooling water			o				o	o					
	minimum number of tube									o	o			
	minimum pressure drop in path of working fluid												o	
	maximum heat transfer rate													o
fixed factors	geometry factors / system layout / direction of mist flow and air flow / inclination of tubes / total number of tubes / number of layers of tubes / number of rows of tubes / tube diameter / tube length / tube pitch / surface geometry / number of path of working fluid					o	o	o	o					
	heat transfer rate	o	o	o	o	o	o	o	o	o	o	o	o	-
	mist flux	o	-		o		o	-	-	o		o	o	o
	air flow rate				o	o								
	flow rate of working fluid	o	o	o	o	o	o	o	o	o	o	o	o	-
	pressure drop in path of mist-air flow	o	o		o				o		o		o	o
	pressure drop in path of working fluid	o	o	o	o	o	o	o	o	o	o	o	-	o
	power		-	o	o	o	-	-	o	o		o		
	condition of atmosphere	o	o	o	o	o	o	o	o	o	o	o	o	o
	condition of working fluid (Pressure level, degree of superheat)	o	o	o	o	o	o	o	o	o	o	o	o	o
	fan									o				o

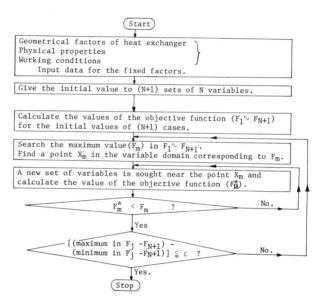

Fig. 10 Flow diagram of optimization program (for searching a minimum of the objective factor)

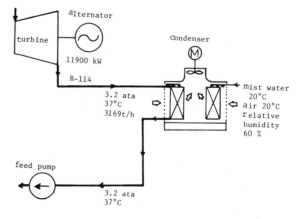

Fig. 11 A typical example of cycle variables before and after of a condenser

increase of wetted area.

(3) Enhancement of condensation heat transfer on the inner surface of tubes is also important for the condenser of a binary cycle plant. For this purpose, small grooves of various depths and pitches were provided on the inner surface. Their rating was determined by the heat transfer performance, the required tube material and the ease of manufacturing. The form of groove judged as the best is the one which has a triangular cross section, a depth of 0.2mm, a circumferential pitch of 0.45mm and the twisting angle relative to the tube axis of 7°.

(4) The data of heat transfer augmentation is useful to design an optimum mist cooled condenser which satisfies various technical as well as economical demands. A new type of condenser is proposed, which incorporates features such as a combination of high-performance tubes placed in three upstream rows to fully utilize the enhancement by mist cooling and conventional high-finned tubes placed in downstream rows where single phase convective heat transfer is the predominant mode. As for the tube-side heat transfer, keeping the vapor quality above 0.2 in the major part of the tubings can help reduce the size of the condenser. Condensate separators will be provided within a network of heat exchanger modules to serve this purpose.

Future works include acquisition of more fundamental data in the laboratory, building of a prototype condenser, and its operation all year round. The present work is sponsored by the Japanese government as one of the technical programs under the Sunshine Project, aimed at the development of alternative energy sources.

ACKNOWLEDGEMENT

The present paper is based on the experimental work carried out by H. Kuwahara and S. Hirasawa of the Mechanical Engineering Research Laboratory, Hitachi Ltd. A full account of the work will be reported in the future with those contributors as co-authors. The authors also wish to thank the Agency of Industrial Science and Technology (MITI) and the Electric Power Development Co. for mainly sponsoring this study.

REFERENCES

1. Kuwahara, Nakayama and Mori, 16th Annual Symposium Heat Transfer Society of Japan, 376-378 (1979).
2. Hodgson, Saterbak and Sunderland, ASME J. of Heat Transfer, 90-4, 457-463 (1968).
3. Mednick and Colver, AIChE J., 15, 357-362 (1969).
4. Sherberg, Wright and Elrod, Progress in Heat Transfer, 6, 739-752 (1972).
5. Simpson, Beggs and Sen, Proceedings of the Symposium of Multi-Phase Flow Systems, Inst. Chem. Engrs. Ser. No. 38, Paper No. H3, 1-22 (1974).
6. Kosky, Int. J. Heat Mass Transfer, 19, 539-543 (1976).
7. Royal and Bergles, ASME J. of Heat Transfer, 100, 17-23 (1978).
8. Reisbig, ASME Paper No. 74-HT-7 (1974).
9. Vrable, Yang and Clark, Vth International Heat Transfer Conference, 3, 250-254 (1974).
10. Kroger, AIChE Symp. Ser. 73-164, 256-260 (1977).
11. Shah, Int. J. Heat Mass Transfer, 22, 547-556 (1979).
12. BOX, Computer Journal, 8, 45-52 (1965).
13. Yang and Clark, Int. J. Heat Transfer, 18, 311-317 (1975).

ROUNDTABLE DISCUSSION

Current Status of Heat Transfer Research on Nuclear Reactor Safety in Japan

S. AOKI
Research Laboratory for Nuclear Reactors, Tokyo Institute
of Technology, Tokyo, Japan

ABSTRACT

This paper describes the current status of the research works concerning the heat transfer problems in the field of nuclear power reactor safety. First, the R & D organization of nuclear reactor safety researches in Japan and next the contents of the following theme are presented introductorily and briefly;
LWR;
1. ROSA-I, II, IV for PWR,
2. ROSA-III for BWR,
3. 2D-3D Reflooding Test for PWR,
4. Full-Scale Mark-II Containment Response,
Test Program;
5. Thermohydraulic Experiment in the Steam Generator,
   (a) CE Type-Steam Generator,
   (b) Fundamental Experiment of Boiling Phenomena in a Crevice under Low Pressure,
   (c) Verification Test on the Reliability of Steam Generator of PWR,
6. Hydrodynamic Stability Tests of Steam Generator for Liquid Metal Fast Breeder Reactor,
   (a) Large Scale Steam Generator Tests,
   (b) Parallel Flow Channel Instability Experiment.

1. INTRODUCTION

As the oil situation in the world becomes more and more severe, the necessity and importance of the peaceful use of nuclear power has been recognized again. However, even before the Three Mile Island accident in the U.S., nuclear reactor safety had become the target of public concern.

Today in Japan, about 20 nuclear power stations are in operation and approximately 10 plants are under construction, which are located on the sea-side. Therefore, as public interests are focused on the safety problem of nuclear reactors, Japanese government is promoting the research projects of nuclear safety. Because of the phenomena in relation to the Three Mile Island accident have been investigated theoretically with simulator and experimentally by simulated experimental facilities.

2. RESEARCH SYSTEM

The safety research in Japan consists of not only the studies on thermal reactors but also on fast breeder reactors (FBR). Fig.1 shows the R & D system in Japan. In the frame of thermal reactor, the safety research involves the heavy-water moderated boiling light-water cooled reactor which has been developed as Japan's national project and named "FUGEN" advanced thermal reactor (ATR). All R & Ds relevant to ATR and FBR are carried out by the Power Reactor and Nuclear Fuel Development Corporation (PNC) at Oharai Engineering Center and co-operating manufacturers supported by PNC. For example, PNC is conducting some large scale experiments on transient critical heat flux, blowdown, instability of boiling flow channels for pressure tube-type re-

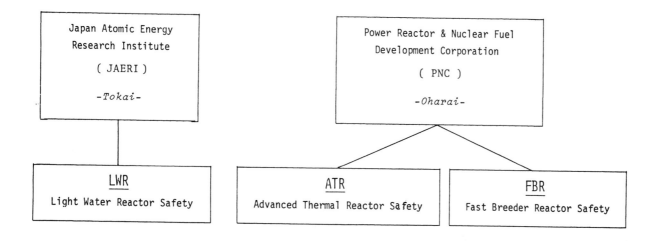

Figure 1   R & D Organization of Nuclear Reactor Safety Research in Japan

actors using 14MW Heat Transfer Loop and other large scale facilities. PNC is also responsible for safety researches for the Fast Breeder Reactor which uses liquid sodium as the coolant. In the case of light water reactors, Japan Atomic Energy Research Institute (JAERI) is responsible for the safety studies. Principally, JAERI carries out large scale experiments and also takes part in the LOFT Project and other international project of PWR Reflooding Test in collaboration with the U.S. and F.R.Germany. Since about 10 years ago, the ROSA Project on blowdown and heat-up of fuel rods has started to study the effect of emergency core cooling systems of PWR. Today, however, the project has moved on to the ROSA-III Program for BWR's emergency core cooling system test. Also ROSA-IV is scheduled for the coming fiscal year, to investigate the effect of small break Loss of Coolant Accident (LOCA) whose importance was recognized in the Three Mile Island accident. In addition, a large scale test facility of BWR Mark-II type containment has been built to make clear the dynamic loads on the suppression pool structures at LOCA. JAERI also has a project for a multipurpose high temperature gas cooled reactor development. Carrying out the detailed design of the reactor, JAERI is now constructing a large scale high temperature helium gas loop (1000°C) named "HENDEL". Also transition from turbulent to laminar has been studied. Some other governmental research institutes are performing nuclear reactor safety researches sponsored by national budgets, and partly by the electric power company groups and fabricators of nuclear power plants themselves are conducting the safety researches. For instance, the Core Counter Flow Limit Test by Hitachi, the Critical Flow test by Toshiba and the Steam Generator Verification Test by Mitsubishi, etc. are a few typical instances. A lot of the basic researches on reactor safety are being carried out at many universities. For example, transient critical heat-flux investigations are done by a couple of universities. In these studies, the critical heat-flux was measured for the cases of loss of pressure, loss of flow and heat generation excursion. The mechanism of reflooding heat transfer and the instability phenomena in boiling flow channels are also being studied.

3. REVIEW OF RESEARCH USING LARGE-SCALE TEST FACILITIES

3.1. ROSA Project

(1) ROSA-I Program /1/
The objectives of the ROSA-I experiments are (a) to measure the pressure and the leak flow transient during blowdown, (b) to obtain data on Departure from Nucleate Boiling during blowdown and (c) to validate computer codes. ROSA-I facility was constructed at the end of 1970 and 61 blowdown tests were conducted by the end of March 1973 for the top and bottom break, with and without a simulated core, a pressurizer and a forced circulation system, and changing the beak size and the initial pressure.

(2) ROSA-II Program /2/
ROSA-II facility consists of a model pressure vessel with the electrically heated core, two simulated primary circulation loops with pumps and steam generators, pressurizer, simulated break units, discharging piping and the emergency core cooling systems. Construction of ROSA-II facility was completed at the beginning of 1974 and eighty-two LOCA tests have been carried out by July 1977.

(3) ROSA-III Program /3/
The ROSA-III facility is a volumetrically scaled (1/424) BWR system with an electrically heated core designed to study the response of a primary core and the emergency core cooling system (ECCS) during the postulated LOCA. The primary objectives of the ROSA-III program are to provide the intergral test required to evaluate the adequacy and improve the analytical methods currently used to predict the LOCA response of large BWRs.

3.2. 2D-3D Program /4/

The 2D-3D program is a partial effect test which covers a part of the refill and reflood phase of a PWR LOCA. This program is now carried out as a three-party, F.R.G., U.S.A. and Japan. JAERI is responsible for experiments using two scaled-down facilities having 2000 simulated fuel rods with full length in cores. One facility is a cylindrical core test facility (CCTF) and another one is a slab core test facility (SCTF).

3.3. Full-Scale Mark-II Containment Response Test Program /5/

Started in fiscal year 1977, the Full-Scale Mark-II CRT (containment Response Test) Program was planned for the period of five years. The objective of the CRT Program is to provide a data base for evaluation of the pressure-suppression-pool hydrodynamic loads associated with a postulated loss-of-coolant accident in the BWR Mark-II containment system. The test facility was designed and constructed from fiscal year 1977 to 1978, and completed in March 1979.

3.4. Thermohydraulic Experiment in Steam Generator

(1) CE Type S.G.
Depending on the systematic measurements, it was confirmed that the Dry & Wet phenomena occur on the tube surrounded by the support-straps. The temperature fluctuations at the Dry & Wet spots were measured. Fig.2 shows an example of temperature fluctuation.

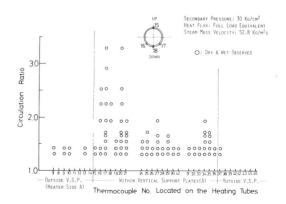

Figure 2 Location of Dry & Wet Observation

(2) Fundamental Experiment of Boiling Phenomena in a Crevice under Low Pressure /6/

The experiments were carried with annular crevices having the gap width of 0.2, 0.3, 0.4, 0.5, 1.0 and 1.5mm for the two cases. (a) In the case of "bottom open", the heat transfer coefficient is improved as the gap width decreases, but it is not affected by the gap length of $40 \leq L \leq 100$mm. (b) In the case of "bottom close", the heat transfer coefficient is not affected by both the gap width and length.

(3) Verification Test on the Reliability of Steam Generator of PWR

The objective of this investigation is to verify the reliability and safety of a steam generator of a pressurized water reactor power plant. And therefore, the following tests have been planned under the actual plant operating conditions, such as, the corrosion test, the thermo-hydraulic test and the tube fracture test. These tests have been conducted based on the time schedule shown in Fig.3.

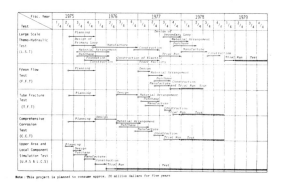

Figure 3 Time Schedule

3.5. Hydrodynamic Stability Tests of Steam Generator for Liquid Metal Fast Breeder Reactor

(1) Large Scale Steam Generator Test /7/

The separated once-through, helical coil type 50MW steam generators and 1MW steam generator were tested by PNC, to understand the instability phenomena in large scale steam generators and to establish criteria for the stable operation.

(2) Parallel Flow Channel Instability Experiment /8/

The mechanisms and characteristics of their instabilities have been studied experimentally, using the parallel channels as simplified as possible, in order to study the above mentioned instability observed in 1MW steam generator with non-insulated downcomer of PNC (we call it "slug excursion instability". Fig. 4 shows two types of flow instability observed in 50MW SG.)

4. CONCLUSION

The experimental facilities for reactor safety research become larger, and the cost of experiment becomes remarkably expensive. For this reason, research work at universities appear to becoming very difficult.   The large scale test can be used only in order to understand the effect of each parameter, and data obtained with such test facilities are very useful to verify the reliability of computer codes. However, as they are not always enough to clarify the effects of geometrical and physical parameter change, other fundamental research are necessary. Such basic research based on the needs of reactor safety should be carried out worldwide   and, if possible, systematically in the frame of international cooperation of universities. Not only the large research establishments, but also universities should promote international collaboration and information exchange more extensively.

5. REFERENCES

/1/. JAERI-M 6318 (1975).
/2/. JAERI-M 6247 (1975), 8287, 6362, 6240, 6241, 6512, 6513, 6709, 6849, 7106, 7236, 7239, 7437, 7505, 7944, 7737 (1978).
/3/. JAERI-M 6703 (1976), 7488, 7712, 7791, 7970, 8185, 8300, 8604, 8588, 8627, 8729, 8723, 8737, 8738, 8728, 8473 (1979).
/4/. 1. Tong, L.S., Status Report of 2D/3D Program, The 7th Water Reactor Safety Research Information Meeting, U.S. Nov. (1979).
   2. Nozawa, M., Japanese Safety Research Program-ROSA, CCTF and SCTF, ibid.
/5/. JAERI-M 8598 (1979), 8780, 8761, 8762, 8763, 8764, 8665, 8765, 8887 (1980).
/6/. Aoki, S., et al., Boiling Heat Transfer in Narrow Gaps, Annual Symposium The 16th, Heat Transfer Society of Japan (1979).
/7/. Tsuchiya, T., et al. (PNC), Hydrodynamic Stability Test for LMFBR Sodium Heated Steam Generators, Proceedings of the Japan-U.S. Seminar on Two-Phase Flow Dynamics, pp.459-482, July (1979).
/8/. Nakanishi, S., Recent Japanese Researches on Two-Phase Flow Instabilities, ibid., pp.55-105, July (1979).

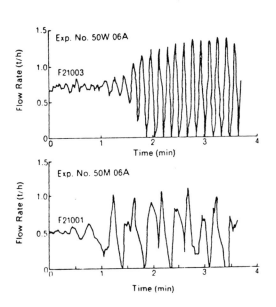

Figure 4 Two-type of Flow Instability

Research on Heat Transfer in Energy Problems in Japan: Additional Topics

TAKASHI SATO
Kyoto University, Kyoto, Japan

At the first session, Prof. F. Ogino (in place of Prof. T. Mizushina) introduced the Japanese present status of the basic researches on energy problems in universities. His explanation was concentrated mainly in the frame of the Grant in Aid for Scientific Research of the Ministry of Education, Science and Culture. At that time, he asked me to report about other research or developing projects, at this round-table discussion.

Since many researches have been carried on in related Ministries, industries and others, it is a much too difficult task for me to cover all of those projects. Therefore, I would like to make a brief introduction only about the Sun-shine Project and the Moon-light project, both of which are planned and proceeded by the Agency of Industrial Science & Technology of the Ministry of International Trade & Industry, with the co-operation of industrial companies. The main items of these Projects, covered with this year's budget, are listed in Prof.Mizushina's report.[1]

The Sun-shine Project is aimed to develop the utilizing techniques of new natural or alternative energy resources instead of crude oil. This plan started in 1974. The target of the Moon-light Project, on the other hand, is to establish the practical technologies and techniques for the saving of energy consumption, more effective use of energy, and the recovery of waste energy or resources. This Project started in 1978. These projects are to be continued for more than 10 years into the future by checking the results & plans, every several years.

Each Project includes several main items as listed [2] and these items are sometimes divided into suitable sub-items. Some are still in the stage of fundamental or planning examination, others are in experimental investigating stages by using the pilot plants or testing equipments, improvement and enlargement of such plants are also carried on, and a few of them are close or have already reached their goals, that is, into commercial base. The utilization of new energy resources is not so easy, from an economical viewpoint, in that it usually will need much more time. It may be suitable here to limit the items only to which are closely connected with heat transfer or thermal engineering problems, and also to which the practical testings are being developed.

In connection with the solar thermal energy, the supplying system of hot water, the heating and cooling system for houses or buildings are almost in the commercial base. And the larger solar systems suitable for industrial factories or larger buildings are being investigated. On the other hand, for the solar power plants, two kinds of model plants of 1,000kW electric output with different types of collector are under construction both at Nio area in Kagawa Prefecture, based on the smaller pilot plants' experiences. The two collector types applied are of the tower collector type in which the inside surface of a cylindrical collector is settled on the top of the tower and which uses concave curved mirrors.

With respect to the geothermal power generation, as Prof. Mori has talked about already, a 10MW model plant using the so-called binary cycle is being planned. It is not a binary cycle in the thermo-dynamical definition, rather it is the single Rankine cycle using organic fluid as the working fluid.

The development of the technologies of MHD power generation and the high efficiency gas turbine proceed on for the purpose of utilizing the energy at a higher temperature level and to increase the total plant cycle efficiency, by combining with the steam cycle or other system using lower level energy. A MHD power generator of 20MWh (100kW times 200hrs) is being built. However both these two items still need a comparatively long lead time to reach the practical use because so many difficult problems involved are to be solved.

In connection with the utilization of waste heat, many kinds of heat exchangers with the high performance heat transfer surfaces including the heat pipe & heat pump systems or the regenerative heat exchangers are proposed and are being investigated to establish the most effective total system, combined with the desirable storage and energy transport systems. Moreover, the pilot plants for the recoveries of energy and also resources from the solid wastes discharged from towns are coming to the testing stage.

Lastly, I would like to add just a few words about the Project Item C [3], Green Energy Project, supported and proceeded by the Ministry of Agriculture, Forest & Fishery. This Project's aims are to develop the practical techniques utilizing the alternative energy or saving energy and chemical fertilizer in such fields.

I have briefly talked about a few items concerning energy problems. Before closing my speech, I would like to point out that it is still neccessary to accumulate much more experiences in each field, and moreover it is important to exchange and to combine the experiences of each item . A closer and more effective co-operation among the researchers and scientists in universities and/or in industries in wider fields is strongly desired in Japan and internationally.

REFERENCES

1. Mizushina, T., First report of this proceeding
2. & 3. ibid p.4 — 5

Performance of an Overall Heat Exchanger Made of Paper

E. NISHIYAMA

Central Research Lab., Mitsubishi Electric Corp., Amagasaki, Japan

ABSTRACT

An overall heat exchanger was made of paper. The structure of the heat exchanger is a cross flow plate fin type and separating plates are treated with hygroscopic solution in order to reduce the gas transmission rate across them. Therefore, in a ventilation apparatus using this heat exchanger, it is possible to exchange both heat and humidity across separating plates without mixing exhausted air and fresh air. The performances of this heat exchanger were calculated numerically and were compared with experimental results. They showed good agreement.

NOMENCLATURE

A = area of sample port for vapor permeation, m^2
G = rate of vapor permeation with paper, kg/m^2h
G' = rate of vapor permeation without paper, kg/m^2h
K_H = overall heat transfer coefficient, $W/m^2°C$
K_M = overall mass transfer coefficient, kg/m^2hcmHg
ℓ_O = thickness of paper, m
T = temperature, K
W = capacity rate, $W/°C$
x,y = distances, m
X,Y = nondimensional distances
$\varepsilon_H, \varepsilon_M$ = efficiencies of heat transfer, mass transfer, respectively
θ = nondimensional temperature

Subscripts

1,2 = fluids 1, 2, respectively

INTRODUCTION

Nowadays, heat recovery ventilation apparatus have been equipped in offices, hospitals and apartments etc. to save energy. There are several types of heat exchangers applied to heat recovery ventilation apparatus; namely rotary regenerators, heat pipe type heat exchangers and plate fin type heat exchangers, etc.
A plate fin type overall heat exchanger made of paper has the following features: The gas transmission rate across the seperating paper is reduced except with water vapor by treating paper with hygroscopic solution. Therefore in a ventilation apparatus using this heat exchanger it is possible to exchange both heat and humidity.
This paper describes the performance of this type of heat exchanger, obtained experimentally and analytically.

STRUCTURE

The structure of the heat exchanger is shown in Fig. 1, which is a cross flow plate fin type heat exchanger. Separating plates are made of Japanese papers treated with hygroscopic solution which

blocks micro pores of papers preventing air from transferring through paper except with water vapor. Fins are made of craft papers.
Since the gas transmission rate across the seperating plates is small except with water vapor, a ventilation unit using this overall heat exchanger has the capability to exchange both heat and humidity without mixing exhausted and intake air.

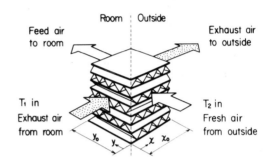

Fig. 1 Construction of an overall heat exchanger element

PERFORMANCE ANALYSIS

Vapor permeability

Vapor permeabilities of papers are measured by the following method (1). A mixture of water and calcium choloride which has a specific vapor pressure is put into an aluminum can with a small port (diameter; 6 mm) covered with a sample paper. The aluminum can is placed in a controlled humidity and transmission rates of water vapor are measured. Permeability P is calculated using the following relation,

$$P = \frac{\ell_O}{A}\left(\frac{1}{G} - \frac{1}{G'}\right)^{-1} / \Delta p \qquad (1)$$

where G and G' are transmitted rates of water vapor with a sample paper and without sample paper respectively and ΔP is a pressure difference between the inside and outside of the can.
An example of experimental permeability of water vapor is shown in Fig. 2. Permeability of water vapor depends on relative humidity. This phenomenon is explained as follows: Water vapor transfers through papers in two processes; namely diffusion and condensation-transmission in liquid-vaporization. The latter process depends on relative humidity and is dominant at a relatively high humidity, on the other hand, a diffusion process is independent of relative humidity. Therefore, permeabilities of water vapor are nearly constant at low relative humidity. Permeabilities of water vapor at a relative humidity of 20% fall in the region of 1.0×10^{-4} - 3.3×10^{-4} kg/mhcmHg depending on the paper and treating method with solution.

On the other hand permeability of carbon dioxide is one tenth less than that of water vapor.

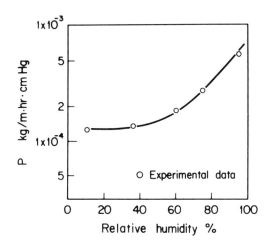

Fig. 2 Humidity dependence of water vapor permeability coefficient through paper

Humidity exchange efficiency

The following assumptions are made in the performance analysis:
1) Heat transfer and mass transfer are independent of each other.
2) Properties concerning heat and mass transfer are constant in a heat exchanger.

The following relation is derived from the heat balance in a differential section in a heat exchanger.

$$K_H(T_1 - T_2) = -\frac{W_1}{y_0}\frac{\partial T_1}{\partial x} = \frac{\partial W_2}{x_0}\frac{\partial T_2}{\partial y} \qquad (2)$$

Using nondimensional quantities, Eq. (2) is modified into the following equation,

$$\frac{\partial^2 \theta_1}{\partial X \partial Y} + NTU_2 \frac{\partial \theta_1}{\partial X} + NTU_1 \frac{\partial \theta_1}{\partial Y} = 0 \qquad (3)$$

where

$$\theta_1 = \frac{T - T_{2in}}{T_{1in} - T_{2in}}, \quad X = \frac{x}{x_0}, \quad Y = \frac{y}{y_0}$$

$$\frac{1}{NTU_1} = \frac{W_1}{K_H x_0 y_0}, \quad \frac{1}{NTU_2} = \frac{W_2}{K_H x_0 y_0}$$

$$W_1 ; W_2 = \text{capacity rate}$$

Boundary conditions are as follows,

$$\theta_1 = \exp(-NTU_1 \times X) \quad \text{at } Y = 0$$
$$\theta_1 = 1 \quad \text{at } X = 0$$

The same equation is derived by substituting humidity H into temperature T, and overall mass transfer coefficient K_M into overall heat transfer coefficient K_H.

Eq. (3) is calculated by numerical method. Thermal efficiency and mass transfer efficiency are obtained from the following equation using the calculated temperature and humidity distributions,

$$\varepsilon = \int_0^1 \theta_2(x,1)dx \qquad (4)$$

Calculated results are compared with experimental results which are shown in Fig. 3. They are found

to coincide each other within 10% deviations. Since the aforementioned vapor permeability depends on relative humidity, Eq. (3) becomes nonlinear considering this effect. In order to consider heat transfer and mass transfer (humidity transfer) one needs to express the equation using enthalpy.

However, in this case it is impossible to modify it into nondimensional form. The difference of these two methods was found small enough in this calculation condition.

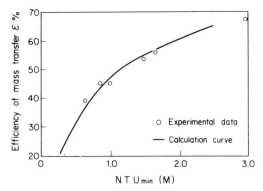

Fig. 3 Efficiency of mass transfer

APPLICATION OF THE HEAT EXCHANGER

An example performance of a heat recovery ventilation unit is shown in Fig. 4. Temperature and humidity of high enthalpy side are 32°C and 55%. Temperature of low enthalpy side is 22°C and humidity is changed in the region of 35%-85%. Thermal efficiency is independent of relative humidity and shows a relative high value of 80%. Mass transfer efficiency changes between 50% and 65%. And enthalpy transfer efficiency falls between the aforementioned two curves.

Test result shows that $\varepsilon_H = 80.6\%$, $\varepsilon_M = 58.9\%$ and $\varepsilon_i = 70.5\%$ respectively at low enthalpy side of relative

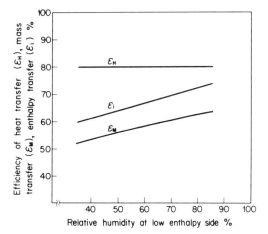

High enthalpy side : 32°C, 55%
Low enthalpy side : 22°C, 35~85%
Air flow rate : 250 m³/hr

Fig. 4 Performance of heat recovery ventilation unit

humidity of 70%.
There are various types of heat recovery ventilation systems using overall heat exchangers. One
duct type ventilation system which is shown in
Fig. 5 is used commonly.

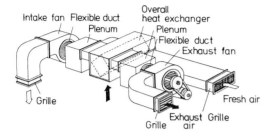

Fig. 5 One duct type ventilation unit

CONCLUSION

The performances of a plate fin type overall heat
exchanger were calculated by ε-NTU method. Calculated thermal efficiency and mass transfer efficiency coincided   with test results very well. A heat
recovery ventilation unit using this overall heat
exchanger was found to have a   heat and mass transfer performance of high efficiency.

ACKNOWLEDGEMENT

This research work was performed by Dr. O. Tanaka,
Mr. H. Kusakawa, Mr. K. Takahashi, Mr. Y. Wakamiya,
Mr. M. Yoshino and Mr. Y. Hashimoto who were members
of the Central Research Lab. and Nakatsugawa works
of the Mitsubishi Electric Corp.
This is the excerpt of their work.

REFERENCE

1. K. Takahashi et al., Kagaku Kogaku Ronbunshu,
    3, 510-513 (1977) (in Japanese)

Heat Transfer in Combustion

H. TSUJI

University of Tokyo, Tokyo, Japan

INTRODUCTION

As is well known, the practical application of the fundamentals of heat transfer from flames to heat sinks covers a wide range of important engineering systems, which includes rocket motors, gas turbine combustors, internal combustion engines, boilers, industrial furnaces and open flames such as fires and flares. In recent years there has been a growing interest in the study of flame radiation for industrial application. This is due to increasing combustion intensities in industrial applications and the need for prediction methods for detailed distribution of radiant heat flux to bounding surfaces [1]. Many studies have been made of the heat transfer from flames to heat sinks, especially the radiative heat transfer from flames. These problems may be considered to be the problem of the heat transfer proper, and in this seminar, these problems were discussed in the session on High-Temperature Heat Transfer - Basic Phenomena.

Recently, however, in connection with the effective use of fuel, especially with the techniques to burn poor fuels and mixtures of very low heat content effectively, some combustion scientists have a greater interest in the problems on heat transfer within the flame, as this heat transfer affects combustion intensity and the rate at which certain types of flames and fires propagate. I would like to make a short comment on this problem.

LIMITING COCENTRATIONS OF FUEL AND OXYGEN

I shall begin with the scientific background to the importance of the heat transfer within the flame. When the calorific value of fuel gas is reduced to a certain limiting value by diluting the fuel with inert gases or combustion products, or when the oxygen concentration in air is reduced to a certain limiting value by adding inert gases or combustion products to air, stable combustion becomes impossible. The study of such concentration limits of fuel and oxidizer gives useful information for the utilization of low calorific value fuel and for the prevention of flame extinction which results from the oxygen shortage in vitiated air.

Recently, the effect of diluting fuel or oxidizer with an inert gas on the extinction of laminar diffusion flame was investigated by Ishizuka and Tsuji, using the counterflow diffusion flame established in the forward stagnation region of a porous cylinder immersed in a uniform oxidizer flow, and the limiting fuel concentration, the limiting oxygen concentration and the limit flame temperatures at these limiting concentrations were examined [2]. The fuels used were methane and hydrogen, and the inert gases were nitrogen, argon and helium.

As the oxidizer-stream velocity is increased or the mixture-ejection velocity is decreased, this counterflow diffusion flame approaches the cylinder surface and finally blows off from the forward stagnation region of a porous cylinder. As the fuel concentration Ω or the oxygen concentration β is decreased, the flame luminosity becomes weak and finally the flame extinction occurs. In this study, we examined the relations between these parameters at the flame extinction and determined the limiting fuel concentration $\Omega_{c,a}$, below which the laminar diffusion flame was never established in various airs ("nitrogen air", "argon air", and "helium air"), and the limiting oxygen concentration $\beta_{c,a}$, below which the laminar diffusion flame of the pure fuel was never established. Also we measured the limit flame temperatures.

The experimental results of the limiting concentrations of fuel and oxygen and the limit flame temperatures are summarized in Tables 1 and 2.

Table 1   Limiting methane concentrations $\Omega_{c,a}$, limiting oxygen concentrations $\beta_{c,a}$ and limit flame temperatures $T_{f,a}$ for methane diffusion flame diluted with various inert gases.

THE LIMITING FUEL CONCENTRATIONS AND THE LIMIT FLAME TEMPERATURES

$\begin{pmatrix}\Omega_{c,a}\\ T_{f,a}°C\end{pmatrix}$		FUEL (CH₄)		
VARIOUS KINDS OF "AIR"		N₂	Ar	He
	N₂	0.165 / 1190	0.112 / 1170	0.33 / 1290
	Ar	0.102 / 1160	0.076 / 1160	0.210 / 1290
	He	0.190 / 1280	0.146 / 1270	0.320 / 1340

THE LIMITING OXYGEN CONCENTRATIONS AND THE LIMIT FLAME TEMPERATURES

$\begin{pmatrix}\beta_{c,a}\\ T_{f,a}°C\end{pmatrix}$		FUEL (CH₄)
VARIOUS KINDS OF "AIR"		
	N₂	0.143 / 1210
	Ar	0.096 / 1170
	He	0.149 / 1350

Table 2   Limiting hydrogen concentrations $\Omega_{c,a}$, limiting oxygen concentrations $\beta_{c,a}$ and limit flame temperatures $T_{f,a}$ for hydrogen diffusion flame diluted with various inert gases.

THE LIMITING FUEL CONCENTRATIONS AND THE LIMIT FLAME TEMPERATURES

$\begin{pmatrix}\Omega_{c,a}\\ T_{f,a}°C\end{pmatrix}$		FUEL (H₂)		
VARIOUS KINDS OF "AIR"		N₂	Ar	He
	N₂	0.114 / 850 (nonuniform)	0.085 / 830 (nonuniform)	0.109 / 790 (uniform)
	Ar	* *	0.081 / 840 (nonuniform)	* *
	He	* *	* *	0.119 / 810 (uniform)

THE LIMITING OXYGEN CONCENTRATIONS AND THE LIMIT FLAME TEMPERATURES

$\begin{pmatrix}\beta_{c,a}\\ T_{f,a}°C\end{pmatrix}$		FUEL (H₂)
VARIOUS KINDS OF "AIR"		
	N₂	0.052 / 740 (uniform)
	Ar	0.036 / 720 (uniform)
	He	0.044 / 810 (uniform)

Table 1 shows the results for the methane flame and Table 2 those for the hydrogen flame. It is found that the limit flame temperature at the limiting fuel concentration is almost equal to the limit flame temperature at the limiting oxygen concentration, except in the case where the non-uniform flame extinction occurs in the hydrogen flame. This suggests that there exists a minimum flame temperature below which the diffusion flame can never be established for the given combination of fuel, oxygen and inert gas, and the controlling factor with diffusion flame under limiting conditions is the limit flame temperature.

The limit flame temperature of diffusion flame for methane diluted with nitrogen is 1,200°C, and that for hydrogen is 740°C, and the limit flame temperatures of these flames are found to be almost equal to the flame temperatures at their lean flammability limits of the premixed gas. Thus, even in the hydrogen flame, unless some instability occurs in the flame zone, the limit flame temperature of the diffusion flame diluted with an inert gas is closely correlated with the flame temperature at the lean flammability limit of the premixed fuel and air. These results suggest that both the lean flammability limit of the premixed flame and the limiting fuel and oxygen concentrations for diffusion flame are primarily controlled by the similar factor.

EXCESS ENTHALPY BURNING

It has been confirmed that there exists a minimum flame temperature below which the flame can never be established for the given combination of fuel, oxygen and inert gas and the controlling factor under limiting conditions is the limit flame temperature. This fact reminds us that, it is necessary to raise the flame temperature above the limit temperature, in order to burn nonflammable, or barely flammable mixtures. For that purpose, it is desirable to preheat the reactants, but preheating the reactants by external heat sources is not suitable for energy saving, and it is necessary to preheat the reactants by their own combustion gases.

In 1971, Weinberg of Imperial College proposed the concept of the excess enthalpy burning to burn poor fuels and mixtures of very low heat content [3]. This concept is based on the idea of recirculating heat without simultaneous dilution - that is raising the temperature in the reaction zone on borrowed heat which is repaid before the gases leave (Fig. 1). Weinberg et al. have investigated the burning systems which recirculate heat without recycling hot products [4]-[10]. Such devices use counterflow heat exchangers, including recirculating particles in shaped fluidized and spouted beds, and are potentially capable of burning the off-gases from various industrial and fermentation processes, from coal seams, ventilation air from mines, fuels of high moisture and inert content such as those expected from solar energy plantations, etc. [9].

The variety of possible, and potentially useful, heat exchangers is very large, and various types of burners have been proposed, but the double spiral burner, namely Swiss roll burner (Fig. 2), is most well known. Experiments on double spiral burners have indicated that the limiting tempera-

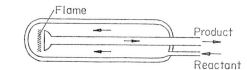

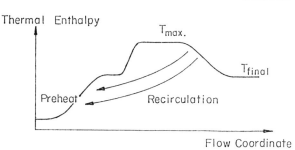

Figure 1 Schematic illustration of excess enthalpy burning.

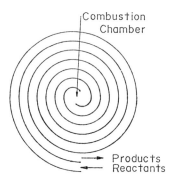

Figure 2 Double spiral ("Swiss roll") burner.

ture for stability in a combustion chamber is virtually constant over a range of velocities and compositions and that the burning system based on this concept is successful to extend the ranges of the flame stability and flammability. The implication is that the power released in such burners can be quite appreciable, in spite of the small chemical energy density in the reactants of these lean mixtures. Such a kind of experimental study has been carried out by Kawamura and Asato of Gifu University, using a simple burner with counter current heat exchanger [11], and it has been shown that increased heat recirculation and decreased heat losses can burn leaner mixtures.

All devices elaborated so far to produce the excess enthalpy use some sort of heat exchangers, which preheat the unburnt mixture before it enters into the flame. This method is indirect, because heat is recirculated outside the flame and the flame structure remains unchanged except for its elevated temperature.

Recently, Takeno and Sato, University of Tokyo, proposed a simpler and more direct method of producing the excess enthalpy flame by changing the internal structure of flame itself [12]-[15]. The idea is just to insert a porous solid into the one-dimensional flame zone (Fig. 3). If the thermal conductivity of the solid is high enough, the heat is recirculated through the solid from the

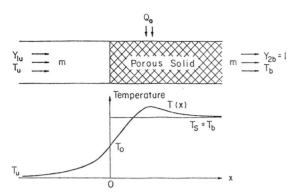

Figure 3   One-dimensional excess enthalpy
flame model.

downstream high temperature region to the upstream
low temperature region. This produces an excess
enthalpy in the upstream region, which may bring
about the favorable characteristic of the flame.
The extent of the produced excess enthalpy may be
controlled by choosing the porosity of the solid
to alter the heat transfer coefficient between
the solid and the gas. The potentiality of the
proposed artificially modified flame was analyzed
on the basis of the simplified one-dimensional
flame theory. The analysis reveals several attractive characteristics of flame and the proposed
idea is promising to burn mixtures of low heat
content in a simple combustion system.

Some experimental studies have also been made
using a specially designed burner. The proposed
idea is very attractive, but a very satisfactory
result of this experiment to support the idea has
not been obtained up to the present.

CONCLUDING REMARKS

I have made a short comment on the topic of heat
transfer within the flame, in connection with the
techniques to burn mixtures of very low heat content effectively. Recently, the heat transfer
problems become more important in combustion science, and I believe that   combustion scientists
must also, take a deep interest in the heat transfer phenomena.

REFERENCES

1. Afgan, N.H. and Beer, J.M., Heat Transfer in
    Flames, Scripta Book Co., Washington, D.C.
    (1974).
2. Ishizuka, S. and Tsuji, H., Eighteenth Symp.
    (Intern.) on Combustion, The Combustion
    Inst., Pittsburgh (1980).
3. Weinberg, F.J., Nature, 233, 239-241 (1971).
4. Hardesty, D.R. and Weinberg, F.J., Comb. Sci.
    and Tech., 8, 201-214 (1974).
5. Lloyd, S.A. and Weinberg, F.J., Nature, 251,
    47-49 (1974).
6. Lloyd, S.A. and Weinberg, F.J., Nature, 257,
    367-370 (1975).
7. Hardesty, D.R. and Weinberg, F.J., Comb. Sci.
    and Tech., 12, 153-157 (1976).
8. Lloyd, S.A. and Weinberg, F.J., Combustion
    and Flame, 27, 391-394 (1976).
9. Jones, A.R., Lloyd, S.A., and Weinberg, F.J.,
    Proc. Roy. Soc., London, A, 360, 97-115
    (1978).
10. Khoshudoodi, M. and Weinberg, F.J., Combus-
    tion and Flame, 33, 11-21 (1978).
11. Kawamura, T., Asato, K., Mazaki, T., and
    Naruse, M., Combustion Research (Proc. Comb.
    Soc. Japan), 51, 33-40 (1979).
12. Takeno, T. and Sato, K., Comb. Sci. and
    Tech., 20, 73-84 (1979).
13. Takeno, T. and Sato, K., Combustion Research
    (Proc. Comb. Soc. Japan), 51, 41-52 (1979).
14. Sato, K., Ph.D. Thesis, University of Tokyo
    (1980).
15. Takeno, T., Sato, K., and Hase, K., Eight-
    eenth Symp. (Intern.) on Combustion, The
    Combustion Inst., Pittsburgh (1981).

Four Specific Heat Transfer Problems Pertinent to Advanced Power Systems

R. VISKANTA

School of Mechanical Engineering, Purdue University,
West Lafayette, Indiana 47907, U.S.A.

This morning in my review of heat transfer problems
in some advanced power systems I cited a large number of areas needing research attention. I would
now like to narrow down this rather lengthy "shop-
ping list" of problems needing solutions and/or
experimental data. A few specific topics of a more
immediate engineering interest, and on which I
would like to focus attention, are the following:

- Combined turbulent convection and radiation
  in cavities

- Optical property determination of high
  temperature combustion products

- Development of realistic and efficient
  models of multidimensional radiation
  transfer models in participating media

- Multi-mode heat in radiation participating media

Combined turbulent convection and radiation in
cavities is relevant to the spread of fires in
buildings and corridors, central solar cavity receivers, and combustion system components. Buoyancy driven or aided turbulence is not fully understood, and both turbulence models and experimental
data for providing guidance to modelers are required. The effects of buoyancy on turbulence
should be accounted for in the numerical prediction
of flow and heat transfer rates in both closed and
open cavities. Such predictions must be verified
in small scale laboratory tests for the purpose of
establishing confidence in the models. This would
then provide a basis for developing relations for
the scaling up from small laboratory models to full
scale systems.

Optical property data of combustion products is
essential for the prediction of radiation heat
transfer in fossil-fuel burning combustion systems
such as coal-fired MHD and utility boilers, gas
turbine combustors, coal conversion and other
combustion systems. In addition to the infrared
radiating gases (CO_2, H_2O, CO, etc.) particulates
such as soot, coal char, coal-ash, and slag, which
form during combustion, also contribute to the
transfer of radiation. The fundamentals of gaseous
radiation are relatively well understood; however,
a need for additional work exists. This work consists of expanding and improving current data on
band parameters at high temperatures and large
optical paths, as well as developing simple analytical characterizations of nonhomogeneous, multi-
species radiation. Particulate radiation includes
both aborption and scattering. The radiation
characteristics (absorption and scattering
coefficients and scattering distribution functions)
depend on the optical constants of the material,
as well as the shape, size and number density
distribution of the particles. An analytical
basis for particle radiation is well developed,

but the actual computation is severely hampered by
the lack of data on the optical constants (soot is
an exception) and information concerning the complex chemistry and physics of particle formation
and agglomeration. Thus, **optical** constants of
particulates need to be measured as a function of
wavelength and temperature as well as the radiation
properties of particulates of the type formed in
combustion systems burning coal and alternate
(derived) fossil fuels, and also the validity of the
analytical models must be verified.

Realistic and efficient models for multidimensional
radiative transfer in participating media are required for the thermal analysis of combustion and
coal conversion systems. Models for scaling radiation heat transfer in coal-fired MHD components such
as the radiant boiler would be particularly valuable
in the MHD development program. For example, the
convective heat transfer coefficient in a tube is
scaled according to the relation, Nu = f(Re,Pr).
Regarding radiation transfer, an important question
is: "What are the relevant scaling parameters and
relations for radiation heat transfer in an enclosure filled with particle-ladden combustion gases?".
It appears that accurate 'bench mark' solutions and
experiments for radiative transfer are needed for
simple physical situations for the purpose of
evaluating various approximate schemes, providing
checkpoints for computer code development, and the
modeling of more complex situations. Radiative
transfer in absorbing and emitting medium is relatively well understood, but models which account for
spectral (e.g., bands of infrared **radiating** gases
and line and continuum radiation by alkali metal
vapors and the continuum radiation by particles
such as soot and coal-ash, need to be developed and
verified. However, when particulates are present
scattering and can become important, and thus the
problem of radiative transfer in a multi-dimensional
system becomes much more complex. Consequently
only planar, one-dimensional problems have been
treated. An extension of the one-dimensional analysis must be considered in order to render application to real systems.

The interaction of radiation with conduction and
convection occurs in coal-fired combustion system
components (i.e., coal-ash slags formed on MHD
channels and radiant boilers), insulations, porous
media, fluidized beds, industrial furnaces, and
other systems of current engineering interest.
Multi-mode heat transfer problems introduce
mathematical complications which are in addition to
those which exist for radiation only heat transfer
problems. Analytical and/or closed form solutions
usually cannot be obtained because of the nonlinear
and integro-differential nature of the governing
equations, and the numerical methods of solution
require rather extensive computational effort. The
concepts of an effective conductivity for combined
conduction and radiation in a participating solid
media is particularly useful for design calcula-

tions. The concept needs to be developed and verified experimentally so that it can be applied with
some confidence. Multi-mode heat transfer in one-
dimensional planar, participating media involving
gases and gas-particle mixtures has been analyzed and a
number of different solution techniques developed.
The extension of these techniques to two or three-
dimensions in enclosures is of fundamental research
interest. Such techniques should be compatible with
current numerical methods for solving fluid flow
problems. The development of simple, yet sufficiently rigorous, analysis which takes into account
the spectral nature and nonisotropic scattering of
radiation is of great concern in many engineering
applications.

Briefly, many essential "building blocks" are still
missing and need to be developed for analyzing
complex multi-mode heat transfer problems involving radiation.

Cryogenic Heat Transfer

EDWARD R. LADY

University of Michigan, Ann Arbor, Michigan, U.S.A.

ABSTRACT

Cryogenic heat transfer research results have been presented in many recent conferences. Principal meetings are listed for reference.

Recent research on brazed aluminum plate-fin heat exchangers is reviewed. Emphasis is being placed on more effective, lower pressure drop exchangers. Rising power costs make increased capital investment in high performance heat exchange surface financially attractive.

Studies continue on heat transfer to HeII, with an effort being made to provide a satisfactory theoretical model for transient and steady state heat flow. Design of very large scale helium cooled cryopanels is noted.

INTRODUCTION

Developments in cryogenic heat transfer research have been reported at two recent conferences: the Joint ASME/AIChE National Heat Transfer Conference, Orlando, Florida, July 27-30, 1980, and the ASME Cryogenic Processes and Equipment Conference, San Francisco, August 19-21, 1980. Prior to these meetings, the LNG-6 Conference held in Kyoto, April 7-10, 1980, and the 1979 Cryogenic Engineering Conference, Madison, Wisconsin, August 21-24, 1979, provided forums for papers on liquefied natural gas (LNG) heat exchange and liquid helium temperature level research. Undoubtedly additional work will be presented at the 1980 Applied Superconductivity Conference taking place this very week, September 29-October 3, 1980.

In addition to the five conferences listed above, which are held annually or biennially, papers on cryogenic heat transfer can be found in the proceedings of other symposia held periodically, such as:
- ASME Winter Meeting
- AIChE Annual Meeting
- International Institute of Refrigeration
- International Conference on Low Temperature Physics
- International Cryogenic Engineering Conference
- Intersociety Energy Conversion Engineering Conference

Numerous periodicals in the fields of cryogenic engineering and low temperature physics also include articles and scholarly papers in the area of cryogenic heat transfer. Clearly this is a field of research where many avenues exist for timely and widespread dissemination of research results. Therefore this discussion will review some of the recent papers presented in Orlando and San Francisco, and make some general observations regarding applications of heat transfer research to the cryogenic industry.

BRAZED ALUMINUM PLATE-FIN HEAT EXCHANGERS

There has been renewed activity in evaluation of brazed aluminum plate-fin heat exchangers used in condensation, boiling, and high performance heat exchange for tonnage oxygen plants. Haseler (1) of AERE, Harwell, measured condensing coefficients of downward-flowing nitrogen in a plate-fin exchanger. Experimental coefficients are significantly higher than predicted by the Nusselt theory, probably due to shear forces exerted on the liquid film by the downward vapor flow, resulting in a thinner, more turbulent film. Robertson (2), also of Harwell, reported on local boiling coefficients of R-11 flowing in a serrated fin test section. These tests complement earlier tests with liquid nitrogen. The Reynolds number of the total flow, using the liquid viscosity, was used to correlate boiling coefficients with fluid quality.

Duncan and Whalen (3) of Trane Company have presented a review of the use of brazed aluminum plate-fin exchangers in the air separation industry. Sizes, performance characteristics, and market developments are given. Current high energy costs are causing designers and manufacturers of these exchangers to optimize life-cycle costs by improving fin design, increasing surface area, and reducing pressure drop. In a 1000 t/day oxygen plant, an additional investment of $500,000 in exchanger surface will result in savings in operating power costs which have a present worth of $1,000,000.

Tsao (4) of Air Products and Chemicals warns that all large scale applications of cryogenic heat transfer surface must be carefully designed to insure uniform distribution of flow in the many parallel circuits. This was emphasized in personal discussion with Dr. J.M. Geist, also of Air Products. For example, flow rates in the 50 to 100 parallel passages of each of 12 cores of a large scale cryogenic system must be balanced to within a few percent for optimum exchanger performance. Heat exchangers must be analyzed by finite difference techniques to include longitudinal and transverse conduction in the aluminum, as well as direct heat flow through the parting wall from the warm fluid to the cold fluid.

ENHANCED SURFACE REBOILER-CONDENSERS

O'Neill and Gottzmann (5) of Linde Division, Union Carbide Corporation, described an advanced air plant main condenser design, using High Flux aluminum tubes with overall condensing nitrogen-boiling oxygen coefficients of 9000-11,000 W/m-m-K (1600-2000 Btu/hr-ft-ft-F). The enhanced condensing surface consists of longitudinal splines or flutes around the outside of a vertically oriented 25mm tube. Surface tension forces drive the condensate into the valleys between splines, resulting in a condensing coefficient 10 times greater than would be calculated by the Nusselt equation for a smooth surface.

The enhanced oxygen boiling surface on the inside of the tube is a layer of aluminum particles bonded to the parent tube wall by brazing. This layer is .3mm thick, with a void fraction of .5. This surface is designed to provide and maintain nucleate boiling over its entire area with temperature differentials in the .2 K range.

A result of this sophisticated, and costly, surface preparation of the reboiler-condenser tubes is a halving of the overall temperature difference from 2 K to 1 K. On a 1000 t/day oxygen plant, with power cost of $.045/kWh, the annual savings resulting from the lower temperature difference of the condenser is about $200,000.

The basic principles of the enhanced surface reboiler-condenser have been known for 30 years. The economics of the higher power costs and improved manufacturing techniques have enabled engineers to put into large scale practice advanced concepts usually restricted to the laboratory.

HEAT TRANSFER AT LIQUID HELIUM TEMPERATURES

In the recent ASME conference in San Francisco, several papers reported on developments in heat transfer to HeII and the use of liquid helium for cryopumping and in cooling superconducting magnets. Van Sciver and Lee (6) of Wisconsin presented experimental and theoretical work on heat transfer from circular cylinders cooled in a bath of HeII. Their work extended into time dependent effects. At this same conference other papers which dealt with superconducting devices did not appear to include much concern or sophistication in the matter of heat transfer to liquid helium. Coils and cryopanels are either flooded with liquid helium, or kept cold by a large forced flow of the helium coolant.

A rather interesting paper by Hood and Bonn (7) of CVI describes the detailed design of the helium cryopanels for the Mirror Fusion Test Facility at Livermore. This is a very large installation currently under construction. The total helium cooled surface is 1100 sq m, giving an estimated pumping speed of 4.5×10^7 L/sec. The panels are cooled by a liquid helium flow of 9.5 L/sec, absorbing an estimated 1200 W. This paper will be of particular interest to those working with rarefied gas dynamics and large high vacuum systems.

Citations 3-7 refer to Cyrogenic Processes and Equipment in Energy Systems, presented at San Francisco, August 19-21, 1980, and published by ASME, New York (1980).

3. Duncan, F.D. and Whalen, M.J., "Large Tonnage Oxygen Plants - Brazed Aluminum Technology for the 80's," p. 29-35.
4. Tsao, T.R., Cassano, A.A. and Tao, J.C., "Acid Gas Removal and Cryogenic Separation of SNG Produced from Coal," p. 11-17.
5. O'Neill, P.S. and Gottzmann, C.F., "Improved Air Plant Main Condenser," p. 37-45.
6. Van Sciver, S.W. and Lee, R.L., "Heat Transfer From Circular Cylinders in HeII," p. 147-154.
7. Hood, C.B. and Bonn, J.W., "Large Scale Cryopumps for Fusion Power Systems," p. 155-161.

REFERENCES

1. Haseler, L., "Condensation of Nitrogen in Brazed Aluminium Plate-Fin Heat Exchangers," ASME Paper 80-HT-57, (1980).
2. Robertson, J.M. and Lovegrove, P.C., "Boiling Heat Transfer with Freon-11 in Brazed-Aluminium Plate-Fin Heat Exchangers," ASME Paper 80-HT-58, (1980).

Index

240